Mechanical Engineering and its Applications

Mechanical Engineering and its Applications

Edited by **Bilroy Lockhart**

CLANRYE
INTERNATIONAL
New Jersey

Published by Clanrye International,
55 Van Reypen Street,
Jersey City, NJ 07306, USA
www.clanryeinternational.com

Mechanical Engineering and its Applications
Edited by Bilroy Lockhart

International Standard Book Number: 978-1-63240-535-7 (Hardback)

Printed in China.

Contents

	Preface	**IX**
Chapter 1	**On Boundaries of Cosmos** Emil Edipovich Lin	**1**
Chapter 2	**Splash Formation Due to a Frog Diving into Water** Yoshihiro Kubota, Osamu Mochizuki	**7**
Chapter 3	**Impact of the Drag Force and the Magnus Effect on the Trajectory of a Baseball** Haiduke Sarafian	**16**
Chapter 4	**Cyclic Elasto-Plastic Fracture Diagram and Some Parameters of Cyclic Crack Growth Resistance for the Plastic Steels** L. A. Sosnovskiy, A. V. Bogdanovich, S. S. Sherbakov	**26**
Chapter 5	**Influence of Soil-Foundation Interaction Properties on Oscillations of the System "Building-Building" and "Building-Stack-Like Structure"** Gaibnazar Khodievich Khojmetov, Alinazar Irismetovich Khodjimetov, Anvar Sayfullaevich Yuvmitov	**32**
Chapter 6	**Nonlinear Free Vibrations of C-C-SS-SS Symmetrically Laminated Carbon Fiber Reinforced Plastic (CFRP) Rectangular Composite Plates** Zakaria Zergoune, Bilal Harras, Rhali Benamar	**43**
Chapter 7	**A Novel Technique for Evaluating Failure Probability of Random Structure** Seifedine Kadry, Abdelkhalak El Hami	**56**
Chapter 8	**Curve Veering in Torsional Systems with Stepped Shafts** Mehdi Eshaghi, Rama Bhat	**68**
Chapter 9	**Joule Heating and Thermal Radiation Effects on MHD Boundary Layer Flow of a Nanofluid over an Exponentially Stretching Sheet in a Porous Medium** Jakkula Anand Rao, Gandamalla Vasumathi, Jakkula Mounica	**76**
Chapter 10	**Thorn-Shape Waveform and Double-Strike Phenomenon Seen in the Impact Force of Soft Materials** Yukio Fujimoto, Chen Liu, Yoshikazu Tanaka, Eiji Shintaku, Toshiki Nakanishi	**90**
Chapter 11	**A Short Vector Solution of the Foucault Pendulum Problem** I. A. Ciureanu, D. Condurache	**101**

Chapter 12 **MHD Mixed Convection Flow of a Casson Fluid over an Exponentially Stretching Surface with the Effects of Soret, Dufour, Thermal Radiation and Chemical Reaction** **114**
Kankanala Sharada, Bandari Shankar

Chapter 13 **A Quaternion Solution of the Motion in a Central Force Field Relative to a Rotating Reference Frame** **127**
Ioan-Adrian Ciureanu, Daniel Condurache

Chapter 14 **Viscosity Transient Phenomenon during Drop Impact Testing and Its Simple Dynamics Model** **136**
Chen Liu, Yoshikazu Tanaka, Yukio Fujimoto

Chapter 15 **Numerical Modeling of Non-Similar Mixed Convection Heat Transfer over a Stretching Surface with Slip Conditions** **145**
A. Subba Rao, V. R. Prasad, N. Nagendra, K. V. N. Murthy, N. Bhaskar Reddy, O. Anwar Beg

Chapter 16 **Numerical Investigation of Aerodynamic Characteristics by a Rotating Thin Plate** **157**
Yoshihiro Kubota, Osamu Mochizuki

Chapter 17 **Aerodynamic Brake for Formula Cars** **163**
Roberto Capata, Leone Martellucci

Chapter 18 **Analytical Model on Steel Tanks Damaged by Corrosion** **179**
Francisco Casanova-del-Angel, Moisés Gaytán-López

Chapter 19 **Inverse Problem of Astrodynamics** **191**
Yuri Menshikov

Chapter 20 **Magnetic Field and Thermal Radiation Effect on Heat and Mass Transfer of Air Flow near a Moving Infinite Plate with a Constant Heat Sink** **199**
S. M. Arifuzzaman, Sajal Kumar Dhali, Md. Abul Kalam Azad, Bibhuti Roy

Chapter 21 **Stability Analysis of a SDOF Mechanical Model with Distinct Critical Points: II. Catastrophe Theory Approach** **213**
Dimitrios S. Sophianopoulos, Vasiliki S. Pantazi

Chapter 22 **Effects of Radiations and Heat Source/Sink on a Casson Fluid Flow over Nonlinear Stretching Sheet** **221**
Chenna Sumalatha, Shankar Bandari

Chapter 23 **Magneto Hydrodynamics Stagnation Point Flow of a Nano Fluid over an Exponentially Stretching Sheet with an Effect of Chemical Reaction, Heat Source and Suction/Injunction** **230**
Ch. Achi Reddy, B. Shankar

Chapter 24 **Rotational Oscillation Effect on Flow Characteristics of a Circular Cylinder at Low Reynolds Number** **241**
Alice Rosa da Silva, Aristeu da Silveira-Neto, Antônio Marcos Gonçalves de Lima

Chapter 25 **Elastic Stress Predictor for Stochastic Finite Element Problems** **256**
Drakos Stefanos

Permissions

List of Contributors

Preface

Mechanics is the scientific and technological study of motion. Mechanical engineering deals with the area of science that studies the changes which occur when a physical body is brought in contact with force and displacement. It applies the laws of physics to comprehend and analyze and create mechanical systems. This book contains some path-breaking studies related to mechanical engineering. It provides detailed information about important theories, its practical applications and other case studies related to this topic. Students, researchers, engineers and all associated with the field of mechanical engineering will benefit alike from this book. It is a complete source of knowledge on the present status of this important field.

This book is a result of research of several months to collate the most relevant data in the field.

When I was approached with the idea of this book and the proposal to edit it, I was overwhelmed. It gave me an opportunity to reach out to all those who share a common interest with me in this field. I had 3 main parameters for editing this text:

1. Accuracy – The data and information provided in this book should be up-to-date and valuable to the readers.
2. Structure – The data must be presented in a structured format for easy understanding and better grasping of the readers.
3. Universal Approach – This book not only targets students but also experts and innovators in the field, thus my aim was to present topics which are of use to all.

Thus, it took me a couple of months to finish the editing of this book.

I would like to make a special mention of my publisher who considered me worthy of this opportunity and also supported me throughout the editing process. I would also like to thank the editing team at the back-end who extended their help whenever required.

Editor

1

On Boundaries of Cosmos

Emil Edipovich Lin

Russian Federal Nuclear Center, All-Russia Research Institute of Experimental Physics, Sarov, Russia
Email: postmaster@ifv.vniief.ru, e.e.lin@ifv.vniief.ru

Abstract

This paper establishes asymptotic time dependences of characteristic sizes of astrophysical and cosmological objects. These dependences are obtained on the basis of uncertainty principle applied in cosmic scales in approximation of spherical symmetry in Euclidean geometry. The proposed analytical approach makes it possible to determine spatial boundaries of the uniformity of matter distribution in the Universe, and a size of cosmic sphere which contains numerous groups of interacting universes.

Keywords

Uncertainty Principle, Cosmos, Astrophysical/Cosmological Objects, Characteristic Sizes, Time

1. Introduction

At present the increasing number of scientific articles is devoted to different problems of astrophysics and cosmology: from formation and evolution of stars and galaxies to formation of the large-scale structure of the Universe under influences of dark matter and dark energy (see, for example, [1]-[10]). The results of these investigations are based on reliable astronomic observations as well as on modern theoretical conceptions about formation and expansion of the Universe after Big bang. Theoretical investigations are connected with the ideas on behavior of gravitational fields and gravitating masses at diversiform possible geometries of cosmic space, in particular, at Euclidean, parabolic and hyperbolic geometries. A stochastic quantum nature of cosmological phenomena is noted.

The most significant cosmological problem is determination of spatial boundaries corresponded to uniformity and isotropy of the Universe, *i.e.*, to Cosmological principle [1]-[3]. The latter signifies that because of insufficient time for growth of gravitational perturbations, the distribution of matter at early stages of formation of cosmological structures does not affect the average value of matter density in up-to-date cosmos, and “heavy heterogeneities and voids are ended at some very large but still unknown distance” [2]. Another significant problem is the following: is cosmos infinite or finite (closed) and where is a boundary of closed cosmic space [1]

[2]?

In accordance with conception [3] about a large-scale structure of the Universe as of non-uniform matter distribution originating from growing adiabatic perturbations of density, there is some distinction between the developed non-linear structures within the scales of less than 10 Megaparsec (Mps) (namely, halo of the galaxy, groups and clusters), and the structures with more regular quasi-linear matter distribution within the scales of up to 100 Mps, *i.e.* superclusters and cosmological "voids". It should be noted that up-to-date the quasi-undisturbed Hubble flow of matter that has been already affected by quantum-gravitational instabilities, still continues.

In this paper, an attempt of qualitative analysis of the above problems is undertaken. With this purpose, extended interpretation of an uncertainty principle applied to the sizes of objects in cosmic scales is proposed. The proposed analytical approach makes it possible to obtain asymptotic time dependences for characteristic sizes of astrophysical and cosmological objects in approximation of spherical symmetry in Euclidean geometry. Globular star clusters, superclusters of galaxies and the Universe itself are considered as such objects. The existence of groups of interacting universes is postulated, and the issue of a size of cosmic sphere involving the great number of such groups is considered.

2. Phenomenological Approach

By analogy with quantum mechanics which considers discrete and continuous energy spectra of microcosm objects it is possible to single out relatively small-scaled regions for cosmic structures, where, as a result of gravitation, mass distribution is non-uniform (analogy is with discrete energy spectrum). In more extended structures, where gravitation "gets smeared" (spreads) in space, mass distribution becomes quasi-uniform (analogy is with continuous energy spectrum).

Such analogy provides precondition for an attempt to extend the uncertainty principle to cosmic scales. The essence of the proposed uncertainty principle in cosmic scales is in the fact that during the time of a single act of gravitational interaction between cosmic structures, their sizes cannot be determined exactly. It is connected with the fact that unless the elementary act is finished, it is impossible to determine to which object each of the interacting surface elements the most closely located to each other is assigned. Stars are elements for globular clusters, galaxies are elements for super-clusters, clusters of galaxies are elements for the Universe.

In accordance with the generally accepted concept about formation of large cosmological objects from smaller ones (embryos) [2], we will consider closed systems whose sizes increase due to interaction of many embryos between themselves and interaction of objects between themselves. A process of irreversible aggregation of objects is described using a concept of distribution density wave $\varphi(a, t)$ in the space of cluster sizes a. The wave propagates with the time t toward an increase in the cluster size. Such one-dimensional approach allows one not to take into account deviation of a geometrical shape of the object from the ideal spherical one.

On the basis of universal relation for half-width of wave packet and half-width of spectral line (resulted from Fourier theorem) $\Delta a \cdot \Delta k \geq 1/4\pi$ (k is wave number), it is possible to write down the following uncertainties relation for a coordinate and a momentum in space of sizes a:

$$|\Delta a| \cdot |\Delta p| \cong \frac{K_c}{2}. \tag{1}$$

Here $\Delta p \sim p = m\,\Delta a/\Delta t$ is uncertainty of momentum, $m = (\pi/6)\,a^3\rho$ is object mass with the density ρ; K_c is 2π-reduced cosmological constant of action; Δt is a typical time period for a single act of object interaction. Uncertainty of momentum by an order of magnitude is equal to momentum itself, since gravitational interaction between objects occurs always. Cosmological constant of action can be determined (by analogy with the Planck mass and length) from consideration of dimensionality as:

$$K_c = a_0 m_{oc} c = \frac{\pi}{6} a_0^4 \rho_c c = \frac{1}{16} a_0^4 \frac{H_0^2}{G} c. \tag{2}$$

Here a_0 is a typical size of embryos from which an object is formed; c is velocity of light in vacuum, $m_{oc} = (\pi/6)\,a_0^3\rho_c$ is conditional mass of an embryo with the critical substance density $\rho_c = 3H_0^2/8\pi G$ [4], above which the Universe becomes closed; H_0 is Hubble constant, G is gravitation constant. Hence it follows that each type of the objects under consideration has own cosmological constant of action: the larger a structure embryo is, the larger K_c is.

Formally, it follows from ratio (1) that $\Delta a \propto \sqrt{\Delta t}$, *i.e.* the instantaneous rate of growth of object size is infinite:

$$\mathrm{d}a/\mathrm{d}t = \lim_{\Delta t\to 0} \Delta a/\Delta t \propto \lim_{\Delta t\to 0} 1/\sqrt{\Delta t} = \infty.$$

That means that formation of a large-scale Universe structure is of pulsed nature, *i.e.* there is development of initial perturbations in matter distribution as a result of instability of cosmological flows leading to appearance of non-uniformities. If perturbations are not large, it is reasonably to consider the mean size $\langle a\rangle$ of objects.

In case of intensive process of object interaction (at close distances or at large flows of dark mass-energy) the value Δt can be determined from the "energy-time" uncertainty relation:

$$\Delta t\cdot|\Delta E| \cong K_c. \tag{3}$$

Here ΔE is energy level width, corresponding to a specific process. Since in relations (1) and (3) we are dealing with object sizes that vary continuously with time, these relations can be considered as differential. From the relations (1) and (3) we obtain the following differential relation:

$$\frac{|\Delta\langle a\rangle|}{\Delta t} \cong \left(\frac{|\Delta E|}{2m}\right)^{1/2}. \tag{4}$$

From here it is obvious that in intensive processes the action constant becomes "virtual".

3. Calculations of Characteristic Sizes of Astrophysical and Cosmological Objects

We consider globular clusters consisting of red giants having an average mass of $m_g = 4M_S$, where M_S is the mass of the Sun [11]. In Newtonian approximation, the gravitational energy (work) necessary for separation of a giant from the surface of a globular cluster or for its capturing from space by this cluster surface, is equal in absolute value to

$$|\Delta E| = G\frac{2m_{bcl}m_g}{\langle a\rangle_{bcl}}. \tag{5}$$

Here m_{bcl} is the mass of globular cluster, $\langle a\rangle_{bcl}$ is its characteristic diameter. Substituting Equation (5) into Equation (4) and solving it in quadratures with regard to zero initial condition, we obtain the following relation between the characteristic diameter of globular cluster and the time:

$$\langle a\rangle_{bcl} \cong \left(\frac{3}{2}\right)^{2/3}\left(Gm_g\right)^{1/3} t^{2/3}. \tag{6}$$

With the Universe life time $t \sim H_0^{-1} \approx 1.4\times10^{10}$ years $= 4.41\times10^{17}$ s [1]-[3], the calculated characteristic diameter of globular clusters is equal to $\langle a\rangle_{bcl} \cong 6\times10^{18}$ m ≈ 200 ps (630 light years). The diameter of the well observed globular cluster Omega Centauri is equal to 600 light years [12].

Thus, the uncertainty principle works well enough within the range of astrophysical scales. This suggests that it might be used in cosmological scales. As Δt we accept a characteristic time of perturbation travel over a cosmological object: $\Delta t = \langle a\rangle/c$. One can expect that during this time interval the matter density in the object remains invariable. Substituting Equation (2) into relation (1) rewritten for the mean size, we obtain the following differential relation:

$$\langle a\rangle^2 \Delta\langle a\rangle = \frac{1}{2^{1/2}}\left(\frac{\rho_c}{\rho}\right)^{1/2} a_0^2 c\Delta t. \tag{7}$$

Here clusters are embryos for superclusters, and superclusters are embryos for the Universe.

Solving Equation (7) in quadratures with regard to the initial condition $\langle a\rangle(t=0)=0$, we obtain the following asymptotic dependence of the mean size of the object on time:

$$\langle a\rangle \cong \left(\frac{9}{2}\right)^{1/6}\left(\frac{\rho_c}{\rho}\right)^{1/6} a_0^{2/3}(ct)^{1/3}. \tag{8}$$

This dependence is valid up to the "moment" of time t_1, from which dark energy begins to prevail in matter density in the Universe. In accordance with generally accepted ideas [3], dark energy prevails in matter density for 3.5 billion of years, which is significantly less than the Universe life time.

Let us calculate consecutively theoretical sizes of the considered structures according with Equation (8) based on the data [1]: $\rho_c/\rho = 33$, $c = 3\times10^8\ \mathrm{m}\cdot\mathrm{c}^{-1}$. In accordance with ideas [2] [3], it is reasonably to accept galactic clusters as "the least element" of large-scaled structure of the Universe. A reliably observed mean size of galactic clusters is approximately equal to $\langle a\rangle_{cl} \approx 1.23\times10^{23}$ m ≈ 4 Mps [11]. Having accepted this value as a_0, we find out that with the time $t_1 \approx 10^{10}$ years $= 3.15\times10^{17}$ s, the calculated mean size of superclusters is approximately equal to $\langle a\rangle_{scl} \approx 2.59\times10^{24}$ m ≈ 84 Mps. It should be noted that the observable sizes of superclusters are in the range 30 - 100 Mps [3] [11]. And void sizes are approximately equal to sizes of superclusters [3]. The value $\langle a\rangle_{scl}$ can be assumed to be "low" in size ranging. From this point on, matter mass distribution in the Universe becomes quasi-uniform. Having taken this value as the size a_0 of embryo of a larger isotropic structure we find out that the calculated characteristic size of the Universe at the time t_1 is equal approximately to $\langle a\rangle_{Un} \approx 1.96\times10^{25}$ m ≈ 640 Mps. One can assume that upper boundary of the Universe is equal to

$$R_{un} = \langle a\rangle_{Un}/2 \approx 0.98\times10^{25}\ \mathrm{m} \approx 320\ \mathrm{Mps}.$$

It should be noted that the size R_{un} is significantly less than the "light" radius $R_{1\mathrm{light}} = ct_1 = 9.45\times10^{25}$ m ≈ 3060 Mps. This circumstance signifies the possibility for interactions within a group of universes, located in the region with the size R_{light}. By analogy with microcosm and mesocosm the exponent 1/3 attached to t in Equation (8) corresponds to Brownian motion of colliding objects, and the final result of collisions consists in irreversible aggregation of objects [13]. In terms of this paper the result of "collisions" between cosmological objects is in their interpenetrations occurring in near-surface regions and in appearance of dark matter and dark energy in these regions. This results in significant increase of expansion rate of the Universe. Substituting $|\Delta E| = mc^2$ into Equation (4), we obtain that starting with t_1 the time dependence of characteristic size of the Universe takes the following form:

$$\langle a\rangle_{Un} \cong \left(\frac{9}{2}\right)^{1/6}\left(\frac{\rho_c}{\rho}\right)^{1/6} a_{scl}^{2/3}\left(ct_1\right)^{1/3} + \frac{\sqrt{2}}{2}c\left(t-t_1\right). \tag{9}$$

The second term in Equation (9) expresses the ultimate ultrarelativistic stage of Universe expansion with the constant velocity $c/\sqrt{2}$. Hence it follows that the upper evaluation of characteristic size of the Universe "at present" is equal to $\langle a\rangle_{Un} \cong 1.96\times10^{25}\ \mathrm{m} + 2.3\times410^{25}\ \mathrm{m} = 4.28\times10^{25}\ \mathrm{m} \approx 1390$ Mps. This value is less than the up-to-date light radius which is equal to $R_{\mathrm{light}} = ct \sim cH_0^{-1} \sim 13.23\times10^{25}$ m ≈ 4290 Mps. This will produce an opportunity for interactions within the group of "closely spaced" universes, whose possible number is evaluated as $\left(R_{\mathrm{light}}/\langle a\rangle_{Un}\right)^3 \approx 30$.

4. Cosmic Sphere

Let us consider an issue about a size of closed cosmic sphere which contains a great number of groups of interacting universes. We will determine the ultimate value of cosmological action constant $K_c^{\lim}$ on the basis of analogy with the Planck mass $m_{pl} = (\hbar c/G)^{1/2}$, which is determined using action constant $\hbar$ in microcosm, gravitation constant G, and velocity of light c. For $K_c^{\lim}$ one can write down the following equation:

$$K_c^{\lim} = \frac{M_c^2 G}{c}. \tag{10}$$

Here, $M_c = m_p N_b\left(\rho_c/\rho\right) \approx 5.5\times10^{54}$ kg is cosmic mass determined as product of baryon mass and the ratio of critical density to baryon density; m_p is proton mass; $N_b \sim 10^{80}$ is number of baryons in the Universe [1]. The value M_c corresponds approximately to the cosmic mass 3.986×10^{54} kg, given in [14]. Reference [14] is devoted to values of physical constants. Substituting expression (10) into Equation (1) and fulfilling the procedure described above, and assuming that the least time period of a single act of interaction between the objects under consideration is equal to time of perturbation travel over galactic cluster (homogeneous structure element) $\Delta t = \langle a\rangle_{cl}/c$, we obtain the following asymptotic equation for cosmic sphere at present "moment" of time:

$$a_{cosm} = \left(\frac{75}{4\pi}\right)^{1/5}\left(\frac{M_c^2 G}{\rho\langle a\rangle_{cl}}\right)^{1/5} t^{2/5}. \tag{11}$$

The exponent 2/5 attached to t corresponds to the analogy with free-molecular flux of Brownian particles, *i.e.*, with irreversible growth of object as a result of successive addition of structure elements. With $\rho = 0.03\rho_c$ [1], $\rho_c \approx 10^{-26}\ \mathrm{kg \cdot m^{-3}}$ [3], we obtain that with the observed cluster size $\langle a \rangle_{cl} = 4$ Mps [11] and with the above stated life time of the Universe, the unknown quantity is equal to $a_{cosm} \approx 9 \times 10^{27}$ m, *i.e.*, a radius of closed cosmic sphere is equal to $R_{cosm} \approx 4.5 \times 10^{27}$ m. This value exceeds the "light" radius R_{light} by almost two orders of magnitude. Hence one can conclude that all interactions within the cosmic sphere can occur only between "neighboring" universes with sizes and distances between them which not exceed the distance travelled by light. The interacting universes are formed independently from each other, for example, as a result of great number of "Big bangs" distributed in space and time randomly inside the sphere (11). Number of universes formed independently inside the cosmic sphere can be evaluated as $\left(a_{cosm}/\langle a \rangle_{Un}\right)^3 \sim 10^7$. And the number of independent groups of interacting universes can be evaluated as $\left(a_{cosm}/R_{\text{light}}\right)^3 \sim 3 \times 10^5$. These values extend the notions [8] on large numbers in quantized cosmos.

5. Discussion of Results

A distinctive feature of this phenomenological approach is in assumption of the existence of many interacting universes in space with spherical Euclidian geometry and in introduction of the concept of cosmological action constant, the value of which depends on a scale of a cosmological object, which is an embryo for a larger structure under consideration. The given results do not contradict to well known ideas [1]-[3] about the sizes of astrophysical and cosmological objects. In addition, the second term in Equation (9) qualitatively reflects the presence of cosmological "repulsion" of objects (antigravitation) at a limiting ultrarelativistic stage of the Universe expansion at a constant velocity. This corresponds to the consensus about nonzero positive cosmological constant $\Lambda > 0$, mentioned in R. Penrose monograph [1].

To describe an intermediate stage of expansion affected by dark energy one can accept the value of mass in Equation (10) for cosmological constant of action as equal to the ultimately possible value $M_c^{\lim} = (\pi/6)\rho_c \langle a \rangle_{Un}^3$. Substituting this equation to relation (1) rewritten for the mean size $\langle a \rangle$, and fulfilling the procedure described in section 3 with $\Delta t = \langle a \rangle / c$, we obtain the following differential relation:

$$\frac{\Delta \langle a \rangle}{\Delta \tau} \cong \left(\frac{\rho_c^2}{\rho} \frac{G}{2} \right)^{1/2} \langle a \rangle. \tag{12}$$

Here $\tau = t - t_1$. Having solved Equation (12) in quadratures with the initial condition

$$\langle a \rangle (\tau = 0) = \left(\frac{9}{2} \right)^{1/6} \left(\frac{\rho_c}{\rho} \right)^{1/6} a_{scl}^{2/3} (ct_1)^{1/3}$$

(see Equation (9)), we obtain the following equation for the Universe size-time relation:

$$\langle a \rangle_{Un} \cong \left(\frac{9}{2} \right)^{1/6} \left(\frac{\rho_c}{\rho} \right)^{1/6} a_{scl}^{2/3} (ct_1)^{1/3} \exp\left(\frac{\rho_c^2}{\rho} \frac{G}{2} \right)^{1/2} (t - t_1). \tag{13}$$

This equation reflects accelerated expansion, and the type of time dependence of characteristic size of the Universe is in compliance with de Sitter cosmological model for Euclidean geometry at $\Lambda > 0$. The type of pre-exponential factor corresponds to Dirac cosmological model for Euclidean geometry at the atomic time scale when $\Lambda = 0$ [15]. In addition, relation (12) with the consideration for the above given ρ_c and ρ is rearranged to the following equation:

$$\frac{\mathrm{d} \langle a \rangle}{\mathrm{d} \tau} \cong 1.41 H_0 \langle a \rangle. \tag{14}$$

Equation (14) corresponds approximately to the Hubble law.

Thus, uncertainty principle and Cosmological principle do not contradict each other and are mutually complementary. The presented results correspond to fundamental concept [10] of stochastic quantum nature of cosmological phenomena.

From Equation (13) we obtain that the universe size at present stage is $\langle a \rangle_{Un} \cong 2.76 \times 10^{25}\ \text{m} \approx 900$ Mps.

Hence we get that the maximum number of universes inside the cosmic sphere is $\left(a_{cosm}/\langle a\rangle_{Un}\right)^3 \sim 3.5\times10^7$, and the maximum number of interacting universes in the region with the size equal to the light radius is equal to $\left(R_{\text{light}}/\langle a\rangle_{Un}\right)^3 \approx 110$. The existence of the great number of mutually interacting universes does not contradict to the statement about existence of a set of images of "unique sample" described by R. O. Bartini in [14].

The above results are obtained in one-dimensional approximation. With more complicated space-time geometry proposed in [14] the value of cosmic radius is $R_{cB} = 5.89\times10^{27}$ m. The discrepancy of the above value $R_{cosm} = 4.5\times10^{27}$ m and R_{cB} is not essential, since both values are much larger than the light radius. If one substitutes theoretical formation time of large-scale structure of the Universe (which is equal to 2.2×10^{10} years [3]) into Equation (11), then the value of the cosmic sphere radius will be $R_{cosm} = 5.39\times10^{27}$ m. The discrepancy with R_{cB} becomes insignificant.

It should be noted that the proposed approach does not assume any prohibitions on existence of great number of independent cosmic spheres (11) in infinite space.

6. Conclusion

Extended interpretation of uncertainty principle applied to the sizes of objects in cosmic scales in approximation of spherical symmetry in Euclidean geometry was proposed. Evidently first phenomenological outer-space constant of action was determined with regard to the sizes and masses of embryos of astrophysical and cosmological objects. The obtained adequate asymptotic time dependences of characteristic sizes of objects under consideration demonstrate that the "ordinary" space-time relations are valid for determination of cosmic boundaries. The presented results indicate the existence of the great number of independent groups of interacting universes. The proposed approach does not contradict to well-known ideas about stochastic quantum nature of cosmological phenomena and completes the existent notions on large numbers in quantized cosmos.

References

[1] Penrose, R. (2004) The Road to Reality. A Complete Guide to the Laws of the Universe. Jonathan Cape, London.

[2] Teerikorpi, P., *et al.* (2008) The Evolving Universe. Springer-Science, New York.

[3] Lukash, V.N., Mikheeva, E.V. and Malinovsky, A.M. (2011) Formation of the Large-Scale Structure of the Universe. *Physics-Uspekhi*, **54**, 983-1005. http://dx.doi.org/10.3367/UFNr.0181.201110a.1017

[4] Doroshkevich, A.G., Lukash, V.N. and Mikheeva, E.V. (2012) A Solution of the Problems of Cusps and Rotation Curves in Dark Matter Halos in the Cosmological Standard Model. *Physics-Uspekhi*, **55**, 3. http://dx.doi.org/10.3367/UFNe.0182.201201a.0003

[5] El Naschie, M. (2014) Cosmic Dark Energy Density from Classical Mechanics and Seemingly Redundant Riemannian Finitely Many Tensor Components of Einstein's General Relativity. *World Journal of Mechanics*, **4**, 153-156. http://dx.doi.org/10.4236/wjm.2014.46017

[6] Bish, R. (2014) Mass Density Distributions in Spiral Galaxies. *World Journal of Mechanics*, **4**, 137-152. http://dx.doi.org/10.4236/wjm.2014.45016

[7] Luo, P.A. (2014) Motion Law of Graviton in Three-Dimensional Space and Its Influence on Star Distribution of Spiral Galaxy. *Journal of Modern Physics*, **5**, 1910-1930. http://dx.doi.org/10.4236/jmp.2014.517186

[8] Ryazantsev, Yan (2013) The Large Numbers in a Quantized Universe. *Journal of Modern Physics*, **4**, 1647-1469. http://dx.doi.org/10.4236/jmp.2013.412205

[9] Berman, M.S. (2013) Heat Flow as Origin of Dark Energy. *Journal of Modern Physics*, **4**, 1468-1654. http://dx.doi.org/10.4236/jmp.2013.410175

[10] Shen, Zhi-Yuan (2013) A New Version of Unified Field Theory—Stochastic Quantum Space Theory on Particle Physics and Cosmology. *Journal of Modern Physics*, **4**, 1213-1380. http://dx.doi.org/10.4236/jmp.2013.410165

[11] Lyubarskii, Yu.E. and Syunyaev, R.A. (1991) Astronomy and Astrophysics. In: Grigor'ev, S. and Meilikhov, E.Z., Eds., *Physical Quantities Handbook*, Energoatomizdat, Moscow, 1223. (In Russian)

[12] Weldon Owen Ltd. (2007) Atlas of Universe. Weldon Owen Inc.

[13] Hellen, E.K. and Alava, M.J. (2002) Persistence in Cluster-Cluster Aggregation. *Physical Review E*, **66**, 026120. http://dx.doi.org/10.1103/PhysRevE.66.026120

[14] Bartini, R.O. (2005) Relations between Physical Constants. *Progress in Physics*, **3**, 34-40.

[15] Zel'manov A.L. (1962) Kosmology. In: Vvedenskii, B.A., Ed., *Physical Enciclopaediacal Dictionary*, Soviet Enciclopaedia, Moscow, Vol. 2, 491-501. (In Russian)

2

Splash Formation Due to a Frog Diving into Water

Yoshihiro Kubota[1*], Osamu Mochizuki[2]

[1]Faculty of Science and Engineering, Toyo University, Saitama, Japan
[2]Department of Biomedical Engineering, Toyo University, Saitama, Japan
Email: [*]kubota548@toyo.jp

Abstract

Herein, we present the results of our experimental investigation of splashes formed by a frog diving into water from the ground or from a leaf and the accompanying sound generated by the impact of the frog on the water. The experiments are performed by visualizing the flow with a high-speed camera. In addition, we used physical models comprising hydrophilic bodies made from hydrogel or acrylic resin to experimentally study how hydrophilicity affects the splash. In these experiments, we use the degree of swelling to define the hydrophilicity degree. The results show that different splashes are caused by the increase in water-film velocity upon an increase in hydrophilicity. For a body with strong hydrophilicity, at a relatively high film velocity, the water film forms when the body impacts the water surface separates from the body surface. In addition, an aircavity forms when the film separates from the body. We empirically examine the relation between the hydrophilicity degree and film velocity. The results indicate that increased hydrophilicity does not reduce the splash. Therefore, we conclude that reducing of the formation of water from the biomimetic point of view is related to the shape of body.

Keywords

Frog Dive, Splash Formation, Flow Visualization

1. Introduction

Herein, we present an experimental investigation of the splash formed by a frog diving into water from the ground or from a leaf and the accompanying sound generated by the impact of the frog on the water. The study focuses on understanding how the hydrophilicity of a body influences splash formation and the subsequent dy-

[*]Corresponding author.

namics, with the aim to understand how splash formation allows a frog to escape predators. The experiments are performed by visualizing the flow with a high-speed camera, with the sound being synchronously recorded. In addition, we experiment with physical objects of varying hydrophilicity to better understand the relation between splash formation and the hydrophilicity degree of a body.

In a related work, in 1929, Karman [1] studied the impact of a seaplane landing on a water surface [1], in which he analyzed the relation between the gliding angle and the landing speed of the seaplane. More recently, Do-Quang *et al.* numerically studied the water splash caused by a spherical body plunging into water [2] and found a relation between wetting phenomena, impact velocity, and splash formation. On the basis of an experimental study in which splash patters were observed for various impact velocities, Duez *et al.* [3] reported a relation between splashes and the accompanying sound. However, they did not discuss this phenomenon in detail. In a previous work [4], we studied the splashes formed by a solid sphere plunging into water [4] and categorized splash patterns according to impact velocity. In addition, we discussed in [4] the characteristics of the film flow generated at the early stages of the splash when a body impacted a water surface. We numerically studied the influence of strength on hydrophilicity in splash formation [5]. We computed the influence of hydrophilicity from the slip of flow.

A water droplet impinging on a solid surface also generates a splash. Yoon *et al.* studied the film-like structure of such splashes and its instability [6]. In a related work, Bejan *et al.* [7] and Bussmann *et al.* [8] used experimental and numerical means respectively to study the deformation of a droplet impinging on a solid surface. Bussmann *et al.* used the Rayleigh-Taylor instability theory to discuss the number of fingers that appeared at the edge of a spreading film. In our previous work [4], we show that instability is related to the number of droplets generated at the edge of the film flow. For a sphere plunging through a water surface, such droplets are generated by the primary splash that appears during the first stage of the splash. This work was related to that of Worthington [9], who investigated the milk crown caused by a droplet-liquid collision and the splash caused by a solid body falling into liquid milk [9]. Akers *et al.* [10] studied the splash of a non-Newtonian fluid [10] and showed how air cavities deform as a function of fluid properties. Ahn *et al.* investigated the influence of surface condition as the strength of hydrophobicity for splash erosion of soil model [11]. Finally, Truscott *et al.* [12] investigated the splash formed when a spinning spherical body plungeed into water, focusing particularly on the formation of underwater air cavities. Eddington *et al.* (2003) reported using the hydrogel as a valve for flow control on micro channels [13].

Thus, although the splash phenomenon is familiar, the detailed mechanisms behind it are not yet well understood. Because frogs dive into water often to escape predators and ensure their own survival, evolutionary reasoning suggests that their splash should be minimized. Thus, we used a high-speed camera to observe the various splashes formed by the diving of two frog species as a function of their habitat. We also recorded the sound generated by the splashes to investigate the relation between splash formation and the dive. In addition, to better understand how the hydrophilicity of a body with a hydrophilic surface (e.g., the body of a frog) affects the splash, we experimented with two homogeneous inanimate physical objects made of materials with different hydrophilicities. On the basis of these experimental results, we explain how the surface of an object affects its splash by modifying the process by which water films form. We also discuss splash formation as a function of the frog-dive technique and how hydrophilicity affects the splash and the film-flow velocity as a frog dives into water. Splashes were visualized using a high-speed CMOS camera. The tested materials are acrylic resin and a hydrogel as the hydrophilic body. These materials result in a distinctive affinity of body and water.

2. Splash Patterns

Figure 1 shows a tree diagram of splash patterns at different time intervals [4] [14] [15]. The evolution of the splash depends on the impact velocity. A splash is represented by the shadowed region in the figure. The three types of splashes considered are primary, secondary, and tertiary splashes. The primary splash appears first in the form of a film flow when the body touches the water surface and tiny droplets are generated from the edge of the film flow. The number of droplets can be determined by applying the instability theory to the film edge. The droplets fly out in a direction tangential to the film at the position where they detach. The secondary splash is a result of the primary splash. If the film flow follows the wall of the spherical object, then the edge of the film recombines at the top point of the object, and the secondary splash takes the form of a mushroom (see **Figure 1**). If the film flow does not follow the wall of the spherical object (*i.e.*, separated flow) but the film recombines

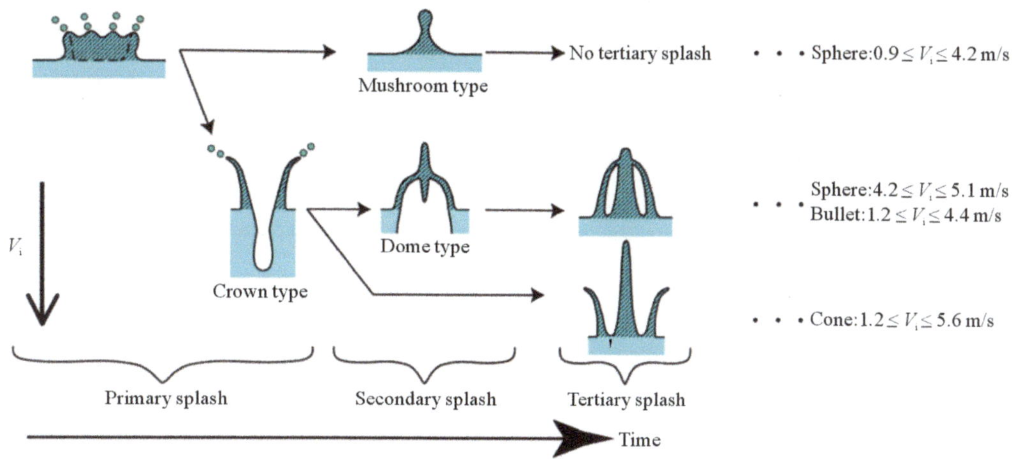

Figure 1. Tree diagram of splash formation for different impact velocities and object shapes.

above the object, the secondary splash takes the form of a dome. If the film does not recombine behind the object, no secondary splash appears. Finally, the tertiary splash appears because of the air cavity that forms when the object enters the water with sufficient speed. If no air cavity forms, no tertiary splash appears. The air cavity is related to the detachment of the film from the surface of the object. Thus, the primary splash strongly affects all subsequent splash processes.

In a previous work [14], we showed how the tail shape of the primary splash affects the secondary splash. More specifically, we reported how the shape of an object and film recombination are related to secondary-splash formation. The tail shape affects the volume of the secondary splash and the time delay between the primary and the secondary splash. We also investigated [15] how the head shape affects the formation of subsequent splash events by studying film formation as a function of the head shape. In that work, we examined the film thickness as a function of the momentum of the body and the film.

3. Experimental Setup

3.1. Splash from Frog Dive

Figure 2 shows that the sketch of experimental setup. The experiments with live frogs were performed in a water tank measuring 240 × 200 × 450 mm^3 and filled with tap water to a depth of 100 mm. Side-view sequential images were acquired by a high-speed CMOS camera(Vision Research Inc., Miro ex4).This camera allows 1200 images to be captured per second (1200 frames per second, or fps). A 500-W halogen lamp was used to illuminate the flow field. The sound due to the frog dives into water was recorded.

Two species of frogs, *Rana porosa* and *Hyla japonica*, were used in this experiment. *R. porosa* lives near flooded paddies, and *H. japonica* lives in trees. The different body shapes and surfaces are useful for understanding frog dives. For both frog species, the body length L (*i.e.*, from nose tip to rear of body) was 50 mm, and the body width W was 20 mm. The initial height of the diving board was 0 W, 2.5 W, and 5 W, where 0 W means that the diving board was at the level of the water surface. The frogs' impact velocity V_i was determined by image analysis and varied from 1.7 to 2.2 m/s. The frog motion is not restricted during the experiments.

3.2. Splash from Inanimate Objects

To observe splashes from a solid inanimate object plunging into water, we used a water tank measuring 300 × 300 × 400 mm^3 and filled it with tap water to a depth of 340 mm. In a previous work [4], we confirmed that contaminants common to tap water do not affect splash formation. To acquire sequential side-view images of splash formation, we used a high-speed CMOS camera (Vision Research Inc., Phantom v7.1) set at 4000 fps. The each frame was exposed for 0.25 ms. A 500-W halogen lamp was used to illuminate the scene.

Spherical objects having the diameter $D = 2r = W = 20$ mm were made from acrylic resin or hydrogel. The hydrogel served as agar gel in our experiment. The wetness of each object was varied to study how the hydrophilicity degree affects the splash. We define wetness (*i.e.*, the amount of water in the object) in terms of the swel-

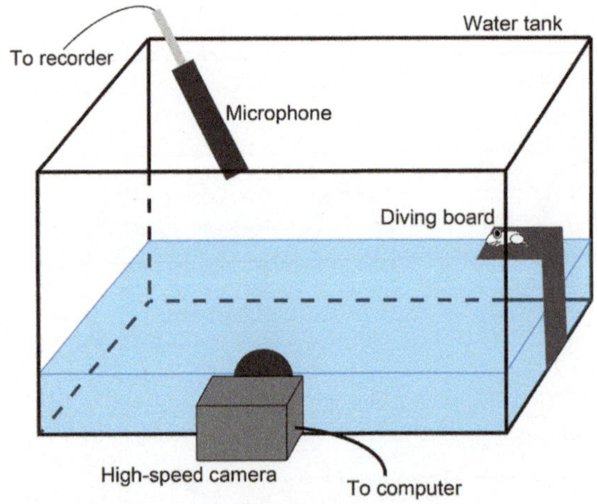

Figure 2. Sketch of experimental setup.

ling degree as follows:

$$s = \left(m_{\text{gel}} + m_{\text{water}}\right) / m_{\text{gel}} , \quad (1)$$

where $m_{\text{gel}}(m_{\text{water}})$ is the mass of gel(water). We consider swelling degrees s = 1, 50, and 100 (s = 1 means no water is contained in the object (e.g., an acrylic object)).

Before each experiment, we carefully wiped the surface of each object with paper towels to remove contaminants such as oil and to ensure similar surface conditions for each experiment. The impact velocity V_i of the object, which is the velocity of the object just before it enters the water, was varied by changing the height h from which it was dropped. Here h is defined as the distance from the water surface to the bottom of the object. The initial height h was varied from $1D$ to $45D$, which resulted in V_i ranging from 0.63 to 4.2 m/s. The value of V_i was calculated using $V_i = (2gh)^{1/2}$, which is based on the conservation of energy. Air drag was neglected in this calculation because of the low air speed involved. A launcher system using suction held the object at the initial height before release, and all objects fell into water without rotating and with no horizontal displacement. Thus, the experiments are all repeatable.

4. Experimental Results

We used three dimensionless parameters to understand splash formation: dimensionless time, Reynolds number, and Weber number. The dimensionless time T is defined as $T = tV_i / D$, where t is the measured time. This approach is useful for comparing how splashes form as a function of V_i. Even if V_i varies, the dimensionless time T = 1 indicates the same dimensionless time interval during which the object moves a distance of one diameter D. The dimensionless time T = 0 indicates the moment when the head of the objects touches the water surface. Further, Reynolds number Re and Weber number We are useful for understanding the relation between the kinetics of the object and the fluid characteristics. The former is defined as $Re = V_i D / v_i$, where v is the kinematic viscosity of water at 25˚C, and the latter is given by $We = \rho V_i^2 D / \sigma$, where ρ is the density and σ is the surface tension of water at 25˚C.

4.1. Splash from Frog Dive

Figure 3 shows consecutive images of the dive of an *R. porosa* frog into the water with an impact velocity of V_i = 1.7 m/s (h/W = 0) and the accompanying audio waveform. **Figure 3(a)** shows that the frog dives into the water head first. **Figure 3(f)** shows the final phase of the dive. The significant event in the dive is the air bubbles separating from the body. The audio waveform accompanying the dive is shown in the bottom part of **Figure 3** and reveals that two significant signal bursts are generated. The first audio signal is emitted during the phenomena shown in **Figure 3(a)** and **Figure 3(b)** and corresponds to a kinematics phenomenon, namely, the sound generated when a body impacts the water surface. The second, larger-amplitude audio signal is due to bubble formation.

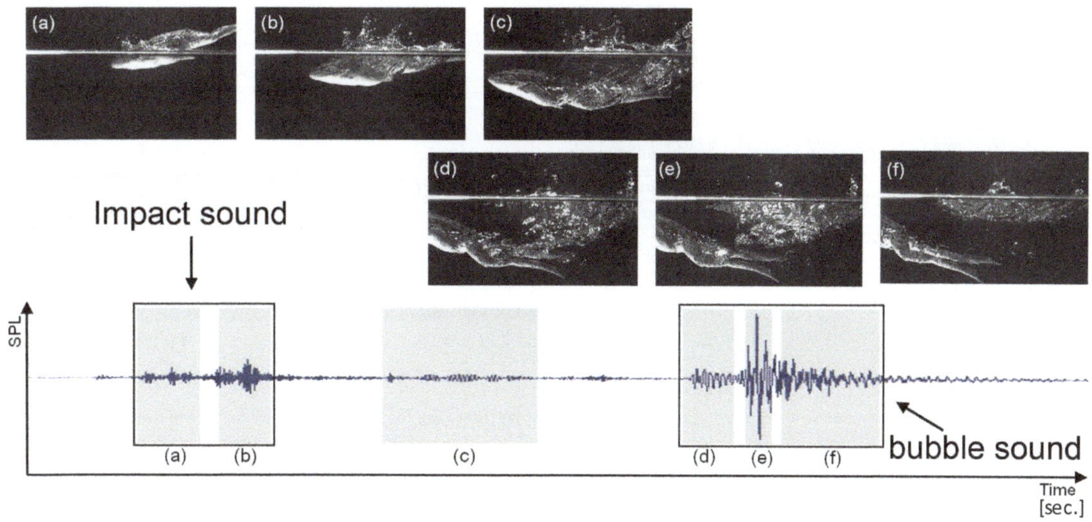

Figure 3. Images showingthe diving sequence of *R. porosa*. Shown at the bottom of the figure is the associated audio waveform.

Figure 4 shows the sequential images of the splash from the dive of an *H. japonica* frog. *H. japonica* enters the water first with its abdomen. Diving abdomen first produces a larger air cavity under the water, and the separation from the body of the cavity generates more small air bubbles than in the case of *R. porosa*. The audio wave form shown at the bottom of the figure again reveals two significant signal bursts, as in the case of the *R. porosa* dive. The initial audio burst occurs when with the body impacts the water and the second audio burst occur when the air bubbles form.

A stronger audio signal occurs at point (c) for *H. japonica* than at the corresponding point in the dive of *R. porosa*. This point corresponds to the formation of air bubbles, as seen clearly in the corresponding image. The different bubble formation for the two dives is due to the diving posture of each frog. Our previous works [12] [13] show that the shape of an object is related to the form of the air cavity created and the splash. Note that *R. porosa* lives near flooded paddies and that *H. japonica* lives in trees. This difference determines in large part how each frog escapes from its predators: *R. porosa* must escape as quietly as possible. Thus, the body posture during the dive influences the formation of air bubbles by determining at which point the air cavity separates from the body. In the next section, we use experiments with inanimate physical objects to discuss how the hydrophilicity of an object's surface affects the splash created when it plunges into water.

4.2. Effect of Splash Formation with a Surface Condition

The hydrophilicity degree of a body is quantified in terms of the swelling s, which is defined in Equation (1). **Figure 5** compares the splash formed by objects having the same shape (spherical) but differing in their degree of swelling. These images correspond to the dimensionless time $T = 2.0$. The acrylic body (**Figure 5(a)**) has $s = 0$ because it contains no water in its body. **Figure 5(b)** and **Figure 5(c)** show the results for the hydrogel objects with $s = 50$ and 100, respectively. For all three objects, a film forms when the object hits the water surface and the droplets are generated at the edge of the film. The splash for $s = 0$ is of the mushroom type, where the film recombines behind the object. A mushroom-type splash forms when no underwater air cavity forms. The splash for the hydrogel for $s = 50$ and 100 is of the dome type. The dome-type splash forms when the body completely penetrates the water surface before the film covers the body. Note that the depth of the air cavity differs for $s = 50$ and $s = 100$. This depth is greater for greater hydrophilicity (*i.e.*, for $s = 100$), which means that the hydrophilicity degree of an object's surface affects the velocity of film flow. To better understand the relation between film velocity and swelling, we measured the film velocity as a function of swelling.

Figure 5(a) shows, as a function of V_i, the maximum film velocity V_{fmax} measured at $T = 0.05$ and for $s = 0$ (red circles). For comparison, the maximum film velocity for $s = 50$ and 100 was measured at $V_i = 3.1$ m/s (triangle, square, and cross). From the result for $s = 0$, we obtain $V_{fmax} = 8.3V_i$. The maximum film velocity for the hydrophilic object at $V_i = 3.1$ m/s is more than twice that for $s = 0$ (*i.e.*, acrylic object), which we attribute to the

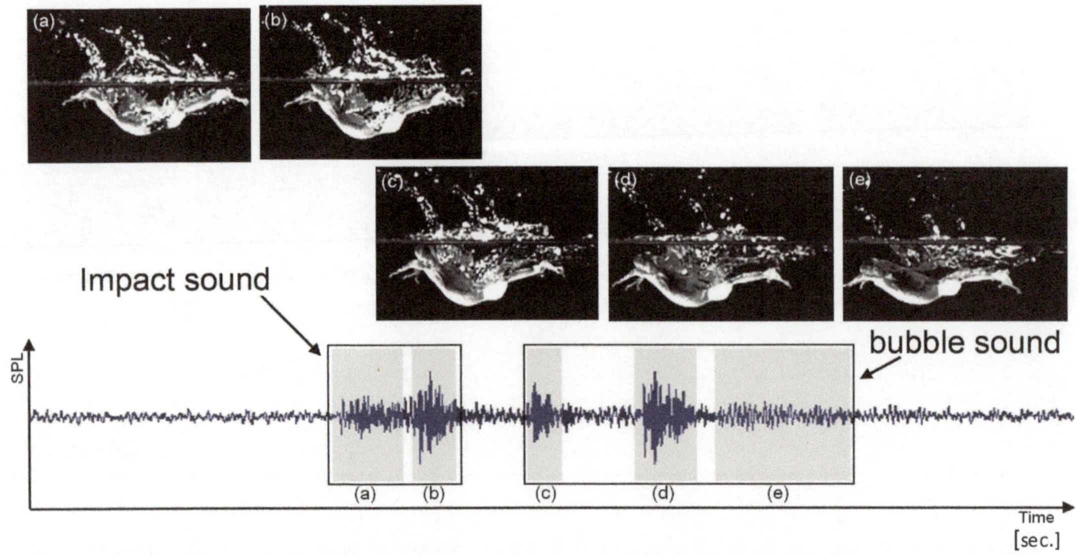

Figure 4. Images showingthe diving sequence of *H. japonica*. Shown at the bottom of the figure is the associated audio wave form.

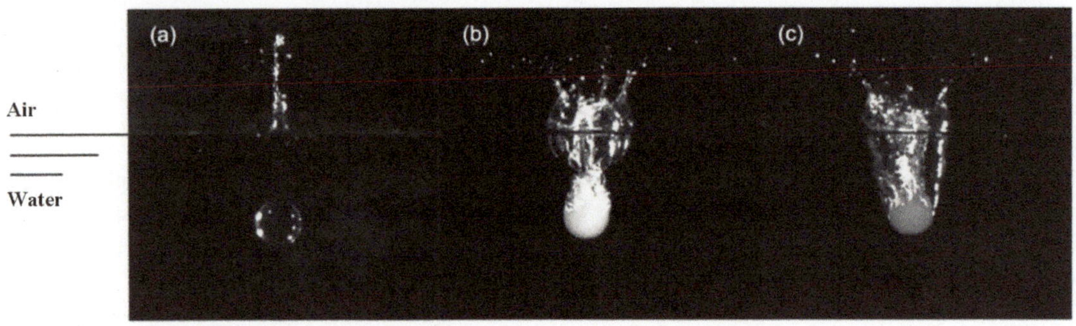

Figure 5. Comparison of splash formation at $V_i = 3.1$ m/s. (a) $s = 0$ (acrylic resin); (b) $s = 50$; and (c) $s = 100$.

film separating from the surface of the object [4]. The film separates when the centrifugal force acting on the film becomes greater than the surface tension of the film [4] [15]. The film velocity increases with swelling, which indicates that hydrophilicity is related to splash formation. Thus, we need to understand the relation between the hydrophilicity degree and splash formation.

Figure 6(b) shows the relation between film velocity and swelling. The red circles show the results of experiments, and the blue line shows the empirical function $V_{fmax} = 0.98\ \ln(s)$. The empirical function is obtained on the basis of the evidence shown in **Figure 5** that film velocity increases with hydrophilicity. **Figure 5** also reveals that the depth of the air cavity increases with an increase in swelling, which means that amore hydrophilic surface leads to a deeper air cavity. Thus, the air cavity must be considered in addition to the impact velocity to determine the potential energy of the system and the film velocity. Therefore, a combination of impact velocity and film velocity affects the tertiary splash, because the tertiary splash forms from the air cavity. In addition, the empirical curve is consistent with the experimental data. Thus, the film velocity of a hydrophilic object can be expressed as a logarithmic function of the swelling of the object.

If the impact velocity is less than 0.9 m/s, then a spire-type splash forms because of the concavity of the water surface [4]. From the relation between film velocity and impact velocity, we infer that the film velocity at a low impact velocity increases. This means that the impact-velocity threshold between spire- and mushroom-type splashes ($V_i < 0.9$ m/s at $s = 0$) changes as a function of the hydrophilicity of an object's surface. To confirm the increase in film velocity, we visualize the splash. **Figure 7(a)** shows that aspire-type splash forms for $s = 0$, and **Figure 7(b)** shows the result for $s = 50$. The spire-type splash in **Figure 7(a)** formed because of a reaction to the concavity of the film. No spire-type splash is seen in **Figure 7(b)**, because the film does not separate from the

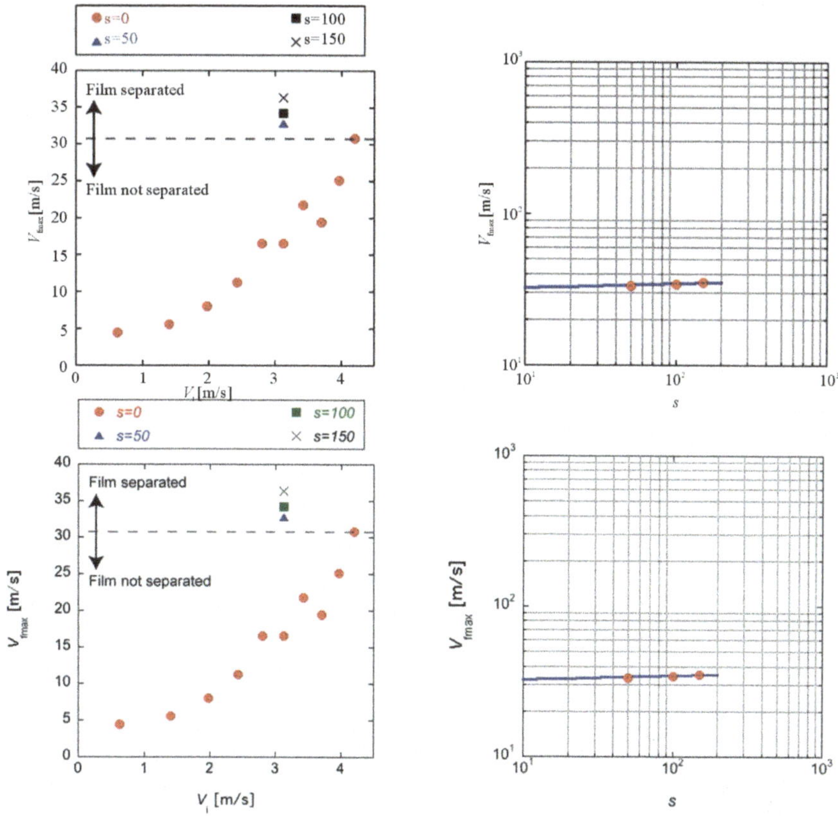

Figure 6. Effect of hydrophilicity on film velocity.

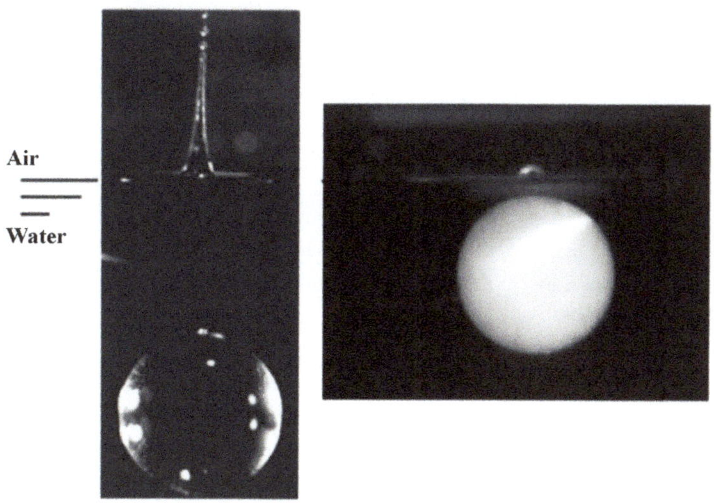

(a) (b)

Figure 7. Splash for V_i = 0.63 m/s and (a) s = 0 (acrylic object), and (b) s = 50 (hydrophilic object).

object before the concavity forms. This result indicates the importance of understanding the relation between film velocity and impact velocity.

5. Concluding Remarks

We observed splash formation for a frog diving into water and for an inanimate solid object falling into water.

The splash for the frog dive was caused by the impact of the frog's body and by the reaction to the separation of the air cavity from the body. The sound generated by the frog dive was strongly related to the formation of air bubbles, which was due to the collapse of the air cavity.

We used a high-speed CMOS camera to observe the splash of a hydrophilic object. The results showed that an object's hydrophilicity affects splash formation. Further, the hydrophilicity degree varied with swelling. We studied splashes from swelling $s = 0$, 50, and 100, where swelling indicated the amount of water inside the object ($s = 0$ for the acrylic object). To study the effect of hydrophilicity, we used objects made of hydrogel. Changing of the objects' surface conditions allowed us to study how they affect the film flow.

The results showed that the secondary splash depended on the hydrophilicity and on whether the film recombines behind the object. The acrylic object ($s = 0$) creates a mushroom-type splash, and the hydrogel objects with $s = 50$ and 100 createdome-type splashes. Whether a mushroom-type splash or a dome-type splash occurrence was determined by whether the film separated from the surface of the object. Film separation led to the formation of an air cavity, which was related to the formation of the tertiary splash. Thus, the hydrogel objects generated an additional splash in the form of a tertiary splash.

The film separated because of excessive film velocity. We obtained an empirical logarithmic expression for film velocity as a function of swelling. Thus, hydrophilicity affected film velocity, which in turn determined whether the film separated from the object's surface. Film separation led to a tertiary splash.

By comparing the results of the frog dive with those of the experiment, we found that the significant bubble entrainment was caused by *H. japonica*, which revealed the importance of shape for experiments on splash formation. In other words, the hydrophilic characteristic of the surface of an object affected splash formation. The splash caused by *H. japonica* would thus be reduced when it adopted an appropriate body position.

References

[1] von Karman, T. (1929) The Impact on Seaplane Floats during Landing. National Advisory Committee for Aeronautics Technical, Note 321.

[2] Minh, D.-Q. and Amberg, G. (2009) The Splash of a Solid Sphere Impacting on a Liquid Surface: Numerical Simulation of the Influence of Wetting. *Physics of Fluid*, **21**, Article ID: 022102. http://dx.doi.org/10.1063/1.3073968

[3] Duez, C.Y., Clanet, C. and Bocquet, L. (2007) Making a Splash with Water Repellency. *Nature Physics*, **3**, 180-183. http://dx.doi.org/10.1038/nphys545

[4] Kubota, Y. and Mochizuki, O. (2009) Splash Formation by a Spherical Body Plunging into Water. *Journal of Visualization*, **12**, 339-345. http://dx.doi.org/10.1007/BF03181877

[5] Yokoyama, M., Kubota, Y., Kikuchi, K., Yagawa, G. and Mochizuki, O. (2014) Some Remarks on Surface Conditions of Solid Body Plunging into Water with Particle Method. *Advanced Modeling and Simulation in Engineering Sciences*, **1**, 1-14. http://dx.doi.org/10.1186/2213-7467-1-9

[6] Yoon, S.S., Jepsen, R.A., Nissen, M.R. and O'Hern, T.J. (2007) Experimental Investigation on Splashing and Nonlinear Fingerlike Instability of Large Water Drops. *Journal of Fluids and Structure*, **23**, 101-115. http://dx.doi.org/10.1016/j.jfluidstructs.2006.08.009

[7] Bejan, A. and Gobin, D. (2006) Constructal Theory of Droplet Impact Theory. *International Journal of Heat and Mass Transfer*, **49**, 2412-2419. http://dx.doi.org/10.1016/j.ijheatmasstransfer.2006.02.001

[8] Bussmann, M., Chandra, S. and Mostaghimi, J. (2000) Modeling the Splash of a Droplet Impacting a Solid Surface. *Physics of Fluids*, **12**, 3121-3132. http://dx.doi.org/10.1063/1.1321258

[9] Worthington, A.M. (1882) On Impact with a Liquid Surface. *Proceedings of the Royal Society of London*, **34**, 217-230. http://dx.doi.org/10.1098/rspl.1882.0035

[10] Akers, B. and Belmote, A. (2006) Impact Dynamics of a Solid Sphere Falling into a Viscoelastic Micellar Fluid. *Journal of Non-Newtonian Fluid Mechanics*, **135**, 97-108. http://dx.doi.org/10.1016/j.jnnfm.2006.01.004

[11] Ahn, S., Doer, S.H., Douglas, P., Bryant, R., Hamlett, C.A.E., McHale, G., Newton, M.I. and Shircliffe, N.J. (2013) Effects of Hydrophobicity on Splash Erosion of Model Soil Particles by a Single Water Drop Impact. *Earth Surface Processes and Landforms*, **38**, 1225-1233. http://dx.doi.org/10.1002/esp.3364

[12] Truscott, T.T. and Techt, A.H. (2006) Cavity Formation in the Wake of Spinning of Sphere Impacting the Free Surface. *Physic of Fluids*, **18**, Article ID: 091113. http://dx.doi.org/10.1063/1.2335903

[13] Eddington, D.T. and Beebe, D.J. (2004) Flow Control with Hydrogels. *Advanced Drug Delivery Reviews*, **56**, 199-210. http://dx.doi.org/10.1016/j.addr.2003.08.013

[14] Kubota, Y. and Mochizuki, O. (2010) Elemental Structure of Splash Generated by a Plunging Solid Body. *Journal of Flow Visualization and Image Processing*, **17**, 359-369. http://dx.doi.org/10.1615/JFlowVisImageProc.v17.i4.70

[15] Kubota, Y. and Mochizuki, O. (2011) Influence of Head Shape of Solid Body Plunging into Water on Splash Formation. *Journal of Visualization*, **14**, 111-119. http://dx.doi.org/10.1007/s12650-011-0071-4

3

Impact of the Drag Force and the Magnus Effect on the Trajectory of a Baseball

Haiduke Sarafian

The Pennsylvania State University, University College, York, PA, USA
Email: has2@psu.edu

Abstract

We consider the impact of drag force and the Magnus effect on the motion of a baseball. Quantitatively we show how the speed-dependent drag coefficient alters the trajectory of the ball. For the Magnus effect we envision a scenario where the rotation of the ball confines the Magnus force to the vertical plane; gravity, drag force and the Magnus force make a trio-planar system. We investigate the interplay of these forces on the trajectories.

Keywords

Drag Force, Magnus Effect, Spinning Ball, Baseball, Nonlinear Physics, *Mathematica*

1. Motivation and Introduction

In introductory physics, engineering and math courses, undergraduate students are traditionally introduced to the concept of projectile motion such that a projectile is thrown at an angle in a vertical plane in a vacuum. The motion in the vacuum is analyzed because in the absence of air, utilizing the Newton's second law yields to a set of two independent second-order linear ODEs. The trivial solution of these equations provides information about the kinematics of the projectile, such as the trajectory, range, time of flight, etc. [1]. If one applies the learned concepts to a real-life situation such as a pitched and/or a batted baseball, one disappointingly would be surprised realizing the gross discrepancies of the two scenarios. For instance, a 45° batted baseball does not maximize the range, and the trajectory of a baseball is not a symmetric parabola.

The motivation of this article stems from the former question, namely "At what angle above the horizontal a baseball should be batted making the range a maximum?" To answer this question, we analyze the problem systematically. In the course of our analysis, we stumbled upon a few interesting related issues. To the author's amazement, the analysis led also to uncharted territories unveiling features of a flying baseball that to date have not been reported in scientific literature. In this article we confirm the previously reported quantitative results and extend the investigation for the "what-if" scenarios.

2. Physics of a Projectile in the Air; The Impact of Air Resistance and the Spin

We consider a ball projected in the air at an arbitrary initial angle above the horizontal. In addition to gravity the ball encounters air resistance. Irrespective of its speed, the drag force (the air resistance) acts in the opposite direction of the motion retarding its movement. In **Figure 1**, a snapshot of a flying ball with its instantaneous velocity and the drag force are shown with $\boldsymbol{v}$ and $\boldsymbol{F}_D$, respectively. In practice a batted ball also spins; it may spin backward (backspin) or forwards (topspin). We quantify the spin by its angular velocity $\boldsymbol{\omega}$. We consider a spinning ball with angular velocity vector perpendicular to the vertical plane. Hence, a back-spin ball orients its $\boldsymbol{\omega}$ parallel to the ground and outward to the vertical plane. Conversely, a top-spun ball orients its $\boldsymbol{\omega}$ parallel and inward to the ground. **Figure 1** displays a back-spinning (counter-clockwise rotating) ball. The dot at the center of the ball represents the head of the $\boldsymbol{\omega}$ arrow emerging the xz-plane. For the chosen scenario, the spinning ball is subject to an additional force; a spin-dependent force, namely the Magnus force [2]. This force is in proportion to $\boldsymbol{F}_M \sim \boldsymbol{\omega} \times \boldsymbol{v}$. Accordingly, this force also lies in the vertical plane making the three active forces namely, $\boldsymbol{w}$, $\boldsymbol{F}_D$ and $\boldsymbol{F}_M$ a trio coplanar. For the given scenario the ball stays in the vertical plane, and therefore, the analysis of the problem becomes two-dimensional.

Figure 1 visually assists distinguishing the impact of each individual force. For instance, it shows the horizontal components of the drag and the Magnus force collectively are acting along the opposite direction of the motion; they have a retarding impact. While the vertical component of the drag force makes the baseball "heavier", the vertical component of the Magnus force acting along the upward direction acts as it makes the baseball "lighter". It is this interplay of the components of the forces that makes the analysis of the problem challenging. The relative quantitative strengths of these components depend on the parameters describing the flying ball and the media. In Section 3 for a chosen set of parameters we address these issues. It needs to be mentioned that for a baseball thrown in a vacuum there are no such challenges; the only active force is gravity!

Here we consider the characteristics of the last two forces. For speeds relevant to the MLB (Main League

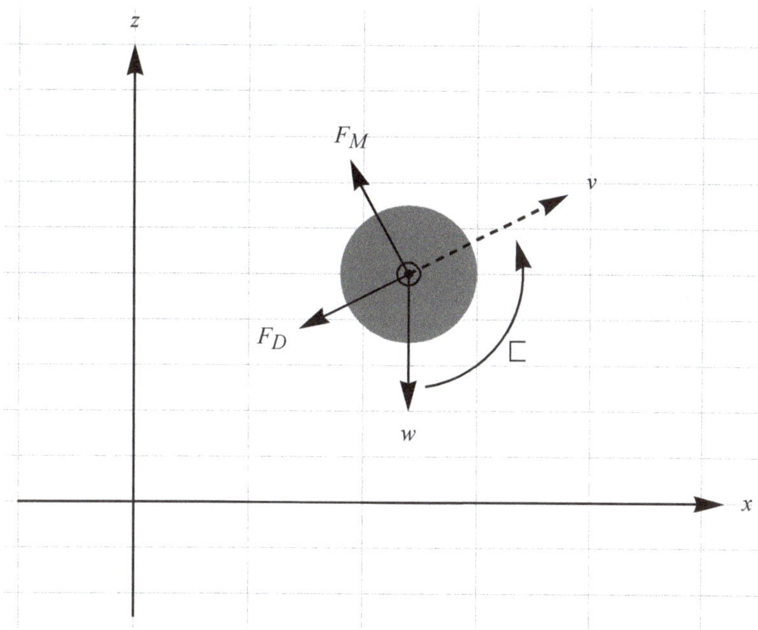

Figure 1. The profile of a flying baseball. Its back-spin (counter-clockwise rotation) is shown with a vector piercing outward and perpendicular to the vertical xz-plane. The head of the angular velocity vector $\boldsymbol{\omega}$ is shown with a black dot at the center of the ball circumvented with a gray circle. The dashed line is the instantaneous velocity of the baseball; the solid arrows are the active forces. From the top in the counter-clockwise direction the lines are: the Magnus force $\boldsymbol{F}_M$, the Drag force $\boldsymbol{F}_D$ and the weight W.

Baseball) the airflow around a baseball is considered as laminar [3]; therefore, the drag force is formatted according to $\boldsymbol{F}_D = \frac{1}{2} C_D \rho A v(-\boldsymbol{v})$. This equation explicitly shows the drag force at any instance is along the opposite direction of motion. The parameters in this equation are C_D, ρ, A, the drag coefficient, the density of the media (air) and the cross sectional area of the ball *i.e.* πR^2, where R is the radius of the ball. Utilizing a Computer Algebra System (CAS) recently this equation has been applied to a "real-life" situation. However, assuming a speed independent coefficient the impact of the drag force has been trivialized leading to inaccurate conclusions [4]. On the other hand, considering speed dependent real-life data, the drag coefficient and therefore the drag force is parameterized leading to a good phenomenological approximation [5]; in our analysis we utilize the latter. In the course of our search we encounter $\frac{1}{m} F_D$ where m is the mass of the baseball. According to [5], this ratio is equated to $\frac{1}{m} F_D = f(v) v^2$, where $f(v) \equiv \frac{1}{2m} C_D \rho A = 0.0039 + \frac{0.0058}{1 + e^{\frac{v - v_d}{\Delta}}}$ with $v_d = 35$ and $\Delta = 5$ both in m/s units. Noting in SI, $f(v)$ has the units of $\left[f(v) \right]_{\mathrm{SI}} = \mathrm{met}^{-1}$. The author being a nuclear physicist realizes that aside from an overall sign this function is somewhat similar to the potential describing certain nuclear reactions [6]. Given this parametrization we solve $f(v)$ for C_D; this yields $C_D = (2m)/(rA) f(v)$. Noting, for slow speeds, C_D approaches ~0.52 *i.e.* $\mathrm{Limit}_{v \to \mathrm{small}} C_D^{\mathrm{TM}} 0.52$. Utilizing the density of the air and the radius of a baseball in SI units, namely $\{\rho, R\} = \{1.23\,\mathrm{kg/m^3}, 3.64 \times 10^{-2}\,\mathrm{m}\}$ **Figure 2** displays how C_D varies vs. the practical range of speed.

According to **Figure 2**, there are three ranges of interest: slow, medium and fast. These correspond to the speeds: $10 < v(\mathrm{m/s}) < 20$, $20 < v(\mathrm{m/s}) < 50$ and $v(\mathrm{m/s}) > 50$, respectively. The corresponding C_D's for the first and the last range are constants and are as high as 0.52 and as low as 0.22, respectively. The middle range is the transition range where the C_D varies from the high value of 0.52 to the low value of 0.22. Consequently, a baseball batted at a typical MLB speed ~88 MPH (=37 m/s) encounters a weak drag force at the beginning, and while traversing along its trajectory continuously slows down encountering a relatively strong drag force. In short, the impact of the drag force is not constant. One intuitively expects the impact of the drag force at the beginning to be less than at the end of the flight. In Section 3, quantitatively we validate our intuition.

The other force is the Magnus force, for a baseball it is parametrized as $\boldsymbol{F}_M = S \boldsymbol{\omega} \times \boldsymbol{v}$, where S is independent of speed and is a constant [5]. The S has a dimension of mass and in SI units is $[S]_{\mathrm{SI}} = \mathrm{kg}$. It is interesting to note in most literature this force is formulated as $F_M \sim \omega v^2$. However, the latter format simply reduces to the former. To see the equivalence of these two formats one may think that in a resting state of the baseball where

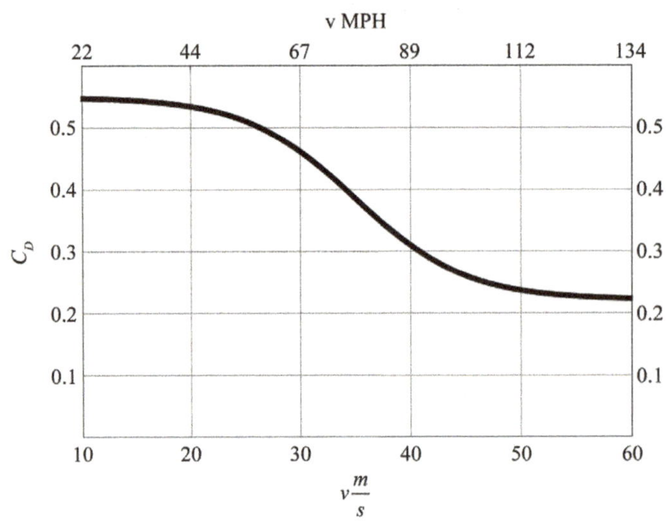

Figure 2. Variation of the drag coefficient for a baseball vs. speed. The lower horizontal axis is in m/s; the upper horizontal axis is the corresponding speed in MPH.

the ball is standing still and the air passes it by with the velocity $\boldsymbol{v}$ from right to left; see **Figure 1**. A point at the top edge of the ball moves with a fast speed, $v+R\omega$, while its counterpoint at the bottom of the baseball moves with a slower speed, $v-R\omega$. The difference between these two speeds according to v^2 format yields $\sim(v+R\omega)^2-(v-R\omega)^2\sim R\omega v$. Therefore, the S is the lump sum of the relevant factors: the lift coefficient, the cross section area of the baseball and the density of the air yielding the former format, namely $F_M=S\omega v$. To determine the orientation of the Magnus force one also may utilize Bernoulli's fluid mechanics law [1]. Paraphrasing, the law states "a fast moving fluid corresponds to a low pressure and vice versa". Since the top points on a ball move at a higher speed vs. the bottom points, the pressure difference would have a tendency to push the ball upward. Consequently, the corresponding force for a back-spinning baseball is $\boldsymbol{F}_D=S\boldsymbol{\omega}\times\boldsymbol{v}$; its direction is depicted in **Figure 1**. In practice, similar to the drag force we encounter the value of $\frac{1}{m}F_M$; this is written as $\frac{1}{m}\boldsymbol{F}_M=B\boldsymbol{\omega}\times\boldsymbol{v}$ where phenomenological unit-less value B for a baseball of mass m = 143 g is $B = 4.1\times10^{-4}$ [5]. The other relevant factor to the Magnus force is the spin value, ω. It is noted [7] that a pitched baseball on its way to a batter spins once per five feet. Knowing the distance between the pitcher's mound, home base and a typical flight time yield to an accepted value ω = 1800 RPM. Furthermore, it is also assumed a pitched or a batted baseball on its trajectory sustains its spin. In other words the value of ω is constant. With this detailed phenomenological information about the parameters describing the drag and the Magnus force in the next section we cast the equations describing the motion of the baseball.

3. Formulation of the Physics of a Batted Baseball

As we explained in the previous section a batted back-spun baseball with its initial angular velocity vector parallel to the horizontal flies and stays in a vertical plane. A vertical plane is the one that contains the initial velocity vector. In other words, the motion of the ball occurs in a 2D space. One such plane with a Cartesian xz coordinate is shown in **Figure 1**. We utilize the active forces depicted in **Figure 1** and compose the equations conducive to the analysis of the motion of the batted ball. Utilizing Newton's second law we write $\boldsymbol{F}_{\text{net}}=m\boldsymbol{a}$. This equation explicitly reads, $m\ddot{\boldsymbol{r}}=\boldsymbol{W}+\boldsymbol{F}_D+\boldsymbol{F}_M$, where the over double-dots means the second order time derivative. We substitute the forces from the previous section yielding $m\ddot{\boldsymbol{r}}=m\boldsymbol{g}+mf(v)v(-\boldsymbol{v})+S\boldsymbol{\omega}\times\boldsymbol{v}$. Since the velocity stays always in the vertical plane we write $\boldsymbol{v}=\{\dot{x},0,\dot{z}\}$. On the other hand, as we discussed the angular velocity sustains its character, meaning, its value and its orientation stay the same. In a right handed coordinate system shown in **Figure 1**, we write $\boldsymbol{\omega}=\{0,-\omega,0\}$. This yields, $\boldsymbol{\omega}\times\boldsymbol{v}=\{0,-\omega,0\}\times\{\dot{x},0,\dot{z}\}=\{-\omega\dot{z},0,\omega\dot{x}\}$. The x and the z components of the force equation yields,

$$\ddot{x}=-f(v)v\dot{x}-B\omega\dot{z} \tag{1}$$

$$\ddot{z}=-g-f(v)v\dot{z}+B\omega\dot{x}, \tag{2}$$

Note the last term of Equations (1), (2) are only operative for a spinning baseball. In other words, the equations associated with a non-spinning baseball are simpler than those given by Equations (1), (2). For a back-spun ball the horizontal components of the drag and the Magnus force are constructively additive, while along the vertical direction they are destructive. Equations (1), (2) describe the motion of the baseball. These are highly coupled non-linear ODEs. The solution of these equations are conducive to a set of explicit time dependent coordinates, namely $\{x(t),z(t)\}$. To highlight the degree of difficulty and the challenge of the task at head, we write these equations explicitly.

$$\ddot{x}=-\left(\alpha+\frac{\beta}{1+\mathrm{e}^{\frac{1}{\Delta}\left(\sqrt{\dot{x}^2+\dot{z}^2}-v_d\right)}}\right)\sqrt{\dot{x}^2+\dot{z}^2}\,\dot{x}-B\omega\dot{z} \tag{3}$$

$$\ddot{z}=-g-\left(\alpha+\frac{\beta}{1+\mathrm{e}^{\frac{1}{\Delta}\left(\sqrt{\dot{x}^2+\dot{z}^2}-v_d\right)}}\right)\sqrt{\dot{x}^2+\dot{z}^2}\,\dot{z}+B\omega\dot{x} \tag{4}$$

where α and β have dimensions of L^{-1} while v_d and Δ have dimensions L T^{-1}; all four parameters are constant. It is obvious that a set of such equations may not have symbolic solutions! Despite of our insistence we were unable to convert these equations to solvable entities. We then apply *Mathematica* symbolic DSolve command, but it also fails to produce analytic solutions. As a last resource we attempt solving them numerically. We introduce a set of meaningful, practical initial conditions and utilize *Mathematica*'s numeric routine, namely, NDSolve. This simple seven letter word successfully produces an output. Utilizing **Figure 2**, we consider a typical initial speed of 88 MPH which corresponds to 37 m/s. Our first objective is to evaluate the initial batted angle that maximizes the range; therefore, in our calculation we treat the initial angle as a variable. Also, we insert two switches in the code. By turning them on and off we include or exclude the drag and the Magnus forces one at a time. The switches are listed as $\{\alpha D, \beta M\}$. This ordered list with possible variations are: $\{0,0\}$, $\{0,1\}$, $\{1,0\}$ and $\{1,1\}$. $\{\text{drag}=0, \text{Magnus}=0\}$ represents the vacuum; $\{\text{drag}=0, \text{Magnus}\neq 0\}$ represents the impact of the pure Magnus effect; $\{\text{drag}\neq 0, \text{Magnus}=0\}$ represents the impact of the pure drag force; and $\{\text{drag}\neq 0, \text{Magnus}\neq 0\}$ represents the impact of the combined drag and the Magnus forces, respectively. The relevant parameters in SI units are:

$$\text{values}=\Big\{\alpha\to 0.0039,\ \beta\to 0.0058,\ v_d\to 35,\ \Delta\to 5,\ \rho\to 1.23,\ R\to 36.410^{-3},$$

$$m\to 0.145,\ g\to 9.8,\ Bb\to 4\times 110^{-4},\ \omega\to 1800\cdot\frac{\pi}{30},\ v_0\to 37\Big\}$$

The initial condition is comprised of the specifications of the initial velocity; on the vertical *xz*-plane this is $v_0=\{v_0\text{Cos}[\theta],0,v_0\text{Sin}[\theta]\}$. For four possible switch settings and for each of the chosen initial angles within the range of {10, 20, 30, 40, 50, 60, 70, 80, 90}° we solve the corresponding equations. Utilizing these solutions we also deduce the time of flight.

The initial angles are discrete, *i.e.* they are 10° apart and so are their corresponding ranges. Utilizing Interpolation command we fill in the gap and generate the needed continuous coordinates. With these continuous coordinates on hand we display the discrete solutions along with the interpolated continuous spectrum. These are shown in **Figure 3**.

With these plots it is easy to draw a few conclusions. For instance, the first and the last plots depict the two extreme scenarios. The first graph corresponds to the vacuum, and the last graph shows the impact of the combined drag and the Magnus effects. The difference between these two scenarios is quite pronounced. Their maximum range not only occurs at two different angles, but their numeric values are almost off by a factor of two. Similarly, one may also compare the second dot to the third plot. Both figures describe the impact of the media; the second plot is with the pure Magnus effect, while the third plot shows the impact of the pure drag force. Although these scenarios do not represent real-life situations they illustrate the impact of the individual forces.

Furthermore, with the interpolated continuous function at hand we search for the initial angle that maximizes the range. The output of the search is tabulated in **Table 1**.

The first column of **Table 1** shows the maximum range occurs at 45°; this is a "classics" knowledge and is an indicative of the accuracy of our analysis. For a real-life scenario, the fourth column shows quite a different result. For a projectile batted at 37 m/s with a 1800 RPM the angle that maximizes the range is 35° *i.e.* it is less by 10° vs. the classic value. The numeric entries in the second and third columns are helpful to draw conclusions about the impact of the individual active forces.

Utilizing the corresponding time of fight in **Figure 4** we display time of flights vs. the departure angles.

These figures are self-explanatory. The headings are the same as in **Figure 3**. These plots clearly show the impact of the forces on the time of flight.

Finally, utilizing the solutions of Equations (1), (2), namely $\{x(t),y(t)\}$ we apply ParametricPlot command and plot the trajectories of the balls. These are shown in **Figure 5**.

Table 1. The first row is the maximum range in meters; the second row shows the corresponding initial angle in degrees. The heading of the matrix explicitly describes the scenario of interest.

	$\{\alpha D,\beta M\}=00$	$\{\alpha D,\beta M\}=01$	$\{\alpha D,\beta M\}=10$	$\{\alpha D,\beta M\}=11$
"max range, m-->"	139.694	162.579	78.2832	84.095
$\theta°$ "-->"	$\theta\to 45$	$\theta\to 38.3973$	$\theta\to 39.4280$	$\theta\to 35.1726$

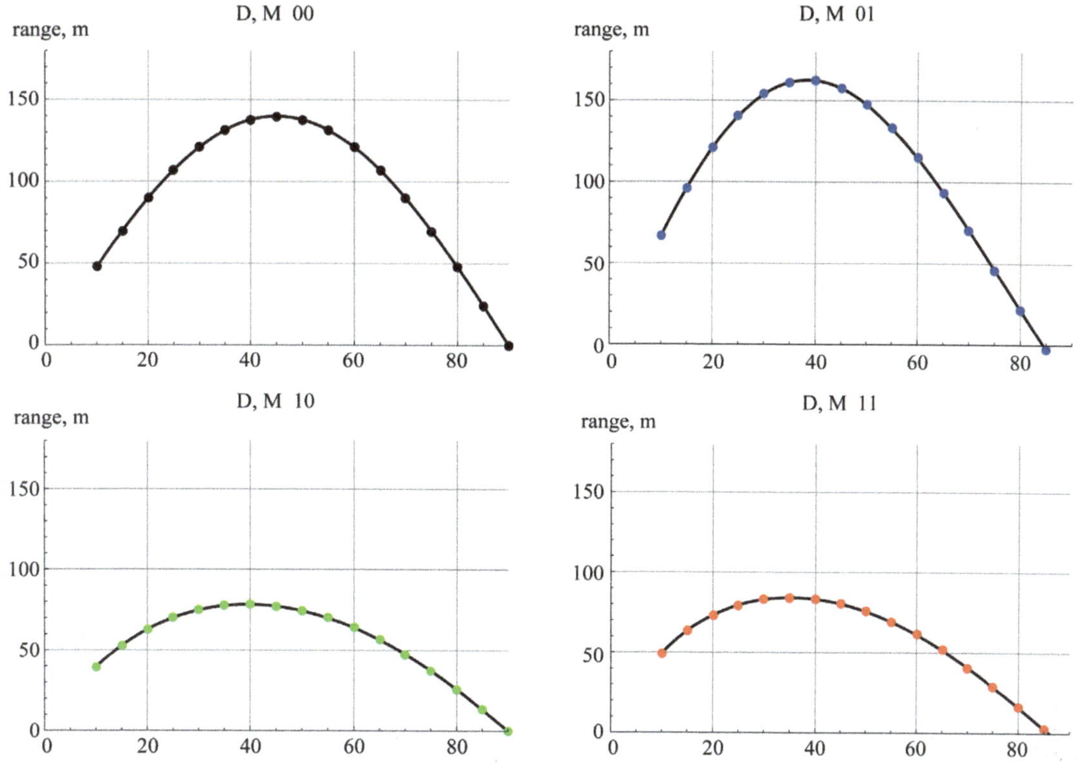

Figure 3. The dots are the numeric solutions of Equations (1), (2) corresponding to the switch setting $\{\alpha D, \beta M\}$. The solid curve in each plot is the corresponding interpolated functions.

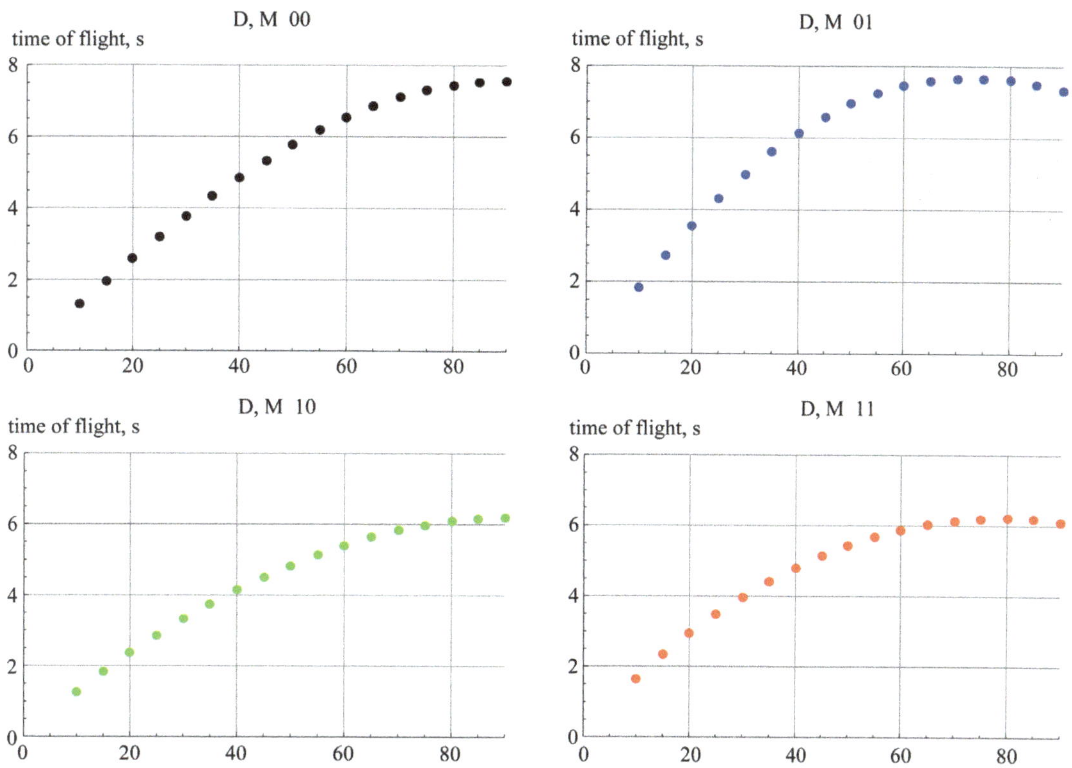

Figure 4. Plots of time of flights vs. the initial angles for four cases of interest.

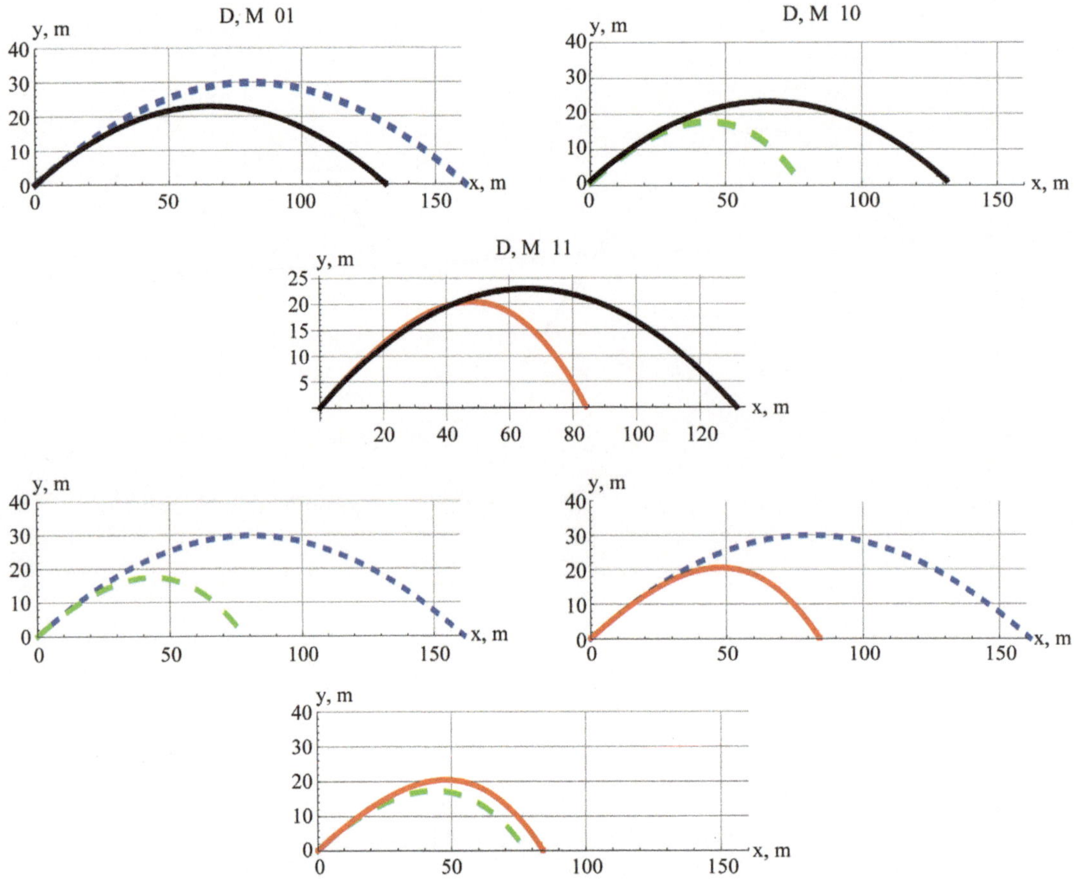

Figure 5. The black curve is the trajectory in a vacuum and is used as a reference. The color code is the same as **Figure 3**.

Trajectories depicted in **Figure 5** are associated with the initial projectile angle 35°. These graphs show the impact of the various scenarios on the trajectory. For instance, the red trajectory is being compared to the black trajectory. The former shows the impact of the combined drag and the Magnus force, while the latter is the trajectory in a vacuum. These two trajectories are indistinguishable for the greater part of the first half of their pathway and deviate more in the second half. After deviating paths, the red curve bends severely, while the black one follows its "natural" curvature. The reasoning for these behaviors are tied to the speed dependent drag coefficient depicted in **Figure 2**. Simply put, initially, a batted baseball flies quickly, and because of the small value of the drag coefficient it encounters a small drag force. Conversely, while it traverses along its trajectory, the loss of its speed is associated with a large drag coefficient and hence it encounters a stronger drag force. The latter is the main cause of the deviation from the counterpart in a vacuum. For the rest of the graphs depicted in **Figure 5** one may easily draw the appropriate conclusions. The author believes these need to be done based on the reader's interest.

We evaluate the maximum height, the results are shown pictorially in **Figure 6**.

Figure 6 contains interesting information. For instance, the blue labels for any chosen initial angle are higher than the rest of the colored labels; this is because the former excludes the drag force. These are also higher than the black labels showing the lifting impact of the Magnus force. The green and the red labels are quite lower than the black ones; this is because the former set contains the drag force and the latter is projected in a vacuum. Also we note the drastic difference between the black and the red labels. That is to say there are huge differences between the vacuum and a real-life scenario. We also note as one should expect intuitively the absolute maximum heights irrespective of the chosen scenario occurs at 90°, *i.e.* an upright projection. A further detailed study of the features displayed in **Figure 6** may also reveal additional detailed information; however, the author leaves them for the interested reader to explore.

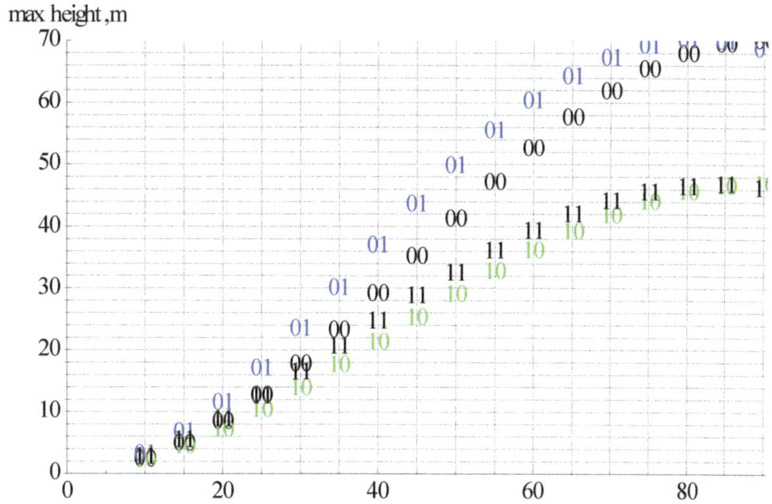

Figure 6. A consolidated plot of the maximum heights vs. the projected angles. The color codes are the same as **Figure 3**.

We continue exploring some more properties of a batted baseball, namely the arc length of a projectile. The length of the projectile comes about from the integration of the arc length element along the trajectory; $d\ell = \sqrt{dx^2 + dz^2}$. However, the coordinates of the projectile are implicit time-dependent functions *i.e.* $\{x(t), z(t)\}$ and hence we write $d\ell = \sqrt{\dot{x}(t)^2 + \dot{z}(t)^2}\,dt$, this yields $\ell = \int_{t_1}^{t_2} \sqrt{\dot{x}(t)^2 + \dot{z}(t)^2}\,dt$. We set $\{t_1, t_2\} \equiv \{0, \text{flighttime}\}$ yielding the length. Results are shown in **Figure 7**.

Having these plots on hand, one may now ask at least two questions. The first questions is "At what angle a ball should be batted to gain the maximum arc length?". The results are tabulated in **Table 2**.

According to **Table 2** the longest trajectory arc length is associated with the second scenario. This makes intuitive sense because in the absence of the drag force the Magnus effect partially "lifts" the ball making it travel longer causing it to gain the longest arc length. Similar self-explanatory information may be deduced for other cases as well. It is reassuring to note that the value of the angle and its associated arc length in a vacuum are exactly what the author reported previously [8].

The second question stems from a careful analysis of the last plot of **Figure 7**. This graph shows that there are a certain *set* of initial projectile angles corresponding to the *same* arc lengths. To make the point, in **Figure 8** we have drawn three horizontal lines. Each of these lines intersects the curve at three points; their abscissa is the values of the initial angles of interest.

Accordingly, this trio angles are {32.51˚, 82.35˚, 87.12˚}. One needs to keep in mind that these angles are specific to the initial speed 37 m/s. In other words, these angles are speed-dependent.

The author would like to comment that although since Newton's era the projectile motion has been visited numerous times no such observation has been reported in scientific literature!

4. Conclusion

Scientific literature is flooded with information about the physics of the baseball. A thorough review of these resources reveals that there is no similarity between the outputs of the presented investigation vs. the historic body of knowledge. In this article we analyzed the impact of the drag force and the Magnus effect on the character of a projectile. Specifically we considered a baseball with a MLB character and included the impact of the speed-dependent drag coefficient. Depending on the choice of factors, four different scenarios were considered. Because of the complexity of the cases we heavily relied on the numeric solutions of the ODEs describing the motion. Results were interpreted and conclusions were reported for individual scenarios. For a comprehensive understanding, the results of each section are depicted graphically. In the last section we reported a new discovered result, namely a set of trio-angles conducive to an equal trajectory arc length.

Table 2. The maximum trajectory arc lengths of a batted baseball at an initial speed of 37 m/s vs. the corresponding initial angles for four scenarios of interest. The headings of the matrix explicitly describes the scenario of interest.

	$\{\alpha D,\beta M\}=00$	$\{\alpha D,\beta M\}=01$	$\{\alpha D,\beta M\}=10$	$\{\alpha D,\beta M\}=11$
"max length, *m*-->"	167.548	184.428	100.646	103.567
θ° "-->"	$\theta\to 56.4556$	$\theta\to 45.744$	$\theta\to 62.166$	$\theta\to 53.8664$

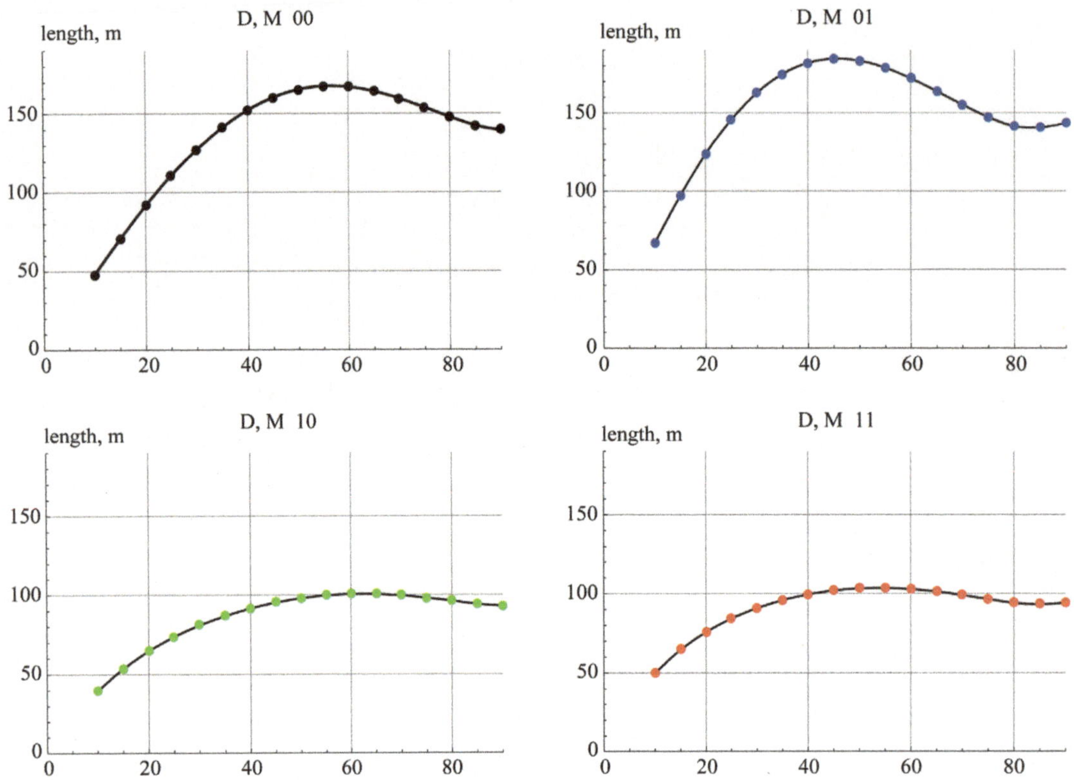

Figure 7. Plots of arc lengths of the trajectory of a baseball batted at 37 m/s vs. the initial projected angle. The title of each plot describes the scenario of interest.

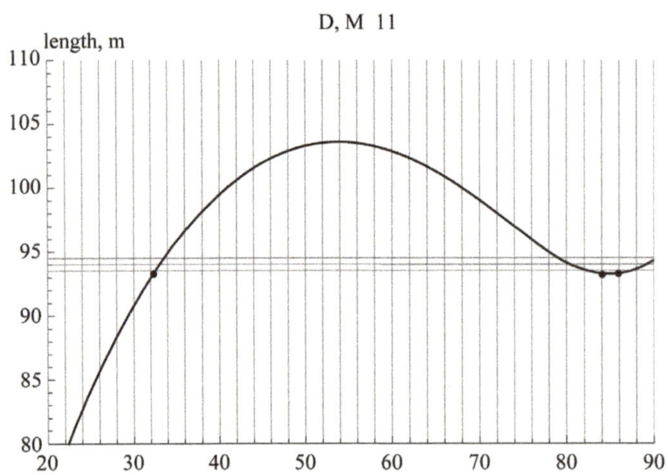

Figure 8. The abscissa of the three dots are the angles corresponding to the initial angles conducive to the same trajectory arc lengths.

Acknowledgements

A version of this investigation was presented at the 11th International Conference on Computer Science and its Applications, Santander/Spain, 2011.

References

[1] Halliday, D., Risnick, R. and Walker, J. (2011) Fundamentals of Physics. 9th Edition, John Wiley & Sons, Inc., Hoboken.

[2] Bauer, W. and Westfall, G. (2011) University Physics. McGraw-Hill, New York.

[3] Tipler, P. (1991) Physics. 3rd Edition, Worth Publisher, New York.

[4] VanWyk, S. (2008) Computer Solutions in Physics. World Scientific, Singapore.

[5] Giordano, N. and Nakanishi, H. (2006) Computational Physics. 2nd Edition, Prentice Hall, Upper Saddle River.

[6] Wolfgang, N. and Weidenmuller, H. (1980) Lecture Notes in Physics, Vol. 51. Springer-Verlag, Berlin, 194.

[7] Adair, R. (2002) The Physics of Baseball. 3rd Edition, Harper-Collins Publishers Inc., New York.

[8] Sarafian, H. (1999) On Projectile Motion. *The Physics Teacher*, **37**, 86-88. http://dx.doi.org/10.1119/1.880184

4

Cyclic Elasto-Plastic Fracture Diagram and Some Parameters of Cyclic Crack Growth Resistance for the Plastic Steels

L. A. Sosnovskiy[1], A. V. Bogdanovich[2], S. S. Sherbakov[2]

[1]S&P Group "TRIBO-FATIGUE", Gomel, Belarus

[2]Department of Theoretical and Applied Mechanics, Belarusian State University, Minsk, Belarus

Email: bogal@tut.by, sherbakovss@mail.ru

Abstract

Method of calculation and experimental estimation of crack growth resistance under cyclic elasto-plastic deformation is proposed. This method is based on measuring of local plastic strain near the crack tip and plotting the cyclic elasto-plastic fracture diagram for a specimen with a crack. Analysis of two types of the cyclic elasto-plastic fracture diagrams and their parameters is made. Experimental D-diagrams of cyclic elasto-plastic fracture for the plastic carbon steel are given.

Keywords

Crack Growth Resistance, Elasto-Plastic Deformation, Compact Specimen, Plastic Steel, Stress Intensity Factor, Contraction

1. Introduction

A significant influence on strength and life of the structures is conditioned by manufacturing defects and operational nature which often become the cause of fatigue cracks. Evaluation of the crack resistance of structures made from plastic steel based on the approaches of nonlinear fracture mechanics is problematic due to non-compliance of the conditions of plane strain. One solution to the problem is in the extrapolation of formulas of linear elastic fracture mechanics for stress intensity factor (SIF) on essentially nonlinear stage of deformation using functions of plasticity amendments. Analytical and experimental method for the estimation of crack growth resistance under cyclic elasto-plastic deformation [1] based on measuring of local plastic strain near the crack tip end is discussed further.

2. Diagram Construction Method

Considering that plastic steel was subjected to test an estimation of applicability of basic formulas of linear elastic fracture mechanics was made. Observance of flat deformation conditions was checked by criteria [2]-[5]:

$$K_I \le K_I^* = \sqrt{t_0 \sigma_{0.2}^2 / 2.5} \quad (1)$$

$$\psi = \frac{t_0 - t_\varphi}{t_0} \times 100\% \le 1.5\% \quad (2)$$

where K_I is stress intensity factor; t_0 is a nominal thickness of the compact specimen; t_φ is a thickness of the compact specimen with the account of elasto-plastic strains; $\sigma_{0.2}$ is yield strength (offset = 0.2%) of a material; ψ is relative contraction of cross-section of the specimen (**Figure 1**).

It has appeared that conditions Equation (1) and Equation (2) are not satisfied for the investigated steel in the upper part of the fatigue crack growth diagram. Formulas of linear elastic fracture mechanics for the estimation of SIF value of the standard compact tension specimen [2]-[4]

$$K_{I\max} = \frac{P_{\max}\sqrt{l}}{t_0 B} \cdot Y\left(\frac{l}{B}\right) \quad (3)$$

where $P_{\max}$ is the maximum load of a cycle; l is the measured length of a crack; t_0, B are the sizes of a dangerous section of the specimen (**Figure 1**); $Y(l/B)$ is the correction function which considers geometry of the specimen and its scheme of loading:

$$Y(l/B) = 29.6 - 185.5(l/B) + 655.7(l/B)^2 - 1017(l/B)^3 + 638.9(l/B)^4 \quad (4)$$

is also correct for elastic deformation under preservation of flat deformation conditions. In order to apply them to elasto-plastic domain it is necessary to correct them for plasticity.

It can be realized by taking into account in Equation (4) the actual sizes of dangerous cross-section of the specimen, *i.e.* those sizes that take place under plastic deformation [1] [6]-[9].

Let us multiply and divide the relation l/B by t_0 value; thus $l/B = (l/B)(t_\varphi/t_0)$. Means, $Y(l/B) = Y[(l/B)(t_\varphi/t_0)]$. At elastic deformation this equality is identical. Taking into account plastic deformation of dangerous cross-section of the specimen in function Y it is necessary to accept the actual thickness $t_\varphi = t_0 - \varphi$ of the specimen, where φ is a lateral component of plastic strain (contraction) of cross-section, *i.e.* we can write [6]:

$$Y\left(\frac{l}{B}\frac{t_\varphi}{t_0}\right) = Y\left(\frac{F_l}{F_0}\right) = Y(\omega_F), \quad (5)$$

where F_0 is the nominal (before deformation) area of dangerous cross-section of the specimen; F_1 is the area damaged by a crack with a length l and defined with the account of the plastic deformation of cross-section. It means that by introduction of Equation (5) into Equation (3) and Equation (4) we obtain a technique of SIF calculation for elasto-plastic domain [1] [6]-[9]:

$$K_{I\max}^F = \frac{P_{\max}}{t_0\sqrt{B}} \omega_F^{1/2} Y(\omega_F); \quad (3a)$$

$$Y(\omega_F) = 29.6 - 185.5(\omega_F) + 655.7(\omega_F)^2 - 1017(\omega_F)^3 + 638.9(\omega_F)^4 \quad (4a)$$

Thus Equation (4a) considers not only geometry of the specimen and its scheme of loading but also integrally the size of plastic strain in dangerous cross-section. And in Equation (3a) the local measure of damage of a specimen with a crack $\omega_F = F_1/F_0$ that has not only geometrical meaning but also physical content is introduced. This measure unambiguously defines the life of an object with a crack [10].

It should also be stressed out that the measure ω_F is defined taking into account plastic strain of dangerous cross-section. According to the developed approach [1] [6]-[9] whole process of elasto-plastic deformation and destruction are described by means of the cyclic elasto-plastic fracture diagram for a specimen with a crack

(CEPF-diagram). This diagram is built in SIF coordinates K_I^F and absolute φ □or relative ψ contraction. Lateral component of plastic strain of the specimen in the zone of crack growth (contraction) is defined as a difference between nominal t_0 and actual t_φ values of thickness of the specimen, *i.e.* $\varphi = t_0 - t_\varphi$ (see **Figure 1(a)**); its relative value is $\psi = \varphi / t_0$. Thus SIF K_I^F is calculated using formulas of linear elastic fracture mechanics, but adjusted for plasticity of the investigated material. For example Equations (3a), (4a) are used for calculation of SIF of the compact specimen (see **Figure 1(a)**).

3. Two Types of CEPF-Diagram

There are two types of CEPF-diagram [8] [9]. If for calculation of K_I^F value we conditionally accept that the maximum load in rupture process remains constant (and it is really possible if the test machine is rigid enough or loading rate is high), then OBCS diagram (see **Figure 1(b)**) is obtained which resembles letter *D* taking into account ordinates axis. Therefore it is named *D*-diagram. If for calculation of K_I^F we consider decrease of loading in rupture process of a specimen (when the test machine has rather low rigidity or the rate of loading is low) diagram $OB_1C_1S_1$ (see **Figure 1(b)**) is obtained. As the form of this diagram reminds letter *Q* it is named *Q*-diagram.

The CEPF-diagram generally consists of two curves: a curve of cyclic elasto-plastic destruction (sections OBC in *D*-diagram and OB_1C_1 in *Q*-diagram) and a curve of quasi-static destruction (rupture) (sections CS in *D*-diagram and C_1S_1 in *Q*-diagram). In points *C* and C_1 the crack reaches the critical size l_c to which limiting contraction φ_c and limiting SIF value—cyclic fracture toughness (values of K_{fc}^F in *D*-diagram and K_{fc}^* in *Q*-diagram) correspond. There is a division of the specimen into two parts in corresponding points *S* and S_1, thus takes place a maximum limiting widening φ_s it's dangerous cross-section on which we define other limiting SIF value—the quasi-static fracture toughness (size K_{sc}^F in the *D*-diagram; $K_I^F = 0$ in this point in *Q*-diagram). Crossing of CS curve with an axis of ordinates gives one more parameter of crack growth resistance K_0^F (see **Figure 1(b)**). The maximum of *Q*-diagram on an SIF axis (point B_1) corresponds to the beginning of

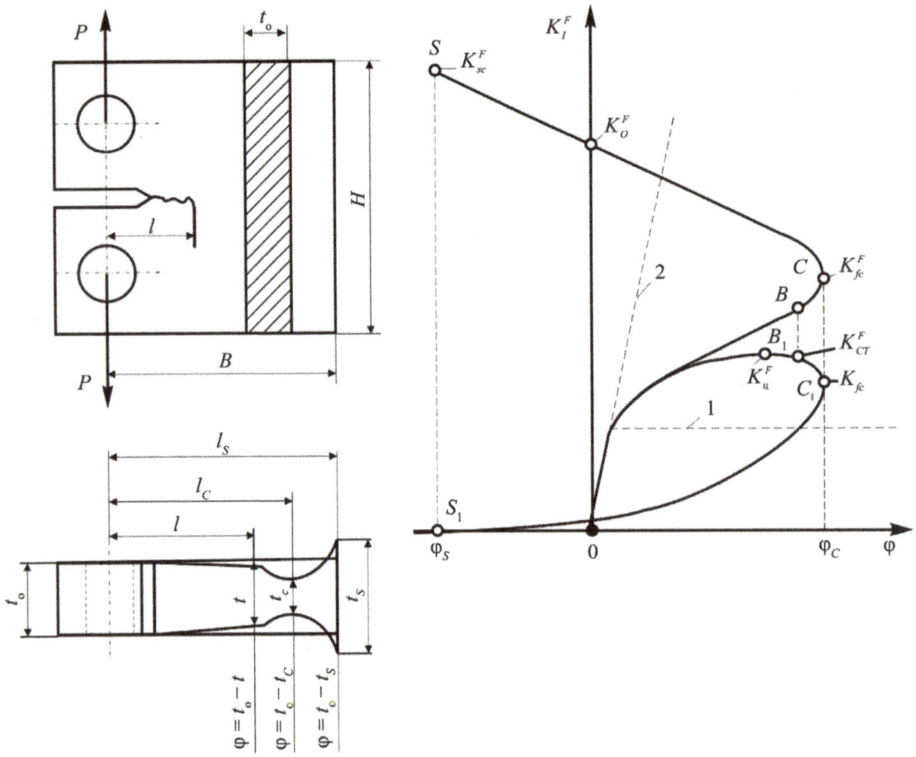

Figure 1. Schemes: test of the compact specimen and measurement of the thickness reduction (a); cyclic elasto-plastic fracture diagram for a specimen with a crack (b).

cyclic rupture and is characterized by parameter $K_{\text{ц}}^F$; parameter $K_{\text{ст}}^F$ in the *Q*-diagram corresponds to the beginning of quasi-static rupture; it is not a characteristic point of this diagram, but it corresponds to the beginning of sharp lifting of curve OBC (a point *B* in the *D*-diagram). In a case of "ideally plastic fracture" the curve of cyclic elasto-plastic destruction is transformed to a straight line 1. In a case of "ideally brittle fracture" $(\varphi = 0)$ this curve coincides with an axis of ordinates. The line 2 divides areas of quasi-brittle and elasto-plastic destructions. Thus the analysis of is viscous-brittle transition, for example, at change of the sizes of a specimen or test temperature is possible by means of CEPF-diagram.

4. Analytical Description of CEPF-Diagram

It is offered three expressions for the analytical description of OBC curve at *D*-diagram [6] [8] [9]. The first is a power equation

$$K_I^F = K_{th}^{\varphi} \cdot \varphi^{m_1} \tag{6}$$

where m_1 is a parameter of cyclic hardening $(0 \le m_1 \le 1)$; K_{th}^{φ} is a plasticity threshold, *i.e.* SIF value below which the plastic strains in a crack top do not influence its value. Parameters m_1 and K_{th}^{φ} are defined on experimental dependence in co-ordinates $\lg K_{I\max}^F - \lg(\varphi/\varphi_{th})$.

The second dependence for the description of a curve of cyclic elasto-plastic destruction OBC looks like:

$$K_I^F = K_w \left(\frac{\psi - \psi_t}{\psi_c - \psi} \right)^{m_2} \quad \text{if } \psi_t < \psi < \psi_c \tag{7}$$

where K_w is the parameter which is subject to definition; m_2 is parameter of hardening; ψ_t is a relative contraction of a specimen, corresponding to the beginning of yield of a material at an axial tension. If $K_I^F = K_w$, $2\psi = \psi_c + \psi_t$ or $\psi = (\psi_c + \psi_t)/2$. Hence parameter K_w is such SIF value which corresponds to relative size of contraction $\psi = (\psi_c + \psi_t)/2$. And as $\psi_t \to 0$ and for plastic materials $\psi_c \gg \psi_t$ so parameter K_w can

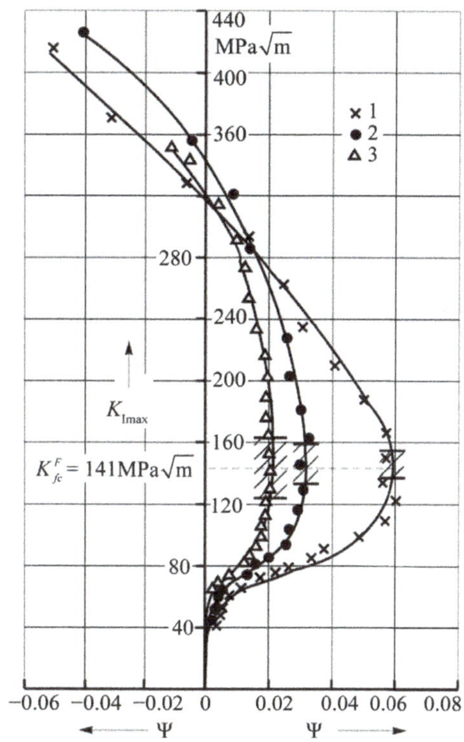

Figure 2. *D*-diagrams of the cyclic elasto-plastic fracture for the carbonic steel constructed by the results of tests of compact samples of 10 (1), 20 (2) and 40 (3) mm thickness.

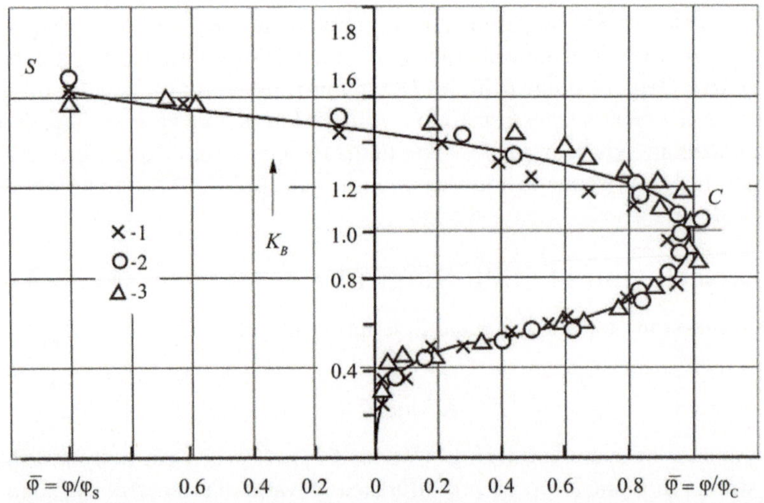

Figure 3. Generalized *D*-diagrams of the cyclic elasto-plastic fracture for the carbonic steel constructed by the results of tests of compact samples of 10 (1), 20 (2) and 40 (3) mm thickness.

be defined for them as such value K_I^F which corresponds to half of limiting contraction $(\psi_c/2)$. Practically value K_w is defined also as value corresponding to value $\lg\left[(\psi-\psi_t)/(\psi_c-\psi)\right]=0$ at representation of OBC curve (see **Figure 1(b)**) in co-ordinates $\lg K_I^F - \lg\left[(\psi-\psi_t)/(\psi_c-\psi)\right]$, and value of parameter m_2 can be found from the same graph as a tangent of an angle of an inclination of the received straight line to an axis of abscises.

For obtaining the third expression it is accepted that experimental points in an average part of OBC part of the *D*-diagram are approximated by a straight line in co-ordinates $\lg K_I^F - \lg(\psi/\psi_t)$. The equation of this straight line at transition to usual co-ordinates is transformed to power dependence of a kind

$$K_I^F = K_t \left(\frac{\psi}{\psi_t}\right)^{m_3} \quad (8)$$

where K_t, m_3 are parameters. Practically value K_t is defined on a point of crossing of the specified straight line with an axis of ordinates in double logarithmic co-ordinates, and value of parameter m_3 is found as a tangent of an angle of an inclination of this straight line to an axis of abscises.

Figure 2 shows CEPF-diagrams (analytical description of which was given above) for compact specimens of different thickness made of the plastic carbonic steel. Influence of specimen sizes on deformation characteristics of crack growth resistance is visible on **Figure 2**. And in **Figure 3** the same diagrams are combined in the form of one dependence SIF-specimen contraction by means of the offered similarity transformation. It is shown how the stated approach can be used for an estimation of pipes survivability.

References

[1] Sosnovskiy, L.A. and Bogdanovich, A.V. (2011) Crack Growth Resistance. BelSUT, Gomel, 169-254.

[2] E-647 (2000) Standard Test Method for Measurement of Fatigue Crack Growth Rates.

[3] E-339 (1976) Standard Test Method for Plane-Strain Fracture Toughness of Materials. Annual Book of ASTM Standards, 471-490.

[4] РД 50-345-82 (1982) Calculations and Tests for Strength. Methods of Mechanical Tests of Metals. Definition of Crack Growth Resistance Characteristics (Viscosity of Destruction) at Cyclic Loading. Methodical Instructions. Moscow.

[5] Drozdovsky, B.A. and Morozov, E.M. (1976) Methods of an Estimation of Viscosity of Destruction. *Zavodskaya Laboratoriya*, **8**, 916-1004.

[6] Sosnovskiy, L.A. (1990) Cyclic Elasto-Plastic Fracture Diagram for a Specimen with a Crack and Its Basic Points. *Vesty AS BSSR, series of phis.-tech. sciences*, **2**, 3-7.

[7] Sosnovskiy, L.A., Hamaza, L.A. and Babich, N.K. (1990) Experimental Research of Cyclic Crack Growth Resistance of Steel 20. *Problems of Strength*, **6**, 34-38.

[8] Sosnovskiy, L.A. and Bogdanovich, A.V. (1995) The Analysis of Cyclic Elasto-Plastic Fracture Diagram for a Specimen with a Crack. Part 1. *Problems of Strength*, **5-6**, 39-51.

[9] Sosnovskiy, L.A. and Bogdanovich, A.V. (1995) The Analysis of Cyclic Elasto-Plastic Fracture Diagram for a Specimen with a Crack. Part 2. *Problems of Strength*, **5-6**, 52-61.

[10] Sosnovskiy, L.A. (1987) Statistical Mechanics of Fatigue Fracture. Science and Technics, Minsk.

5

Influence of Soil-Foundation Interaction Properties on Oscillations of the System "Building-Building" and "Building-Stack-Like Structure"

Gaibnazar Khodievich Khojmetov, Alinazar Irismetovich Khodjimetov, Anvar Sayfullaevich Yuvmitov

Tashkent Automobile and Road Institute, Tashkent Architectural-Building Institute, Branch of I. M. Gubkin Russian State University of Oil-Gas, Tashkent, The Republic of Uzbekistan

Email: anvar.sayfullaevich@mail.ru, ahodjimetov@mail.ru, diyorbek_84@mail.ru

Abstract

Seismic oscillations of the "building-building" system which is interconnected buildings built close to each other, and "building-stack-like structure" system which is adjacent and connected in different ways to existing building are considered in the paper. Different types of connections, such as dampers, including the ones suggested by the authors, are studied. Seismic impact is given as a harmonic function and various existing accelerograms, including synthesized ones. Distinctive feature of this paper from previously published ones [1] [2] is the fact that the emphasis falls on the influence of soil-foundation interaction properties, which are described using various models of load-displacement connections. Calculation results are compared in the case of representation of the building as concentrated masses and spatial systems. Ways to reduce seismic response of buildings during the earthquakes are pointed out. Results of experimental studies are given in the paper and are compared with calculations.

Keywords

Interaction Properties of Soil-Foundation, The Rigid Foundation, Kelvin-Voigt and Maxwell Models, Damper, the Coulomb's Force, Modeling, Similarity Multipliers, Dynamic Characteristics, Oscillations, Sensors, Accelerogram, Absorption Factor

1. Introduction

The difficulty to predict where, when and with what magnitude an earthquake will strike leads to the strengthening of the necessity to build earthquake-resistant buildings and structures and to develop seismic engineering. One of the ways to improve earthquake resistance of buildings is to provide constructive solutions, which absorb seismic energy. Usually, such solutions are carried out in the process of design, calculations and construction of buildings. Efficient use of land resources in modern construction and an expansion of cities nowadays lead to a trend to construct adjacent buildings—close to each other.

In addition, there is a tendency to expand the living space, building an attachment in the form of stack units to the main building. Stack-like structures of different height may be observed in the reconstruction of a number of administrative buildings. Connections of "building-building" system are usually carried out using various types of oscillations dampers. As for the system "building-stack", the parts may be connected rigidly or with dampers. Structures of connecting elements are numerous; some of them are given in [3] [4].

The study of oscillations of the buildings, connected with different types of dampers, is presented in the papers [5] [6]. They give results of calculations, and define dynamic characteristics of adjacent buildings and standing apart ones. In these works, the foundation is considered to be rigidly fixed in soil.

2. Statement of the Problem

It is assumed, that the building has an n floors. Under seismic effect it oscillates in a plane of the least rigidity, *i.e.* it oscillates in horizontal transverse direction only. At the beginning, the building is modeled by adopted design scheme in the form of a console with concentrated masses on the floors. Viscous-elastic characteristics of the building itself, and foundation displacement relative to soil are considered here. Both buildings (**Figure 1**) in the "building-building" system, interconnected on floor levels with dampers are subjected to shear waves, and in the "building-stack" system the building is subjected to shear vibrations and the stack–to bending ones.

Oscillation equation of the system "building-building" is written in the following form [7]:

$$
\begin{cases}
k_n\left(y_{n-1}-y_n\right)-\mu_n\dot{y}_n=m_n\left(\ddot{y}_{00}+\ddot{y}_n\right)\\
k_{n-1}\left(y_{n-2}-y_{n-1}\right)-k_n\left(y_{n-1}-y_n\right)-\mu_{n-1}\dot{y}_{n-1}=m_{n-1}\left(\ddot{y}_{00}+\ddot{y}_{n-1}\right)\\
\cdots\\
k_1((y_0-y_1)-k_2\left(y_1-y_2\right)-\mu_1\dot{y}_1=m_1\left(\ddot{y}_{00}+\ddot{y}_1\right)\\
-k_1\left(y_0-y_1\right)-R=m_0\left(\ddot{y}_{00}+\ddot{y}_0\right)\\
\\
\bar{k}_n\left(\bar{y}_{n-1}-\bar{y}_n\right)-\bar{\mu}_n\dot{\bar{y}}_n=\bar{m}_n\left(\ddot{y}_{00}+\ddot{\bar{y}}_n\right)\\
\bar{k}_{n-1}\left(\bar{y}_{n-2}-\bar{y}_{n-1}\right)-\bar{k}_n\left(\bar{y}_{n-1}-\bar{y}_n\right)-\bar{\mu}_{n-1}\dot{\bar{y}}_{n-1}=\bar{m}_{n-1}\left(\ddot{y}_{00}+\ddot{\bar{y}}_{n-1}\right)\\
\cdots\\
\bar{k}_1\left(\bar{y}_0-\bar{y}_1\right)-\bar{k}_2\left(\bar{y}_1-\bar{y}_2\right)-\bar{\mu}_1\dot{\bar{y}}_1=\bar{m}_1\left(\ddot{y}_{00}+\ddot{\bar{y}}_1\right)\\
-\bar{k}_1\left(\bar{y}_0-\bar{y}_1\right)-\bar{R}=\bar{m}_0\left(\ddot{y}_{00}+\ddot{\bar{y}}_0\right)
\end{cases}
\tag{1}
$$

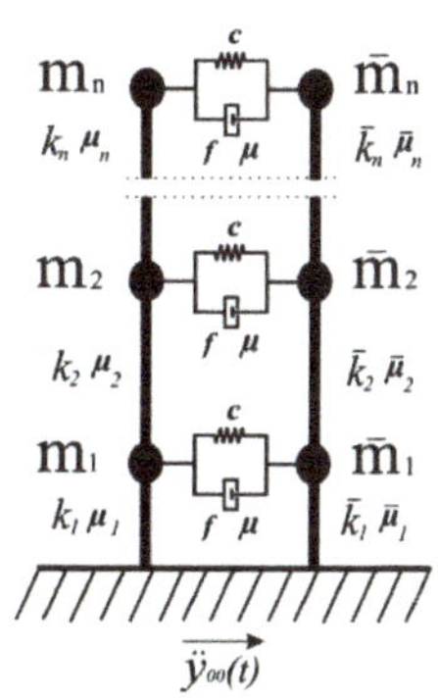

Figure 1. Design scheme of the system connected with a damper.

And for the system "building-stack" equation of stack oscillations is written in the form [8]:

$$\begin{cases} \bar{y}_1 - \bar{y}_0 = -\delta_{11}\left(\bar{m}_1\left(\ddot{y}_{00} + \ddot{\bar{y}}_1\right) + \bar{\mu}_1\dot{\bar{y}}_1\right) \cdots - \delta_{1n}\left(\bar{m}_n\left(\ddot{y}_{00} + \ddot{\bar{y}}_n\right) + \bar{\mu}_n\dot{\bar{y}}_n\right) \\ \cdots \\ \bar{y}_n - \bar{y}_0 = -\delta_{1n}\left(\bar{m}_1\left(\ddot{y}_{00} + \ddot{\bar{y}}_1\right) + \bar{\mu}_1\dot{\bar{y}}_1\right) \cdots - \delta_{nn}\left(\bar{m}_n\left(\ddot{y}_{00} + \ddot{\bar{y}}_n\right) + \bar{\mu}_n\dot{\bar{y}}_n\right) \\ -\bar{R} = \bar{m}_0\left(\ddot{y}_{00} + \ddot{\bar{y}}_0\right) \end{cases} \tag{2}$$

where k_i, $\bar{k}_i$ $(i = 1, 2, 3, \cdots, n-1, n)$ are floor rigidities on shear; y_i, $\bar{y}_i$ are floor deflections; $\dot{y}_i$, $\dot{\bar{y}}_i$ are floor velocities; $\ddot{y}_i$, $\ddot{\bar{y}}_i$ are floor accelerations; m_i, $\bar{m}_i$ are floor masses; δ_{ij} is floor displacements from unit force; μ_i, $\bar{\mu}_i$ are floor viscosity coefficients.

In (1) and (2), $R(y_0)$ is a force of the effect between foundation and soil; it may be written in different ways. When this connection obeys to elastic law, it has the following form:

$$R = K_x \cdot y_0 \tag{3}$$

where K_x is foundation-soil interaction coefficient; y_0 is foundation displacement relative to soil.

When load-displacement connection obeys to elastic-plastic law, it has the form [9]:

$$R = K_x \cdot \left[1 - \omega\left(y_0\right)\right] \cdot y_0 \tag{4}$$

where $\omega\left(y_0\right)$ is some experimental function, which characterizes non-linear properties of interaction. The form and the value of the function $\omega\left(y_0\right)$ obtained experimentally at foundation shear relative to soil, are presented in [10].

Besides, the force R on the basis of hereditary theory of viscous-elasticity may be taken also in the form [11]:

$$R = \int_0^t r(\tau)\,\mathrm{d}u(\tau) \tag{5}$$

Here $r(\tau)$ is a relaxation function.

Calculation results, obtained when viscous-elastic properties of interaction are taken according to Kelvin-Voigt model, are given in the paper:

$$R = K_x \cdot y_0 + \mu \cdot \dot{y}_0 \tag{6}$$

or according to Maxwell model

$$n \cdot \dot{\tau} + \tau = K \cdot \dot{y}_0 \tag{7}$$

here $n = \dfrac{K}{E}$ is a constant called relaxation time; E is elasticity modulus of material; K is viscosity coefficient or coefficient of internal resistance to element strain.

Writing $R = \tau \cdot S$ (τ—tangential stress in foundation-soil contact; S—contact area) and giving foundation displacement relative to soil in the form $y_0 = A\mathrm{e}^{-\beta t} \cdot \sin \omega_0 t$, after some computation we will get:

$$\tau = -\frac{\mathrm{e}^{-\left(m + \frac{1}{n}\right)t}}{m^2 + \omega_0^2}\left(\left(am - b\omega_0\right)\sin \omega_0 t + \left(a\omega_0 + bm\right)\cos \omega_0 t\right) + \frac{a\omega_0 + bm}{m^2 + \omega_0^2} \tag{8}$$

where $a = -\dfrac{AK\beta}{n}$; $b = \dfrac{AK\omega_0}{n}$ and $m = \beta - \dfrac{1}{n}$ are symbols at transformation, β is coefficient of soil damping; ω_0 is circular frequency.

When solving Equations (1) and (2), the forces of response of connecting element (damper) are added to inertia forces for the first building with positive sign, for the second—with opposite one.

It is supposed that the buildings are interconnected with oscillations damper [12], its structure shown in **Figure 2**. It consists of ferroconcrete column base 1, sand 2, compensator of steel sheet 3, reinforcement as an elastic connection 4 and steel plates 5. In the base of this structure lays the damper, offered by Klgatyan [13].

Here, response forces of connecting element are written in the following form:

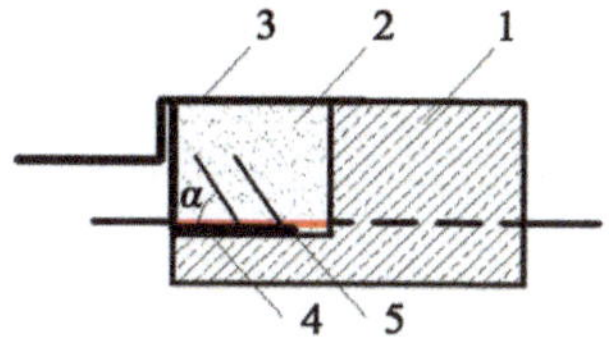

Figure 2. General view of the damper.

$$R = c\left(y_i - \overline{y}_i\right) + \mu\left(\dot{y}_i - \dot{\overline{y}}_i\right) + F_k \tag{9}$$

where c is a rigidity of elastic element; μ is viscosity coefficient of a damper; F_k is Coulomb's force, which (between the plate and sand) is written in the form:

$$F_k = nf\frac{y_n - \overline{y}_n}{\ell_0}ES_{pl}\frac{\vartheta_r}{|\vartheta_r|}\cos\alpha \tag{10}$$

here n is a number of plates; f is coefficient of dry friction of sand with steel; y_n, $\overline{y}_n$ are displacements of the left and right ends of compensator, respectively; ℓ_0 is a length of compensator; E is deformation modulus of sand; S_{pl} is plate area; $\frac{\vartheta_r}{|\vartheta_r|} = \pm 1$ is unit vector, which points the direction of system motion; α is an angle of plate inclination. The structure of connecting element (**Figure 2**) allows us to select the most rational value of the force F_κ for the system "building-stack" by varying a number of plates n and an angle of its inclination α.

3. Experimental Studies of the System "Building-Building" and "Building-Stack-Like Structure" under Micro-Seismic Effects

It is known that field experiments require vast capital investments. Besides, it is almost impossible to conduct twin-tests in field conditions. Therefore, to solve this problem the tests have been conducted with the models using the principles of physical simulation, which are described in details in [14]. The most acceptable principle for the solution of this problem is a simple similarity [15]. In modeling we may use the same material as in field conditions. In this case if field condition presents heavy concrete of high grade, for the model we may take light concrete of low grade. The aim of the experiment is to determine dynamic characteristics of the building, assuming that strains are elastic ones only. So material of a framework may not be modeled. To solve this problem a section of four-storey large-panel building with attached stack-like structure has been modeled (**Figure 3**).

The processes of static and dynamic strain of rigid bodies are characterized by the following values: linear dimension ℓ, stress σ, relative strain ε, material density ρ and time t. Conditions between field experiments and a model depend on their material.

In this work the scale of modeling (similarity factor) ia taken as equal to $\alpha = 1/5$. To prevent cantledge of the model (to compensate the lack of volume forces) accelerations of full-scale experiment and a model are taken as similar ones.

Strength characteristics of field structure and a model, taken for the material of field structure and calculated for a model, are presented in **Table 1**, where R is a cubic strength of concrete; R_{pr} is prism strength of concrete; E_0 is elasticity modulus of concrete; R_a is yield strength of a reinforcement; E_a is elasticity modulus of a reinforcement; asterisk refers to a model.

Design similarity factors of discussed structure for these values are (**Table 2**).

In fabrication of structure elements of the section of large-panel building, steel annealed rods with respective diameters have been used as reinforcement. Their yield stress and elastic modulus, obtained from laboratory test results are 16% and 22% different from design values, as shown in **Table 1**. It is assumed that these differences effect insignificantly on dynamic characteristics of the building. Expanded-clay concrete was taken as a light concrete. Brand of cement in concrete mortar is 400 M, and it is added to a mortar according to concrete grade.

Figure 4 shows the model of building section A and the model of attached stack B.

Physical-mechanical properties of soil are the following: density −16.7 kN/m^3; porosity coefficient −0.595; an angle of internal friction −29.1˚; cohesion coefficient −21.4 kPa; strain modulus −9.6 MPa.

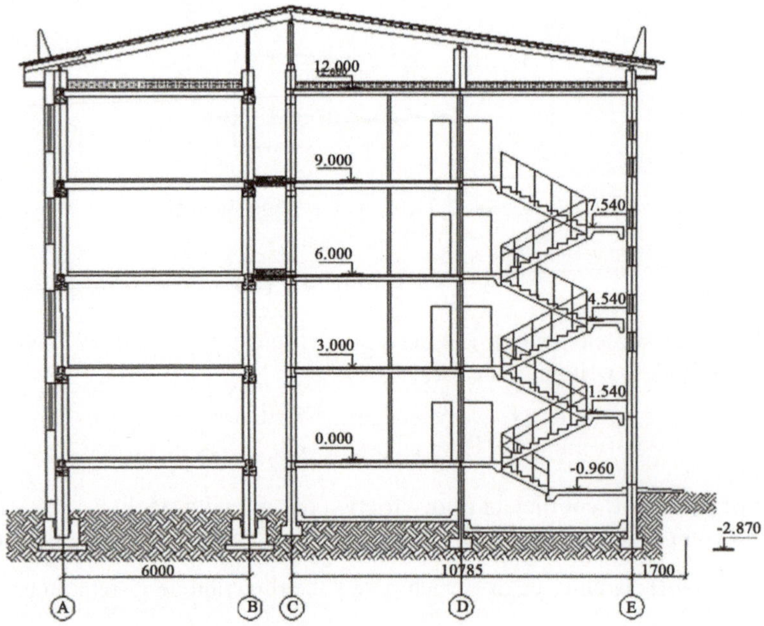

Figure 3. Section of four-storey large-panel building with attached stack-like structure.

(a) (b)

Figure 4. Model of the section of the building **(a)** and model of built stack-like structure **(b).**

Table 1. Strength characteristics of building material of full-scale field structure and a model.

Name of material	Value and dimension units	Ferroconcrete structures, foundations, floor plates, cores, seismic belts, wall panels brickwork	
		Original	Model
Ordinary heavy concrete	R, R^* MPa R_{pr}, R^*_{pr} MPa E_0, E^*_0 MPa λ_r, λ^*_r	20 14.5 26,500 0.6	4 2.9 8748.6 0.6
Reinforcement	R_a, R^*_a MPa E_a, E^*_a MPa	300 210,000	60 41,500

Table 2. Similarity factors.

Linear dimensions and stresses	$\alpha = \beta = 0.2$	Linear loads	$\alpha^2\delta = 0.04$
Strains	$\gamma = 1$	Concentrated loads	$\alpha^3\delta = 0.008$
Time	$\xi = 0.45$	Force moments	$\alpha^4\delta = 0.0016$
Acceleration	$\alpha\gamma\xi^{-2} = 1$	Deflections and displacements	$\alpha\gamma = 0.2$

During the tests dynamic characteristics of the model of the building and the stack have been determined separately and then connected with offered damper. Oscillations of the model of the building have been carried out by two ways: by tension with wire and weight dropping. During the tests, accelerations in different points along the height of the building and the stack have been recorded.

Oscillations have been recorded by DYTRON (USA) detecting elements (sensors) with piezoelectric elements. The sensors have been manufactured to record the signals of one or three components. The signals in the form of stresses via convertor enter the computer. Computer software transforms stress signals into accelerations. The sensors may be adjusted to different frequency ranges. During the tests the computer was adjusted to the frequencies most often observed under micro-seismic oscillations.

Experimental results have been translated to full-scale buildings, using the scale of modeling and built spectral densities of "Matlab" program. Before starting serial experiment, preliminary experiments have been conducted, where all floors (except the first) have been connected by connecting elements in different combinations of them. These preliminary experiments have shown that the greatest effect is achieved when connecting the third and the fourth floors by viscous-elastic coupling. Below are the results of experiments when only the third and the fourth floors are connected by connecting element of "building-stack" system.

According to accelerations records and built spectral densities, floor displacements of "building-stack" system, connected by proposed connecting element have been determined. For this purpose, the graphs have been built, based on calculated values of maximal amplitudes of oscillations, of dependences of displacements of standing apart building and stack-like structure, and after their connection, that is, "building-stack" system, schematized in the form of a cantilever beam with concentrated masses (**Figure 1**).

4. Results of Calculations and Experimental Investigations and Their Analysis

Seismic effect in calculations have been given in the form of damping sinusoid:

$$y_{00} = Ae^{-\alpha t}\sin\omega t \tag{11}$$

Accepted parameters of these effects are the following: amplitude of soil oscillation $A = 0.15$ m; damping ratio $\alpha = 0.15$; period of oscillation $T = 0.5$ s; $\omega = 2\pi/T$.

Calculations have been carried out to the effect of accelerogram, taken in the form of synthesized accelerogram CA-482, which generalizes the properties of all types of accelerograms, in particular, of low frequency (Bucharest, 1977), high frequency (Gazli, 1976) and mid-frequency (El-Centro, 1940) earthquakes. This accelerogram has been digitized by the step 0.005 s, with duration 4.3 s.

Floor masses and rigidities are taken as the ones of 4-storey large-panel building of 76-017CA/53 series.

Calculated parameters of existing building based on project data are the following: $m_1 = 258$; $m_2=213$; $m_3 = 213$; $m_4 = 213$; $m_5 = 190$; (10^3 N·s^2/m); $k_1 = 35.04 \times 10^6$; $k_2 = 26.846 \times 10^6$; $k_3 = 26.846 \times 10^6$; $k_4 = 26.846 \times 10^6$; $k_5 = 26.846 \times 10^6$ (10^3 N/m).

The ones accepted by design data for a stack-like structure are $m_1^* = 149.8$; $m_2^* = 161.8$; $m_3^* = 161.8$; $m_4^* = 161.8$; $m_5^* = 127.34$ (10^3 N·s^2/m); $\delta_{11} = 0.155\times10^{-7}$; $\delta_{12} = \delta_{21} = 0.155\times10^{-7}$; $\delta_{13} = \delta_{31} = 0.155\times10^{-7}$; $\delta_{14} = \delta_{41} = 0.155\times10^{-7}$; $\delta_{15} = \delta_{51} = 0.155\times10^{-7}$; $\delta_{23} = \delta_{32} = 0.98\times10^{-7}$; $\delta_{24} = \delta_{42} = 0.98\times10^{-7}$; $\delta_{25} = \delta_{52} = 0.98\times10^{-7}$; $\delta_{34} = \delta_{43} = 1.97\times10^{-7}$; $\delta_{35} = \delta_{53} = 1.97\times10^{-7}$; $\delta_{45} = \delta_{54} = 2.95\times10^{-7}$; $\delta_{22} = 0.98\times10^{-7}$; $\delta_{33} = 1.97\times10^{-7}$; $\delta_{44} = 2.95\times10^{-7}$; $\delta_{55} = 3.93\times10^{-7}$ (10^{-3} m/N).

With account of floor resistance force, resistance coefficient has been determined by the following expression:

$$\mu_i = \frac{\psi}{2\pi}\sqrt{m_i k_i} = \frac{\psi}{2\pi}\sqrt{m_i\frac{1}{\delta_{ni}}} \tag{12}$$

where ψ – is absorption coefficient, for practical use taken as equal to 0.6 [16].

Figure 5 and **Figure 6** show deflections of the buildings considered as a console beam with concentrated masses on the floors (A) and as a spatial system (B). In these calculations the foundation is considered to be rigidly fixed in soil, and seismic effect is taken in the form (11). Floor viscosity has been neglected. As seen from **Figure 5** and **Figure 6**, the difference in deflection values is null, except for middle floors. Similar conclusions may be drawn from results of calculations of 9-storey and 16-storey buildings (**Figure 6**). Based on above mentioned, calculation results by console scheme only are presented (floor viscosity is taken into consideration in all cases).

To compare experiment results with design data on displacements values from Building Code [17] seismic forces (shear forces and bending moments), generated on the floor, have been determined. Results of experiments and calculations have shown that the most effective is the connection of a buildings and a stack by connecting element, having two plates. Below we compare the results obtained from experiments and calculations. **Figure 7** shows the results of experiments to determine the displacements in "building-stack" system; **Figure 8** shows results of calculations. Comparison of **Figure 7** and **Figure 8** shows that the displacements of upper floor of a building and a stack differ on 59% and 48%, respectively. In experiments displacements values are greater. This can be explained by the following. Calculations take into account real rigidity of the building. It is known that in construction of large-panel buildings, butt joints of the panels and floor slabs after welding are filled with concrete. This increases the rigidity of the whole building. In manufacturing the model of the building their panels and floor slabs have been welded together, but to fill them with concrete appeared impossible. Therefore, it was impossible to achieve the rigidity of the model corresponding to the one of a full-scale building. To our opinion this is the cause of significant differences of experimental and calculated data. In addition, this difference is also affected

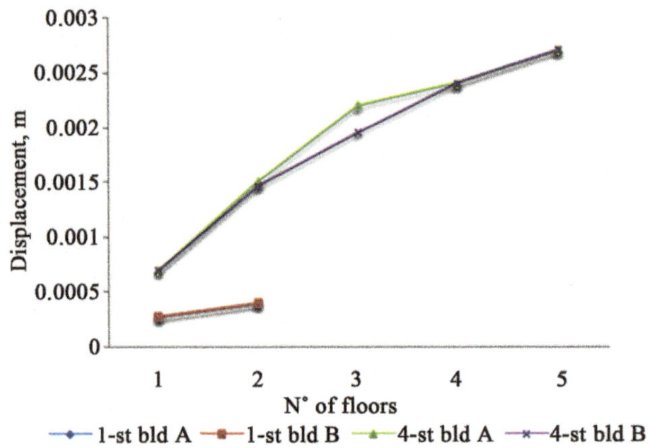

Figure 5. Displacements of one-storey and four-storey buildings. A: design of the bar with concentrated masses; B: design of the building as a spatial box-like structure.

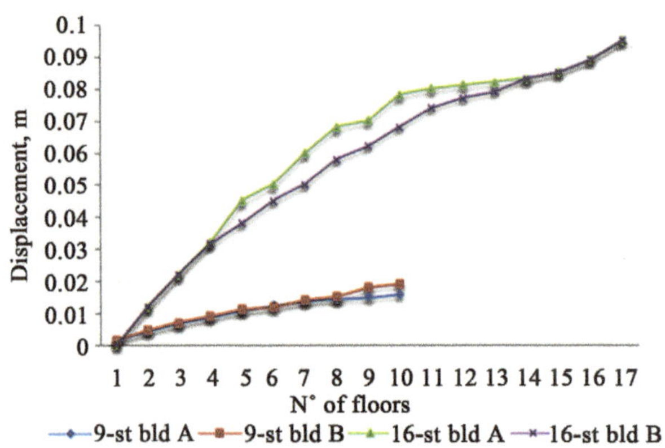

Figure 6. Displacements of nine-storey and sixteen-storey buildings. A: design of the bar with concentrated masses; B: design of the building as a spatial box-like structure.

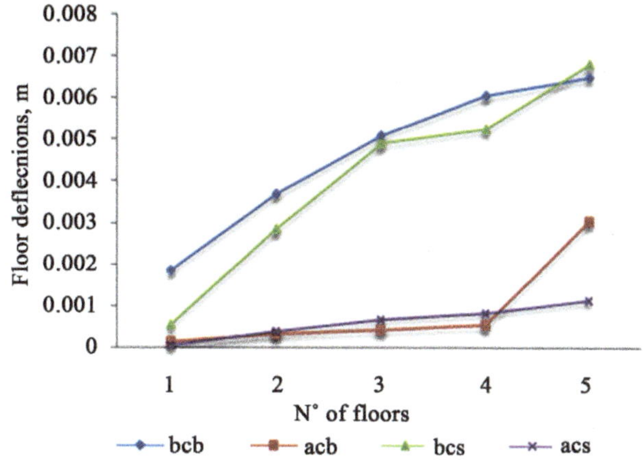

Figure 7. Floor deflections: Results of experiments of the system. bcb and acb: before and after connection of a building with a damper; bcs and acs: before and after connection of a stack with a damper.

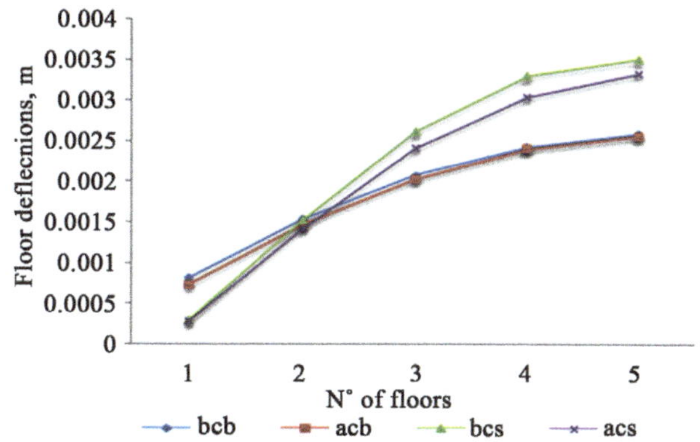

Figure 8. Floor deflections: Results of calculations of the system. bcb and acb: before and after connection of a building with a damper; bcs and acs: before and after connection of a stack with a damper.

by the properties of material used and the technology of model construction. Nevertheless, the above factors (shortcomings) have not effected significantly on the goal achievement, because the experiments before connecting the building with a stack and after their connection have been conducted under identical conditions.

The difference of test data from calculated ones after connection of the building with a stack-like structure with a damper on the upper floor is 15% for a building and 36% for a stack. In tests the damper reduces building displacement to 47%, and stack displacement to 83%. Similar conclusions may be drawn for lower floors.

The use of proposed connecting element in the system "building-stack" reduces force factors (shear forces and bending moments), according to tests and calculation data.

A design for the first case has been conducted to compare displacements and force factors occurring at rigid connection of the building and stack-like structure with a damper. The building and stack-like structures are considered as a single system. In determining the displacements it was assumed that the system performs shear oscillations only. A comparison of numerical results shows that the displacement at rigid connection is 16% greater, than at "building-stack" connection. Load on the building at rigid connection is 3% less, and with the use of a damper it is reduced by 12%. Therefore, it can be stated that the connection of "building-stack" system with a damper increases seismic stability of the building by 15%, as compared to rigid connection.

The use of proposed connecting element gives a good result at stack-like structure attachment to a building of less rigidity (such as brick ones) than the large-panel ones. The decrease in the deflection at the upper floor is 66% and at the lower floor-51%. However, the deflection of the upper and lower floors of stack-like structure shows more than 2 times rise. This increase slightly effects on "building-stack" displacement on the whole.

Figure 9(a) and **Figure 9(b)** show calculation results with the model of soil-foundation rigid connection, elastic (3), visco-elastic Kelvin-Voigt (4) and Maxwell models (5). In these calculations $K = 4 * 10^7$ N/m^3 and η

= 0.152 * 10^5 Ns/m^2 are taken; they corresponds to loamy soil. The character of soil-foundation interaction has no practical effect on displacement of a standing apart building and stack-like structure. The influence of this factor when connecting them with a damper is significant. Depending on accepted model the values of building displacements are reduced to 54%, and stack-like structure-to 3%. Similar calculations have been conducted at K = 3 * 10^7 N/m^3 and η = 0.32 * 10^5 N s/m^2 and K = 5 * 10^7 N/m^3 and η = 0.78 * 10^5 N s/m^2, they correspond to sand and loess, respectively. Conclusions made on the basis of these calculation results are practically the same as the ones for loamy soil.

Calculations have also been performed at altering K or η, ceteris paribus (**Figure 10** and **Figure 11**). With ac-

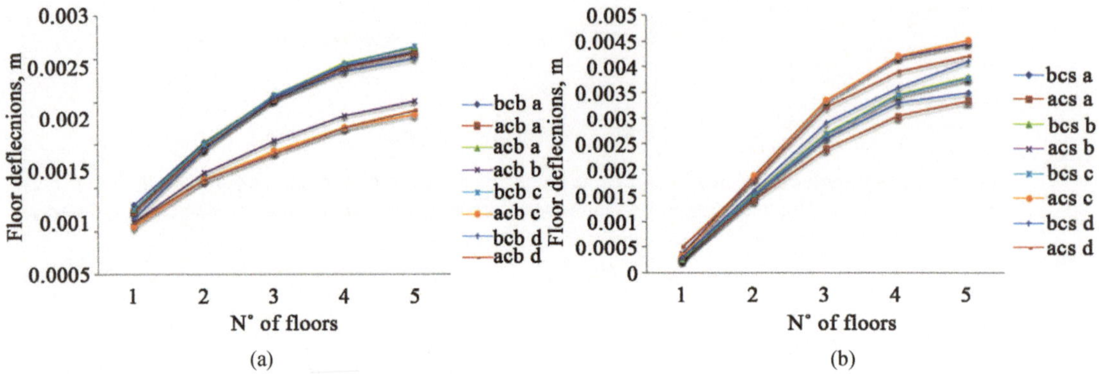

Figure 9. Results of calculations of the system: a—Building and b—Stack-like structure, before and after connection with damper. (a) Rigid foundation; (b) Elastic foundation, c—Elastic-viscous foundation by Kelvin-Voigt model, d—Elastic-viscous foundation by Maxwell model.

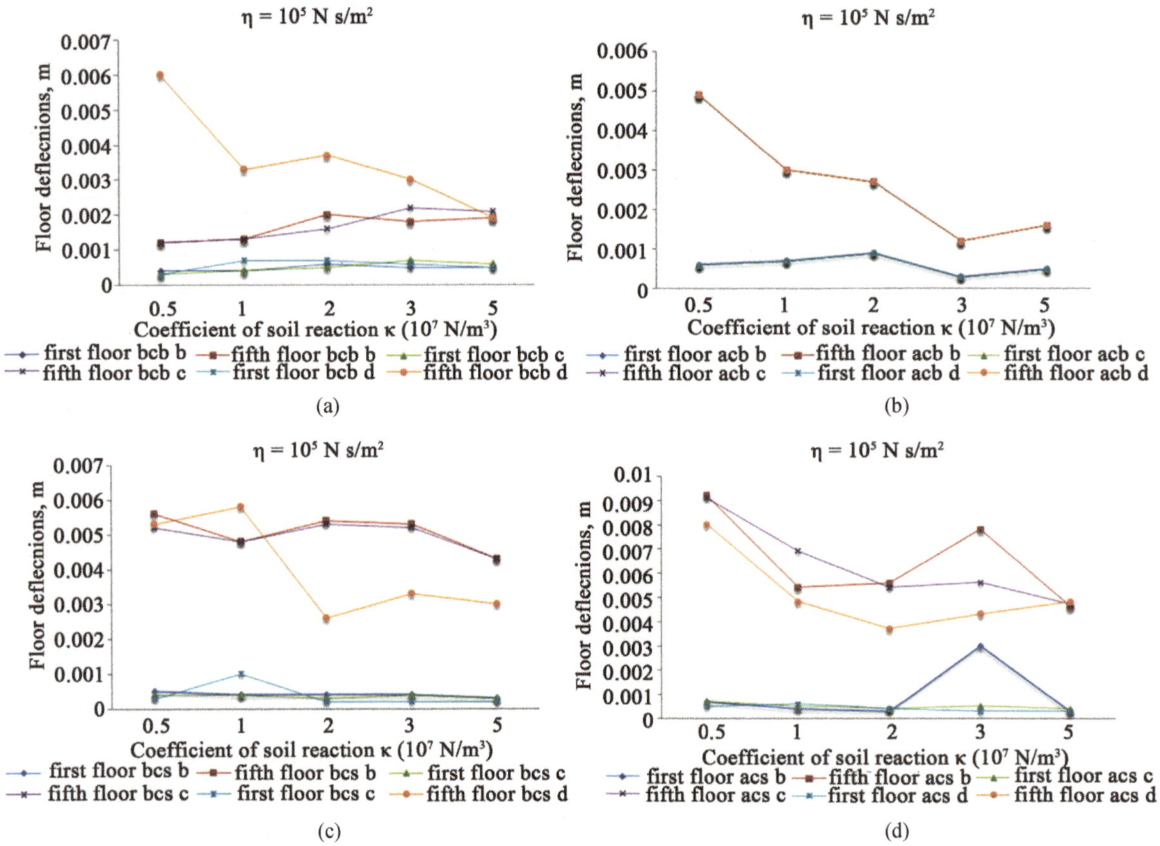

Figure 10. Results of calculation of the system at altering value of K and constant value of з. Building before (a) and after (b) connection with damper; Stack-like structure before (c) and after (d) connection with damper.

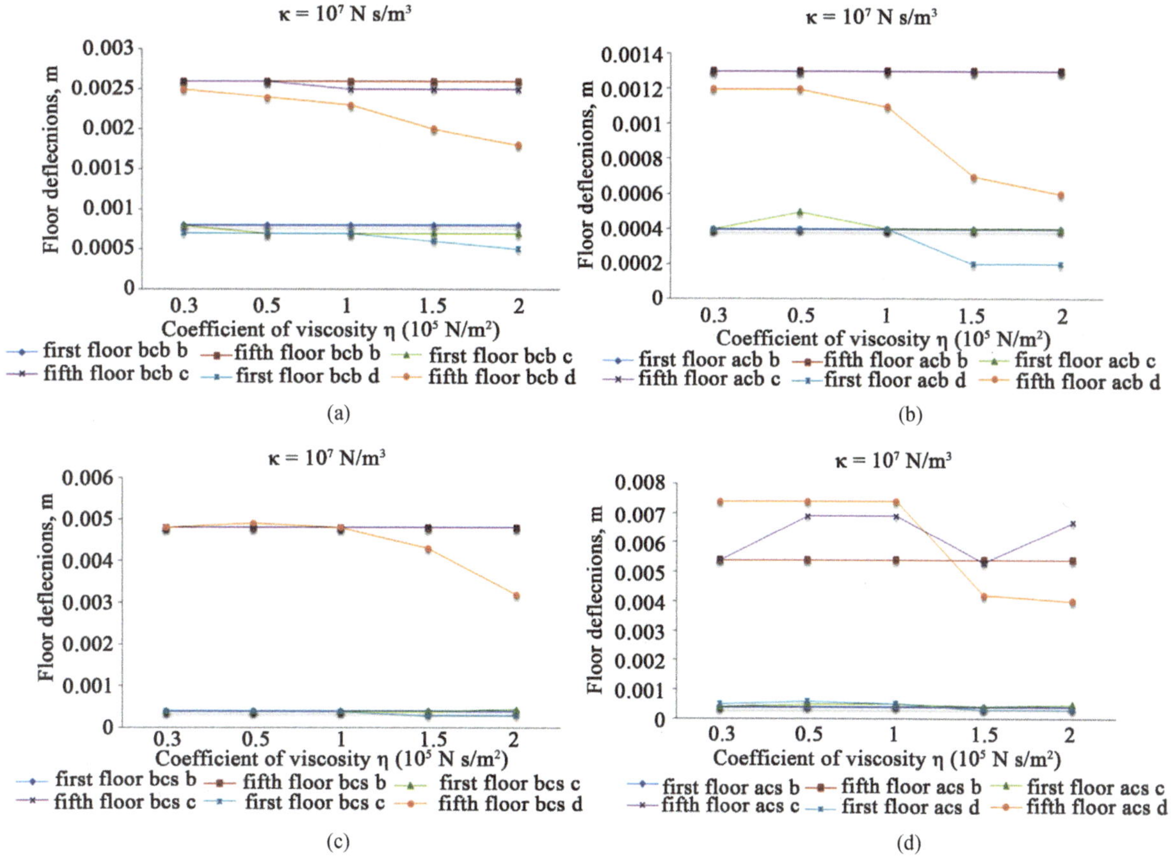

Figure 11. Results of calculation of the system at altering value of η and constant value of K.

cepted in calculations values of K its alteration from 0.5 to 5 (10^7 H/m^3) at constant value of η effects insignificantly on deflections of the building before and after its connection with stack-like structure with a damper. This conclusion is valid for any accepted model of soil-foundation interaction, except for Maxwell model. The increase of η in any case decreases the values of deflection both before connecting with a damper and after it. Calculations (of 9 and 16-storey buildings) show that the higher the building, the greater is the effect of foundation yielding.

5. Conclusions

- Calculations show that the deflections in calculation of the building as a bar with concentrated masses on the floors and as a spatial box are practically the same, except for middle floors, where the difference is insignificant (5% - 6%). This conclusion is also applied to buildings interconnected by means of dampers.
- Connection of buildings with dampers on all floors is optional; it is sufficient to connect them in two places along the height: at a height (0.35 ÷ 0.4) h and (0.6 ÷ 0.7) h (h is a total height of the building).
- Accepted models of soil-foundation interaction affect slightly buildings of low height. This effect increases with the height of the building.
- The effect of the dampers is greater when brick building (less rigid) is attached to stack-like structure (more rigid).

References

[1] Khojmetov, G.Kh. and Yuvmitov, A.S. (2012) Vibrations of Buildings Connected with Dampers under Seismic Effects. *Uzbek Journal "Problems of Mechanics"*, **4**, 36-39.

[2] Khojmetov, G.Kh. and Yuvmitov, A.S. (2012) Seismic Vibrations of Buildings with Attached Structures Connected with Vibration Dampers. *Proceedings of International Scientific-Practical Conference Problems of Geomechanics and*

Natural Disciplines Teaching, Almaty, 18 October 2012, 278-280.

[3] Korenev, B.G. and Reznikov, L.M. (1988) Dynamic Dampers of Vibrations (Theory and Technical Propositions). Nauka, Moscow, 108 p.

[4] Aizenberg, Ya.M. (1983) Seismic Insulation and Adaptive Systems of Seismic Protection. Nauka, Moscow, 37-40, 139.

[5] Patel, C.C. and Jangid, R.S. (2010) Seismic Response of Adjacent Structures Connected with Maxwell Dampers. *Asian Journal of Civil Engineering*, **11**, 585-603.

[6] He, Q. and Xu, Y.L. (1998) Dynamic Analysis of Damper-Connected Adjacent Buildings under Earthquake. *Transaction of Tianjin University*, **4**, 128-133.

[7] Korchinskiy, I.L., Borodin, L.A., Grossman, A.B., Preobrajenskiy, V.S., Rjevskiy, V.A., Tsipenyuk, I.F. and Shepelev V.F. (1971) Quakeproof Construction of the Buildings. Edited by I. L. Korchinskiy, High School, Moscow, 125-130.

[8] Nazarov, A.G. (1959) Method of the Engineering Analysis of the Seismic Forces. Academy of the Sciences of the Armenian Republic, Erevan, 285 p.

[9] Rashidov, T.R. and Khojmetov, G.Kh. (1973) Viscous-Elastic Characteristics of Soil-Structure Interaction. No. 2, Mechanics of Polymers, Tashkent.

[10] Arif, M.Y. (1986) Seismic Stability of Structures with Non-linear Viscous-Elastic Properties of Material and Soil-Foundation Interaction. Dissertation to Award Scientific Degree of Candidate of Sciences, Tashkent, 120.

[11] Rjanitsin, A.R. (1986) Theory of Yielding. Stroyizdat, Moscow, 11-14.

[12] Khojmetov, G.K. and Yuvmitov, A.S. (2014) Coupling Unit of the System "Building-Adjacent Structure". Patent of the Republic of Uzbekistan on Effective Model, No. FAP 00973 03.11.2014.

[13] Klgatyan, Z.M. (1999) Seismic Connection for Multi-Storey Buildings. Report of Restoration of 5-Storey Residential House in Gyumri, Erevan.

[14] Nazarov, A.G. (1965) On Mechanical Similarity of Rigid Deformable Bodies. Armenian Academy of Sciences, Erevan, 218 p.

[15] Zavriev, K.S., Nazarov, A.G., Aizenberg, Ya.M., Darbinyan, S.S., Kartsivadze, G.N., Rasskazovskiy, V.T., Khachiyan E.E. and Shaghinyan, S.A. (1970) Bases of the Theory of Seismic Stability of Buildings and Structures. Edited by the Publishing House of Academician of Georgian Ac. of Sci. K.S.Zavriev, Academician of Armenian Ac. of Sci. A. G. Nazarov and Candidate of Tech. Sci. G. N. Kartsivadze, Moscow, 171.

[16] Korchinskiy, I.L., Polyakov, S.V., Bykhovskiy, V.A., Duzinkevich, S.Y. and Pavlik, V.S. (1961) Bases of Projecting of Buildings in Seismic Areas. Gosstroyizdat, Moscow, 319.

[17] Building Code 2.01.03-96 Engineering in Seismic Areas. Tashkent, 1997.

6

Nonlinear Free Vibrations of C-C-SS-SS Symmetrically Laminated Carbon Fiber Reinforced Plastic (CFRP) Rectangular Composite Plates

Zakaria Zergoune[1], Bilal Harras[1], Rhali Benamar[2]

[1]Laboratoire de Génie Mécanique, Faculté des Sciences et Techniques, Fez, Morocco
[2]Laboratoire d'Études et de Recherches en Simulation, Université Med V Rabat, École Mohammadia d'Ingénieurs, Rabat, Morocco
Email: zergoune.uni@gmail.com, harrasbilal@yahoo.fr, rhali.benamar@gmail.com

Abstract

The purpose of this paper is to apply the theoretical model developed in References [1]-[6] in order to analyze the geometrically nonlinear free dynamic response of C-C-SS-SS rectangular CFRP symmetrically laminated plates so as to investigate the effect of nonlinearity on the nonlinear resonance frequencies, the nonlinear fundamental mode shape and associated bending stress patterns at large vibration amplitudes. Various values of the plate aspect ratio and the amplitude of vibrations will be considered, and useful numerical data also are provided.

Keywords

CFRP, C-C-SS-SS Rectangular Plates, Mode Shape, Nonlinear Free Vibration, Laminate Theory

1. Introduction

The geometrically nonlinear vibration of structures is of continuing interest, due to the tendency to build more reliable structures, with high strength, high stiffness, and low weight such as composites. Such new materials have a more accentuated nonlinear behavior [1]-[6]. It is then necessary to develop new design concepts, taking into account the nonlinear behavior induced by large vibration amplitudes, which may occur for example in severe environments. The large displacement amplitudes induce geometrical nonlinearity that is mainly caused by the development of in-plane membrane stresses. In such cases, one of the basic assumptions of the linear vibration theory, *i.e.* small displacement amplitudes compared to the plate thickness is violated. Therefore, it is of crucial theoretical and practical interest, to develop new approaches which take into account the membrane stress effect, neglected in the classical treatment of plate vibrations. Even though the vibration of plates overall has interested researchers for a long time, the exact linear solutions are known only for the six cases having two opposite edges simply supported corresponding to a square plate or a rectangular plate [7] [8]. Also, as has been mentioned in the study made by Leissa in 1973 [7], it was pointed out that until 1954, when Warburton derived his formulae based on a single-term representation of the deflection shapes for the natural frequencies of plates with various boundary conditions, no solution, even approximate, was known for six boundary condition cases. The general accuracy of Warburton's formulae is discussed in references [7] [8].

A great deal of studies and a wide program of experimental work carried out by White and his co-workers [9]-[18] has been developed in order to understand the dynamic behavior of homogeneous and CFRP beams and plates. In Chia's book [19], an extensive information is given on the nonlinear analysis of plates, with presentation of a variety of geometrically nonlinear static and dynamic problems. Also, a survey was presented of the literature on the geometrical nonlinear analysis of laminated composite elastic plates in reference [20]. In Sathyamoorthy's work [21], a review was presented which deals with analytical, numerical and experimental methods used in the geometrically nonlinear dynamic analysis of plates.

In recent works carried out by F. Alijani and M. Amabili [22] [23] in 2013, nonlinear vibrations of completely free laminated and sandwich rectangular plates are investigated using a multi-modal energy approach based on Lagrange equations and by using classical and higher-order shear deformation theories [24] with von Karman type nonlinearities [25].

In several engineering applications, large vibration amplitudes of plate-type structures are encountered especially in the aerospace sector. The plates are assumed to be whether fully or partially free, simply supported, or clamped. The clamped boundary conditions assume that both displacements and rotations are prevented. In exchange, the simply supported boundary conditions assume that displacements are allowed, but rotations are not. In practice, it is extremely hard to achieve both cases [26], because the real plate boundaries are neither completely clamped nor simply supported. So, in most cases, the real boundary conditions are a combination of clamped, simply supported, or free conditions. As a real illustrative example of the combination of the boundary conditions, aircraft wing panels may be mentioned [27].

The purpose of this paper is to apply the theoretical model developed in references [1]-[3] in order to analyze the geometrically nonlinear free dynamic response of C-C-SS-SS rectangular CFRP symmetrically laminated plates so as to investigate the effect of nonlinearity on the nonlinear resonance frequencies, the nonlinear fundamental mode shape and associated bending stress patterns at large vibration amplitudes. Various values of the plate aspect ratio and the amplitude of vibration will be considered, and useful numerical data will be provided.

2. General Formulation of the Geometrically Nonlinear Free Vibration of C-C-SS-SS Symmetrically Laminated Rectangular Composite Plates

2.1. Constitutive Equation at Large Deflections

Consider the transverse vibration W of the plate of dimensions a, b and H shown in **Figure 1** with a coordinate system taken such that the xy plane coincides with the mid-plane of the plate.

The strain-displacement relationships for large deflections, considering the classical plate laminated theory [28], are given by the expressions taken from reference [29] which assume that (a) the plate is thin (the thickness H is much smaller than the typical plate dimension, H << a, b), (b) the magnitude of the transverse deflection

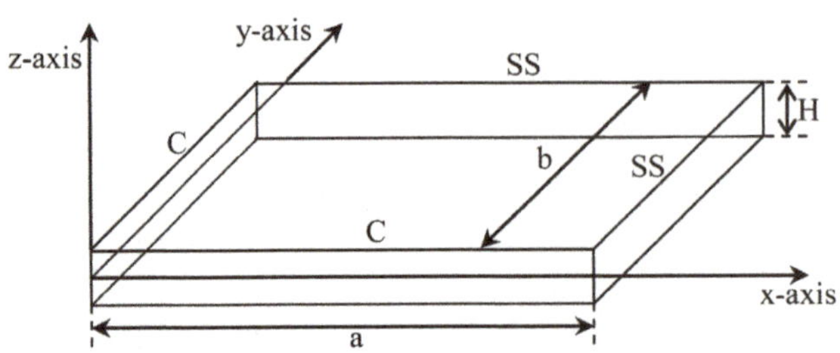

Figure 1. The plate notation.

W is of the same order as the thickness of the plate, and (c) the in-plane displacements U and V are linear functions of the z-coordinate:

$$\{\varepsilon\} = \{\varepsilon^0\} + z\{k\} + \{\lambda^0\}. \tag{1}$$

where $\{\varepsilon^0\}$, $\{k\}$ and $\{\lambda^0\}$ are given by [1]-[5]:

$$\{\varepsilon^0\} = \begin{Bmatrix} \varepsilon_x^0 \\ \varepsilon_y^0 \\ \varepsilon_z^0 \end{Bmatrix} = \begin{Bmatrix} \dfrac{\partial U}{\partial x} \\ \dfrac{\partial V}{\partial y} \\ \dfrac{\partial U}{\partial y} + \dfrac{\partial V}{\partial x} \end{Bmatrix}, \quad \{k\} = \begin{Bmatrix} k_x \\ k_y \\ k_{xy} \end{Bmatrix} = \begin{Bmatrix} -\dfrac{\partial^2 W}{\partial x^2} \\ -\dfrac{\partial^2 W}{\partial y^2} \\ \dfrac{\partial^2 W}{\partial xy} \end{Bmatrix}, \quad \text{and} \quad \{\lambda\} = \begin{Bmatrix} \lambda_x^0 \\ \lambda_y^0 \\ \lambda_z^0 \end{Bmatrix} = \begin{Bmatrix} \dfrac{1}{2}\left(\dfrac{\partial W}{\partial x}\right)^2 \\ \dfrac{1}{2}\left(\dfrac{\partial W}{\partial y}\right)^2 \\ \dfrac{\partial W}{\partial x}\dfrac{\partial W}{\partial y} \end{Bmatrix} \tag{2)(3)(4}$$

where U, V, and W are the displacements of the plate mid-plane, in the x, y and z directions respectively. In the development of the nonlinear energy expression, the transverse displacement W of the plate mid-plane has been taken into consideration while the in-plane displacements U and V have been neglected. Therefore, the strain energy (5, 6) of an elastic solid is written in Cartesian coordinates as follows [1]-[5]:

$$V_b = \frac{1}{2}\iint\left[D_{11}\left(\frac{\partial^2 W}{\partial x^2}\right)^2 + 2D_{12}\frac{\partial^2 W}{\partial x^2}\frac{\partial^2 W}{\partial y^2} + D_{22}\left(\frac{\partial^2 W}{\partial y^2}\right)^2 + 4D_{16}\frac{\partial^2 W}{\partial x^2}\frac{\partial^2 W}{\partial xy} + 4D_{26}\frac{\partial^2 W}{\partial x^2}\frac{\partial^2 W}{\partial xy} + 4D_{66}\left(\frac{\partial^2 W}{\partial xy}\right)^2\right]\mathrm{d}x\mathrm{d}y \tag{5}$$

$$V_a = \frac{1}{2}\iint\left[\frac{A_{11}}{4}\left(\frac{\partial W}{\partial x}\right)^4 + \frac{A_{22}}{4}\left(\frac{\partial W}{\partial y}\right)^4 + \left(\frac{A_{22}}{2} + A_{66}\right)\left(\frac{\partial W}{\partial y}\right)^2\left(\frac{\partial W}{\partial x}\right)^2 A_{16}\left(\frac{\partial W}{\partial x}\right)^3\frac{\partial W}{\partial y} + A_{26}\left(\frac{\partial W}{\partial y}\right)^3\frac{\partial W}{\partial x}\right]\mathrm{d}x\mathrm{d}y \tag{6}$$

The kinetic energy of the plate may be written as [1]-[5]:

$$T = \frac{1}{2}\rho H\iint\left(\frac{\partial W}{\partial t}\right)^2 \mathrm{d}x\mathrm{d}y \tag{7}$$

In which the in-plane and rotary inertia terms are neglected.

2.2. Numerical Model for the Nonlinear Mode Shapes and Resonance Frequencies of C-C-SS-SS Rectangular CFRP Symmetrically Laminated Plates

The transverse displacement function W may be written as in references [29] [30] in the form of a double series as follows:

$$W(x,y,t) = w(x,y)\sin\omega t = \sum_{i=1}^{N}\sum_{j=1}^{N} a_{ij}w_{ij}(x,y)\sin\omega t \tag{8}$$

where ij is the contribution coefficient of the function obtained as product of the ith clamped simply-supported beam mode shape in the x-direction, with the jth clamped simply-supported beam mode shape in the y-direction for the case of C-C-SS-SS rectangular plates, and vice versa.

The discretization of the strain and kinetic energy expressions (5)-(7) can be carried out leading to [1]-[5]:

$$V_b = \frac{1}{2}a_i a_j k_{ij}\sin^2\omega t, \quad V_a = \frac{1}{2}a_i a_j b_{ijkl}\sin^4\omega t \quad \text{and} \quad T = \frac{1}{2}a_i a_j \omega^2 m_{ij}\cos^2\omega t \tag{9)(10)(11}$$

where m_{ij}, k_{ij}, and b_{ijkl} are respectively the mass tensor, the rigidity tensor and the geometrical nonlinearity tensor. These tensors are related to the corresponding nondimensional ones by:

$$m_{ij} = \rho H^5 abm_{ij}^*, \quad k_{ij} = \frac{aH^5E}{b^3}k_{ij}^* \quad \text{and} \quad b_{ijkl} = \frac{aH^5E}{b^3}b_{ijkl}^* \tag{12)(13)(14}$$

where the nondimensional tensors m_{ij}^*, k_{ij}^*, and b_{ijkl}^* have the below form:

$$m_{ij}^{*} = \iint w_i^* w_j^* \mathrm{d}x^* \mathrm{d}y^* \tag{15}$$

$$\begin{aligned} k_{ij}^{*} = \iint \Bigg[& D_{11}^{*}\alpha^4 \left(\frac{\partial^2 w_i^*}{\partial x^{*2}} \frac{\partial^2 w_j^*}{\partial x^{*2}} \right) + D_{12}^{*}\alpha^2 \left(\frac{\partial^2 w_i^*}{\partial x^{*2}} \frac{\partial^2 w_j^*}{\partial y^{*2}} + \frac{\partial^2 w_i^*}{\partial y^{*2}} \frac{\partial^2 w_j^*}{\partial x^{*2}} \right) + D_{22}^{*} \left(\frac{\partial^2 W_i^*}{\partial y^{*2}} \frac{\partial^2 W_j^*}{\partial y^{*2}} \right) \\ & + 2D_{16}^{*}\alpha^3 \left(\frac{\partial^2 w_i^*}{\partial x^{*2}} \frac{\partial^2 w_j^*}{\partial x^* y^*} + \frac{\partial^2 w_i^*}{\partial x^* y^*} \frac{\partial^2 w_j^*}{\partial x^{*2}} \right) + 2D_{26}^{*}\alpha \left(\frac{\partial^2 w_i^*}{\partial y^{*2}} \frac{\partial^2 w_j^*}{\partial x^* y^*} + \frac{\partial^2 w_i^*}{\partial x^* y^*} \frac{\partial^2 w_j^*}{\partial y^{*2}} \right) \\ & + 4D_{66}^{*}\alpha^2 \left(\frac{\partial^2 w_i^*}{\partial x^* y^*} \frac{\partial^2 w_j^*}{\partial x^* y^*} \right)^2 \Bigg] \mathrm{d}x^* \mathrm{d}y^*. \end{aligned} \tag{16}$$

$$\begin{aligned} b_{ijkl}^{*} = \iint \Bigg[& \frac{A_{11}^{*}}{4}\alpha^4 \left(\frac{\partial w_i^*}{\partial x^*} \frac{\partial w_j^*}{\partial x^*} \frac{\partial w_k^*}{\partial x^*} \frac{\partial w_l^*}{\partial x^*} \right) + \frac{A_{22}^{*}}{4} \left(\frac{\partial w_i^*}{\partial y^*} \frac{\partial w_j^*}{\partial y^*} \frac{\partial w_k^*}{\partial y^*} \frac{\partial w_l^*}{\partial y^*} \right) \\ & + \left(\frac{A_{12}^{*}}{4} + \frac{A_{66}^{*}}{2} \right)\alpha^2 \left\{ \left(\frac{\partial w_i^*}{\partial y^*} \frac{\partial w_j^*}{\partial y^*} \frac{\partial w_k^*}{\partial x^*} \frac{\partial w_l^*}{\partial x^*} \right) + \left(\frac{\partial w_i^*}{\partial x^*} \frac{\partial w_j^*}{\partial x^*} \frac{\partial w_k^*}{\partial y^*} \frac{\partial w_l^*}{\partial y^*} \right) \right\} \\ & + \frac{A_{16}^{*}}{2}\alpha^3 \left\{ \left(\frac{\partial w_i^*}{\partial x^*} \frac{\partial w_j^*}{\partial x^*} \frac{\partial w_k^*}{\partial x^*} \frac{\partial w_l^*}{\partial y^*} \right) + \left(\frac{\partial w_i^*}{\partial x^*} \frac{\partial w_j^*}{\partial x^*} \frac{\partial w_k^*}{\partial y^*} \frac{\partial w_l^*}{\partial x^*} \right) \right\} \\ & + \frac{A_{26}^{*}}{2}\alpha \left\{ \left(\frac{\partial w_i^*}{\partial x^*} \frac{\partial w_j^*}{\partial y^*} \frac{\partial w_k^*}{\partial y^*} \frac{\partial w_l^*}{\partial y^*} \right) + \left(\frac{\partial w_i^*}{\partial y^*} \frac{\partial w_j^*}{\partial x^*} \frac{\partial w_k^*}{\partial y^*} \frac{\partial w_l^*}{\partial y^*} \right) \right\} \Bigg] \mathrm{d}x\mathrm{d}y. \end{aligned} \tag{17}$$

Applying Hamilton's principle to the vibration problem gives:

$$3a_i a_j a_k b_{ijkr}^{*} + 2a_i k_{ir}^{*} - 2\omega^{*2} a_i m_{ir}^{*} = 0, \quad r = 1, \cdots, n \tag{18}$$

In matrix form, the above expression (18) may be rewritten as:

$$3\left[B^*(A)\right]\{A\} + 2\left[K^*\right]\{A\} - 2\omega^{*2}\left[M^*\right]\{A\} = \{0\} \tag{19}$$

where ω^* is the nondimensional nonlinear frequency parameter, the expression of which can be obtained by pre-multiplying Equation (18) by $[A]^{\mathrm{T}}$ from the left-hand side, which leads to the following equation:

$$\omega^{*2} = \frac{a_i a_j k_{ij}^{*} + \frac{3}{2} a_i a_j a_k a_l b_{ijkl}^{*}}{a_i a_j m_{ij}^{*}} \tag{20}$$

The set of nonlinear algebraic Equations (20), in which the parameters b_{ijkl}^*, k_{ij}^* and m_{ij}^* were computed numerically by using Simpson's rule, has been solved numerically by using the Harwell library routine NS01A. This routine is based on a hybrid method combining the steepest descent and Newton's methods, to obtain the numerical results presented for the first nonlinear mode of a C-C-SS-SS anisotropic rectangular plate.

3. Applications to C-C-SS-SS Rectangular Laminated CFRP Composite Panels

Table 1, given below, shows the geometrical and material characteristics of the CFRP rectangular composite plates examined in the present work, taken from Reference [31].

Table 1. Geometric and material properties of the rectangular plate.

Number of layers	Orientation of principal axes	a (mm)	b (mm)	H (mm)
8	(90, 45, −45, 0) sym	480	320	1
E_x (Gpa)	E_y (Gpa)	G_{xy} (Gpa)	ν_{xy}	ρ (kg/m^3)
120.5	9.63	3.58	0.32	1540

In the present work, the set of admissible functions concerning C-C-SS-SS composite rectangular plates are obtained as products of linear clamped simply-supported beam mode shapes in the *x*- and *y*-directions. Such beam functions have been used previously to build plate functions and shown to be appropriate in both linear and nonlinear vibration studies [2] [5].

The beam functions, whose analytical expressions are given in Appendix, used in the present work lead to plate-functions which satisfy all of the C-C-SS-SS theoretical boundary conditions, *i.e.* zero displacement along the four edges, zero slope along the two clamped edges, and zero moment with respect to x and y along the two simply supported edges.

3.1. Linear Analysis

In order to verify the accuracy of the results obtained in the present work corresponding to the C-C-SS-SS rectangular plate, we will compare the solution of Equation (21), that is considered as an eigen problem which will be solved by using MATLAB software, with the previous published results.

$$a_j k_{ir}^* = a_i \omega^{*2} m_{ir}^*, \quad r = 1, \cdots, n \tag{21}$$

The calculation was made by using 36 basic functions obtained as products of the first six clamped simply-supported beam functions, leading to square mass and rigidity matrices of dimension 25.

In **Table 2**, the nonlinear frequency parameters obtained from the nonlinear analysis at very small vibration amplitudes (a_{11} = 0.001 and 0.15) corresponding to various values of the plate aspect ratio, are compared with results obtained from linear analysis. Consequently, the results show the convergence of the nonlinear model to the linear approach at very small vibration amplitudes.

3.2. Nonlinear Analysis

Table 3 gives a summary of the first nonlinear mode shape of C-C-SS-SS rectangular plate results, corresponding to $\alpha = 0.667$ for a_{11} = 0.001 and 0.15, which are calculated here using 25 basic functions.

The qualitative comparison that is done in **Figure 2** for the nonlinear frequency parameters of the first

Table 2. Comparison of nondimensional frequency parameters.

	Aspect ratio α				
	0.4	0.5	0.667	1.0	1.5
ω_1^*	9.12141	11.462242	17.89957	46.447142	166.25531
ω_{nl}^* (a_{11} = 0.001)	9.121334	11.462156	17.89945	46.446766	166.25356
ω_{nl}^* (a_{11} = 0.15)	9.8864237	12.339166	19.123762	49.411693	176.96215

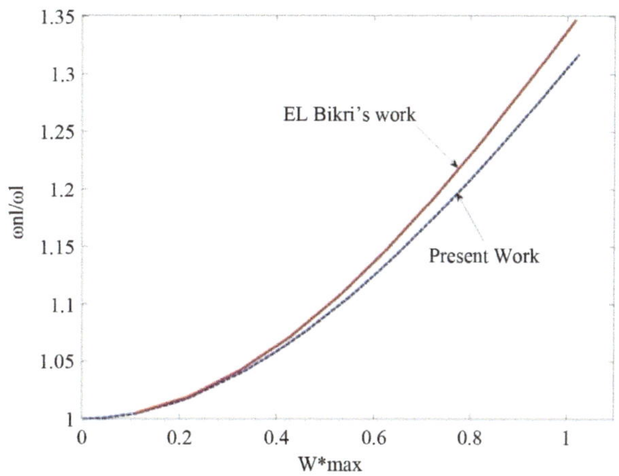

Figure 2. Comparison of the nonlinear frequency parameters of the first nonlinear mode of C-C-SS-SS rectangular plate. -: present work, ---: value taken from [31] corresponding to $\alpha = 0.4$.

Table 3. Comparison of nondimensional frequency parameters.

		$W_{max} = 0.11173319E-01$	$W_{max} = 0.26809814$
	(a) $\omega_l^* = 17.89957$	(b) $\omega_{nl}^* = 17.89945$	(c) $\omega_{nl}^* = 19.123762$
a_{11}	1	0.001	0.15
a_{12}	0.034640933	0.34693649E-04	0.71160884E-02
a_{13}	0.007849452	0.78612035E-05	0.10368977E-02
a_{14}	0.002492575	0.24964054E-05	0.61552283E-03
a_{15}	0.000990265	0.99177280E-06	0.21810887E-03
a_{21}	0.040354981	0.40415664E-04	0.62294442E-02
a_{22}	−0.001785123	−0.17877778E-05	−0.14783568E-03
a_{23}	−0.001032162	−0.10337164E-05	−0.14807638E-03
a_{24}	−0.000468162	−0.46886538E-06	−0.63046742E-04
a_{25}	−0.00021805	−0.21837845E-06	−0.32070459E-04
a_{31}	0.009393579	0.94078473E-05	0.18552141E-02
a_{32}	−0.001170972	−0.11727421E-05	−0.18827547E-03
a_{33}	−0.000763724	−0.76482717E-06	0.33978506E-04
a_{34}	−0.000396506	−0.39711893E-06	−0.10327654E-03
a_{35}	−0.000203788	−0.20409761E-06	−0.44565495E-04
a_{41}	0.003032144	0.30366964E-05	0.42418279E-03
a_{42}	−0.000565483	−0.56632875E-06	−0.67252538E-04
a_{43}	−0.000425984	−0.42663248E-06	−0.87193658E-04
a_{44}	−0.000250132	−0.25050635E-06	−0.31356367E-04
a_{45}	−0.000140975	−0.14118626E-06	−0.19236005E-04
a_{51}	0.001214007	0.12158334E-05	0.18055087E-03
a_{52}	−0.000273669	−0.27407930E-06	−0.34861000E-04
a_{53}	−0.000229392	−0.22973835E-06	−0.36839827E-04
a_{54}	−0.000148152	−0.14837451E-06	−0.21774339E-04
a_{55}	−9.01E-05	−0.90258064E-07	−0.13305140E-04

(a) Linear results calculated here; (b) present results obtained from nonlinear analysis for $a_{11} = 0.001$; (c) present results obtained from nonlinear analysis for $a_{11} = 0.15$.

nonlinear mode, between the results obtained here and those obtained in Reference [31] for an isotropic square plate with similar boundary conditions showed that the curve (continuous line) obtained in Bikri's work increases more than the curve (dashed) corresponding to the present work. On the other hand, the comparison done with the work of B. Harras *et al.* in **Figure 3** shows that the fully clamped anisotropic rectangular plate exhibits less nonlinearity than that examined in the present work.

In **Figure 4** are plotted the nonlinear frequency versus the maximum nondimensional amplitude for the first non-linear mode shape of C-C-SS-SS rectangular plate for various values of the plate aspect ratio. All curves are similar to those mentioned in References [5] [31].

The curves in **Figure 5** and **Figure 6**, obtained via the present model, are plotted respectively for the values of the maximum nondimensional amplitudes $W_{max}^*(\alpha = 1.0) = 0.223E-02,\ 1.360,\ 2.560$, and

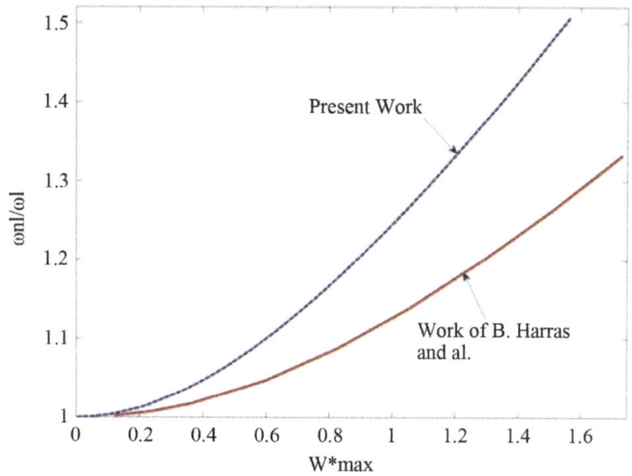

Figure 3. Comparison of the nonlinear frequency parameters of the first nonlinear mode of anisotropic C-C-SS-SS rectangular plate (present work) with anisotropic fully clamped rectangular plate (work of B. Harras *et al.* values taken from [5]) corresponding to $\alpha = 0.667$.

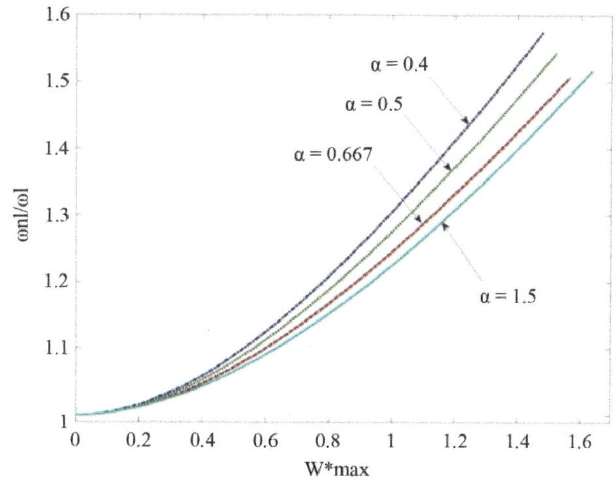

Figure 4. Comparison of the change frequency of the first mode for $\alpha = 0.4$, 0.5, 0.667, and 1.5.

$W^*_{\max}(\alpha = 1.5) = 0.222\text{E}-02,\ 1.386,\ 2.604$, along the y-direction of the nonlinear fundamental mode shape of a rectangular plate ($\alpha = 1.0$ and 0.5) corresponding to $X^*_{\max}$. They show the amplitude dependence of the nonlinear mode shape and an increase of curvatures near to the clamped edges. On the other hand, no changes are noticed close to the simply supported ones, thereby, we suppose that the flexural stresses will increase nonlinearly near to the clamped edges accompanying increase of the vibration amplitudes.

Figure 7 and **Figure 8** represent the nondimensional flexural stress distributions associated to the first non linear mode shape of a C-C-SS-SS rectangular plate, which show the effect of large vibration amplitudes along the x and y directions, corresponding to $Y^* = 0.5$ and $X^* = 0.25$ for $\alpha = 0.667$. Besides, all curves show that the flexural stress is equal to zero close to the edge where the rectangular plate is simply supported, but on the other side the flexural stress exhibit a higher increase which has been proved in the linear vibration approach.

4. Conclusions

The nonlinear free vibrations of C-C-SS-SS rectangular symmetrically laminated plates are studied in this paper

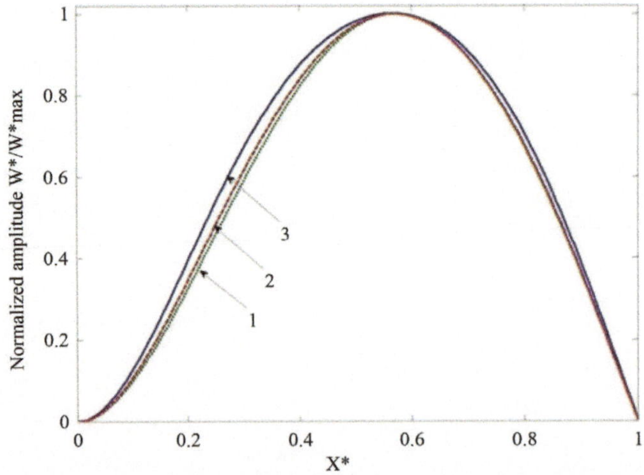

Figure 5. Normalized first nonlinear mode of a C-C-SS-SS rectangular plate corresponding to X^*_{max} and $\alpha = 1.0$. Curve 1, lowest amplitude; Curve 3, highest amplitude.

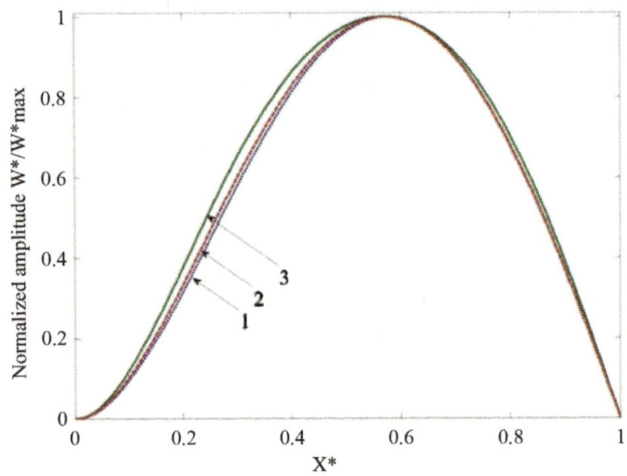

Figure 6. Normalized first nonlinear mode of a C-C-SS-SS rectangular plate for X^*_{max} and $\alpha = 1.5$. Curve 1, lowest amplitude; Curve 3, highest amplitude.

using a theoretical model developed previously in References [4]-[6] for geometrically nonlinear free vibrations of thin elastic structures. The objective was to determine the effects of large vibration amplitudes on the first nonlinear mode shape, the fundamental natural frequency, and the associated flexural stress distribution. The model, based on Hamilton's principle, reduces the nonlinear free vibration problem to solution of a set of nonlinear algebraic equations involving the classical rigidity and mass tensors and a fourth order tensor due to the geometrical nonlinearity.

The study of the first nonlinear mode shape for the CFRP plates considered has shown that the curvature near the edge increases very rapidly with increase of the vibration amplitude. In addition, the mode shape close to the center of the plate becomes flatter when the amplitude increases. It was also noticed that the deformation of the mode shape, for a given value of the normalized amplitude of vibration, increases as the aspect ratio α increases. The nondimensional bending stress distribution associated with the CFRP rectangular plate first nonlinear mode shows an increase with increasing a_{11} which is much higher than that obtained for CFRP plates of identical aspect ratio. As a consequence of the deformation of the mode shape, a higher rate of increase was noticed in the induced bending stresses in a region close to the clamped edges. Furthermore, the nondimensional bending stress distribution associated with the CFRP rectangular plate first nonlinear mode for different amplitudes-

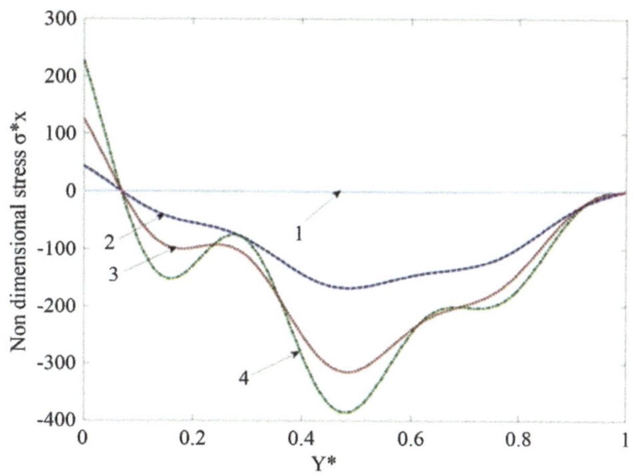

Figure 7. Nondimensional bending stress distribution associated with C-C-SS-SS rectangular plate first nonlinear mode along $Y^* = 0.5$ for $\alpha = 0.667$. Curve 1, lowest amplitude; Curve 4 highest amplitude.

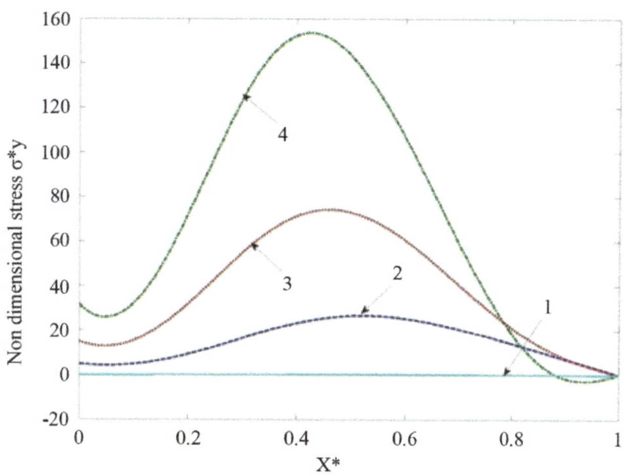

Figure 8. Nondimensional bending stress distribution associated with C-C-SS-SS rectangular plate first nonlinear mode along $X^* = 0.25$ for $\alpha = 0.667$. Curve 1, lowest amplitude; Curve 4 highest amplitude.

of vibration show a nonsymmetry along the length of the section considered because of the influence of the fiber orientation. It is worth noting here that further investigations are needed in order to check the distribution of the nonlinear bending stresses obtained at very high amplitudes of vibration.

At last, this present work enabled us to find out how the thin elastic structures behave at the large vibration amplitude by plotting different nonlinear dynamic characteristics. In addition, the qualitative comparison done with the results of an isotropic rectangular plate with similar boundary conditions obtained in reference [31], and whith an anisotropic rectangular plate examined in reference [5] showed good and satisfactory results.

References

[1] Benamar, R., Bennouna, M.M.K. and White, R.G. (1991) The Effects of Large Vibration Amplitudes on the Fundamental Mode Shape of Thin Elastic Structures. Part I: Simply Supported and Clamped Clamped Beams. *Journal of Sound and Vibration*, **149**, 179-195. http://dx.doi.org/10.1016/0022-460X(91)90630-3

[2] Benamar, R., Bennouna, M.M.K. and White, R.G. (1993) The Effects of Large Vibration Amplitudes on the Fundamental Mode Shape of Thin Elastic Structures. Part II: Fully Clamped Rectangular Isotropic Plates. *Journal of Sound*

and Vibration, **164**, 295-316. http://dx.doi.org/10.1006/jsvi.1993.1215

[3] Benamar, R., Bennouna, M.M.K. and White, R.G. (1990) The Effects of Large Vibration Amplitudes on the Fundamental Mode Shape of a Fully Clamped, Symmetrically Laminated Rectangular Plate. *Proceedings of the* 4*th International Conference on Recent Advances in Structural Dynamics*, Southampton.

[4] Harras, B., Benamar, R. and White, R.G. (2002) Experimental and Theoretical Investigation of the Linear and Non-Linear Dynamic Behavior of a Glare 3 Hybrid Composite Panel. *Journal of Sound and Vibration*, **2**, 281-315. http://dx.doi.org/10.1006/jsvi.2001.3962

[5] Harras, B., Benamar, R. and White, R.G. (2002) Geometrically Non-Linear Free Vibration of Fully Clamped Symmetrically Laminated Rectangular Composite Plates. *Journal of Sound and Vibration*, **4**, 579-619. http://dx.doi.org/10.1006/jsvi.2001.3713

[6] Harras, B., Benamar, R. and White, R.G. (2002) Investigation of Non-Linear Free Vibrations of Fully Clamped Symmetrically Laminated Carbon-Fibre-Treinforced PEEK (AS4/APC2) Rectangular Composite Panels. *Composite Science and Technology*, **62**, 719-727. http://dx.doi.org/10.1016/S0266-3538(01)00133-6

[7] Leissa, A.W. (1973) Free Vibrations of Rectangular Plate. *Journal of Sound and Vibration*, **31**, 257-293. http://dx.doi.org/10.1016/S0022-460X(73)80371-2

[8] Leissa, A.W. (1969) Vibration of Plates. NASA-SP-160. US Government Printing Office, Washington DC.

[9] White, R.G. (1978) A Comparison of Some Statistical Properties of the Responses of Aluminum and CFRP Plates to Acoustic Excitation. *Composites*, **9**, 251-258. http://dx.doi.org/10.1016/0010-4361(78)90178-7

[10] Wolfe, H. (1995) An Experimental Investigation of Non-Linear Behaviour of Beams and Plates Excited to High Levels of Dynamic Response. Ph.D. Thesis, University of Southampton, Southampton.

[11] White, R.G. (1988) The Acoustic Excitation and Fatigue of Composite Plates. *Proceedings of the* 29*th AIAA/ASME Structures, Structural Dynamics and Materials Conference*, Williamsburg, Paper 88-2242, 253-260.

[12] Bennouna, M.M.K. (1981) Non-Linear Behaviour of a Clamped-Clamped Beam with Consideration of Fatigue Life. Ph.D. Thesis, Institute of Sound and Vibration Research, Southampton.

[13] Benamar, R. (1990) Non-Linear Dynamic Behaviour of Fully Clamped Beams and Rectangular Isotropic and Laminated Plates. Ph.D. Thesis, Institute of Sound and Vibration Research, Southampton.

[14] White, R.G. (1971) Effects of Non-Linearity Due to Large Deflections in the Resonance Testing of Structures. *Journal of Sound and Vibration*, **16**, 255-267. http://dx.doi.org/10.1016/0022-460X(71)90486-X

[15] White, R.G. (1975) Some Measurements of the Dynamic Mechanical Properties of Mixed, Carbon Fiber Reinforced Plastic Beams and Plates. *Aeronautical Journal*, **79**, 318-325.

[16] White, R.G. and Abdin, E.M.Y. (1985) Dynamic Properties of Aligned Short Carbon Fiber-Reinforced Plastics in Flexure and Torsion. *Composites*, **16**, 293-306. http://dx.doi.org/10.1016/0010-4361(85)90282-4

[17] White, R.G. and Mousley, R.F. (1987) Dynamic Response of CFRP Plates under the Action of Random Acoustic loading. *Proceedings of the Fourth International Conference on Composite Structures*, Paisley College of Technology, 1519-1535.

[18] Ng, C.F. and White, R.G. (1988) Dynamic Behaviour of Postbuckled Isotropic Plates under In-Plane Compression. *Journal of Sound and Vibration*, **120**, 1-18. http://dx.doi.org/10.1016/0022-460X(88)90331-8

[19] Chia, C.Y. (1980) Nonlinear Analysis of Plates. McGraw-Hill, New York.

[20] Chia, C.Y. (1988) Geometrically Nonlinear Behavior of Composite Plates: A Review. *Applied Mechanics Reviews*, **41**, 439-451. http://dx.doi.org/10.1115/1.3151873

[21] Sathyamoorthy, M. (1987) Nonlinear Vibration Analysis of Plates: A Review and Survey of Current Developments. *Applied Mechanics Reviews*, **40**, 1553-1561. http://dx.doi.org/10.1115/1.3149544

[22] Alijani, F. and Amabili, M. (2013) Nonlinear Vibrations of Laminated and Sandwich Rectangular Plates with Free Edges. Part 1: Theory and Numerical Simulations. *Composite Structures*, **105**, 422-436. http://dx.doi.org/10.1016/j.compstruct.2013.05.034

[23] Alijani, F., Amabili, M., Ferrari, G. and D'Alessandro, V. (2013) Nonlinear Vibrations of Laminated and Sandwich Rectangular Plates with Free Edges. Part 2: Experiments & Comparisons. *Composite Structures*, **105**, 437-445. http://dx.doi.org/10.1016/j.compstruct.2013.05.020

[24] Reddy, J.N. (1996) Mechanics of Laminated Composite Plates: Theory and Analysis. CRC Press, Boca Raton.

[25] Nayfeh, A.H. and Frank Pai, P. (2004) Linear and Nonlinear Structural Mechanics. John Wiley & Sons, Inc., Hoboken. http://dx.doi.org/10.1002/9783527617562

[26] Abrate, S. and Schoeppner, G. (1997) Identification of Support Conditions for Composite Beams and Plates from Natural Frequencies. 29 *International SAMPE* (*Society for the Advancement of Materials and Process Engineering*) *Tech-*

nical Conference, Orlando, 28 October-1 November, 725-732.

[27] Teh, C.E. (1989) Dynamic Behaviour and Acoustic Fatigue of Isotropic and Anisotropic Panels under Combined Acoustic Excitation and Static in Plane Compression. Ph.D. Thesis, Institute of Sound and Vibration Research, Southampton.

[28] Ashton, J.E. and Whitney, J.M. (1970) Theory of Laminated Plates. Technomic Publication, Stanford.

[29] Timoshenko, S. (1959) Theory of Plates and Shells. McGraw-Hill, New York, 416.

[30] Timoshenko, S., Weinsowsky-Krieger, S. and Jones, R.M. (1975) Mechanics of Composite Materials. International Student Edition, McGraw-Hill Kogakusha, Ltd., Tokyo, 51.

[31] El Bikri, K., Benamar, R. and Bennouna, M. (2003) Geometrically Non-Linear Free Vibrations of Clamped Simply Supported Rectangular Plates. Part I: The Effects of Large Vibration Amplitudes on the Fundamental Mode Shape. *Journal of Sound and Vibration*, **81**, 2029-2043.

Nomenclature

$\{\varepsilon\}$	column matrix of total strains;
$\{\varepsilon^0\}$	column matrix of strains due to the in-plane displacements U, V, W;
x, y, z	point co-ordinates;
$\{K\}$	column matrix of bending or twisting;
$\{\lambda^0\}$	column matrix of strains induced by large displacements W;
$\varepsilon_x, \varepsilon_y, \gamma_{xy}$	strain tensor components;
$\sigma_x^{(k)}, \sigma_y^{(k)}, \sigma_{xy}^{(k)}$	stresses in the kth layer;
$[\bar{Q}]$	6 × 6 matrix of transformed stiffness;
$[\bar{Q}]^k$	6 × 6 matrix of transformed stiffness for the kth layer;
$N_x^{(k)}, N_y^{(k)}, N_{x,y}^{(k)}$	force resultant for the kth layer;
$M_x^{(k)}, M_y^{(k)}, M_{x,y}^{(k)}$	moment resultant for the kth layer;
h_k	distance from the mid-plane to the layer surface of the kth layer;
A_{ij}, B_{ij}, D_{ij}	extensional, coupling and bending stiffness coefficients for the laminated plate;
A_{ij}^*, D_{ij}^*	nondimensional extensional and bending stiffness coefficients;
$\{A_k\}^T$	matrix of coefficients corresponding to the kth harmonic;
$\{W\}^T$	basic spatial functions matrix;
a_{ij}	contribution coefficient of the plate deflection function obtained as a product of the ith and j^{th} beam mode shapes in the x and y directions respectively;
a, b	length, width of the plate;
E	Young's modulus;
H	plate thickness;
K_{ij}, m_{ij}, b_{ijkl}.	general term of the rigidity tensor, the mass tensor and the nonlinearity tensor respectively;
$K_{ij}^*, m_{ij}^*, b_{ijkl}^*$	general term of the nondimensional rigidity tensor, the mass tensor and the nonlinearity tensor respectively;
$U(x,y,t)$, $V(x,y,t)$	in-plane displacements at point (x,y) of the plate;
V_b, V_a and V	bending, axial and total strain energy respectively;
T	kinetic energy;
$W(x,y,t)$	transverse displacement at point (x,y) of the plate midplane;
$W^*(x,y,t)$	nondimensional transverse displacement at point (x,y) of the plate midplane;
$W_{\max}^*$	maximum of the nondimensional transverse displacement;
α	nondimensional parameter (plate aspect ratio) given by $\alpha = b/a$;
ν_{xy}	major Poisson's ratio;
E_x	longitudinal modulus of elasticity;
E_y	transverse modulus of elasticity;
G_{xy}	shear modulus;
ρ	mass density per unit volume of the plate;
ω, ω^*	Frequency and nondimensional frequency parameter respectively;
ω_l, ω_{nl}	linear frequency and nonlinear frequency respectively;
σ_{xy}, σ_{yx}	dimensional bending stresses;
$\sigma_{xy}^*, \sigma_{yx}^*$	nondimensional bending stresses.

Appendix

The clamped simply-supported beam functions $P_i^*(x)$ are defined as follows (**Table 4** and **Figure 9**):

$$P_i^*(x) = \left[\cos\left(\frac{\lambda_i x}{a}\right) - \cosh\left(\frac{\lambda_i x}{a}\right)\right] - \gamma_i\left[\sin\left(\frac{\lambda_i x}{a}\right) - \sinh\left(\frac{\lambda_i x}{a}\right)\right] \tag{22}$$

where

$$\gamma_i = \frac{\mathrm{ch}(\lambda_i) - \cos(\lambda_i)}{\mathrm{sh}(\lambda_i) - \sin(\lambda_i)} \tag{23}$$

Table 4. The eigenvalue parameters for a C-C-SS-SS beam.

i	λ_i	i	λ_i
1	3.926990817	9	29.059732046
2	7.068583471	10	32.201324700
3	10.210176124	11	35.342917353
4	13.351768778	12	38.484510006
5	16.493361431	13	41.626102660
6	19.634954085	14	44.767695314
7	22.776546739	15	47.909287967
8	25.918139392	16	51.050880621

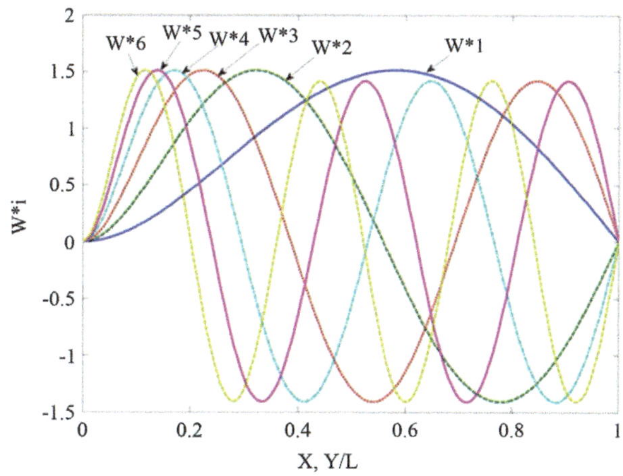

Figure 9. Clamped simply-supported beam functions for i = 1, 2, 3, 4, 5, and 6.

In which λ_i for $i = 1, 2, \cdots$ are the eigen value parameters for a clamped simply-supported beam. Besides the values of the parameters λ_i were computed solving numerically the transcendental equation $\tanh\lambda_i - \tan\lambda_i = 0$ by using Newton's method.

$$w_i^* = \frac{1}{G} P_{\alpha i}^* \left(x^*\right) P_{\beta i}^* \left(x\right) \tag{24}$$

where

$$x^* = \frac{x}{a}, \quad y^* = \frac{y}{b}, \quad w_i^* \left(x^*, y^*\right) = \frac{w_i}{H\left(x^*, y^*\right)} \tag{25}$$

G is a normalization scaling factor given by:

$$G = \sqrt{\int\int \left(X_{\alpha i}^* \left(x^*\right) Y_{\beta i}^* \left(x\right)\right)^2 \mathrm{d}x^* \mathrm{d}y^*} \tag{26}$$

7

A Novel Technique for Evaluating Failure Probability of Random Structure

Seifedine Kadry[1], Abdelkhalak El Hami[2]

[1]American University of the Middle East, Egaila, Kuwait
[2]INSA of Rouen Institute, Normandy University, St. Etienne du Rouvray, France
Email: skadry@gmail.com, abdelkhalak.elhami@insa-rouen.fr

Abstract

The purpose of this article is to develop a new methodology to evaluate the statistical characteristic of the response of structures subjecting to random excitation, by combining the Finite Element Method (FEM) with the Transforming Density Function (TDF). Uncertainty modeling of structure with random variables encourages the coupling of advanced TDF for reliability analysis to analyze problems of stochastic mechanical systems. The TDF is enthusiastically applicable in the situation where the relationship between input and output of structures is available in explicit analytical form. However, the situation is much more involved when it is necessary to perform the evaluation of implicit expression between input and output of structures through numerical models. For this aim, we propose a new technique that combines the FEM software, and the TDF method to evaluate the most important statistical parameter the Probability Density Function (PDF) of the response where the expression between input and output of structures is implicit. Once the PDF is evaluated, all other statistical parameters are derived easily. This technique is based on the numerical simulations of the FEM and the TDF by making a middleware between Finite Element software and Matlab. Some problems, range from simple to complex, of structures are analyzed using our proposed technique. Its accuracy is validated through Monte-Carlo simulation.

Keywords

Random Parameter, Uncertainty, Transforming Density Function, Finite Element Method, Finite Element Software, Matlab, Structures

1. Introduction

Many applications in mechanics require consideration of stochastic properties of materials, geometry or loads.

The problem of reliability analysis of stochastic mechanical systems is of central importance in the safety assessment of structures. In a stochastic system, a large number of random variables influence the performance of the system, e.g. young's modulus, external loads... The performance of the system is evaluated by a *best-estimate* code. Consider a performance criterion Y of the system depending on the input variables $X_1, X_2, \cdots, X_n$. the function $Y = g(X_1, X_2, \cdots, X_n)$ is a random variable to be determined. In order to get the information about the uncertainty of Y, a number of *FE* runs have to be performed. For each of these runs, all identified uncertain parameters are varied simultaneously. According to the exploitation of the result of these studies, the uncertainty on the response can be evaluated either in the form of an uncertainty range, or in the form of a probability density function.

Reliability methods focus on the calculation of the probability of failure associated with a limit-state function. First-order and second-order reliability approximations methods (FORM and SORM) and various simulation methods are commonly used in reliability analysis [1]. Because of the typically high level of reliability of civil structures, the failure probability is usually small (in the order of 102 - 106). To account for the spatial variability of uncertain quantities (for instance, that of a material property), a characterization in terms of a random field is usually employed. Through a process of discretization, it is possible to represent the random field by a vector of random variables. One of the methods mentioned above may then be used to carry out second-moment or reliability analysis. The spectral stochastic finite element method (SSFEM) proposed by Ghanem and Spanos [2]- [5] is an approach well suited to analysis involving random fields. It is based on two discretization of the system of (stochastic) partial differential equations governing the problem under consideration, one in the spatial domain and one in the probabilistic domain. The response (e.g. the random vector of nodal displacements) is cast as a series expansion in standard normal random variables. This can be interpreted as an "intrinsic" representation of the random response, from which quantities such as statistical moments can be computed by post- processing, either analytically or by simulation.

In this paper, a proposed technique: the Finite Element Method (FEM) software coupled with the Transforming Density Function (TDF) program is applied in order to evaluate numerically the probabilistic and statistical characteristics of the response of stochastic mechanical system. To demonstrate the efficiency of the proposed technique, we have carried out different applications to cover several structural problems

2. Reliability Analysis

Reliability methods aim at evaluating the probability of failure of a system whose modeling takes into account randomness. The design of structures and the prediction of their good functioning lead to the verification of a certain number of rules resulting from the knowledge of physical and mechanical experience of designers and constructors. These rules traduce the necessity to limit the loading effects such as stresses and displacements. Each rule represents an elementary event and the occurrence of several events leads to a failure scenario. The objective is then to evaluate the failure probability corresponding to the occurrence of critical failure modes. In addition to the vector of deterministic variables x to be used in the system control and optimization, the uncertainties are modeled by a vector of stochastic physical variables Y affecting the failure scenario. The knowledge of these variables is not, at best, more than statistical information and we admit a representation in the form of random variables. For a given design rule, the basic random variables are defined by their probability distribution associated with some expected parameters; the vector of random variables is noted herein Y whose realizations are written y. The safety is the state where the structure is able to fulfill all the functioning requirements: mechanical and serviceability, for which it is designed. To evaluate the failure probability with respect to a chosen failure scenario, a limit state function $G(x, y)$ is defined by the condition of good functioning of the structure. The limit between the *state of failure* $G(x, y) \le 0$ and the *state of safety* $G(x, y) > 0$ is known as the *limit state surface* $G(x, y) = 0$ (**Figure 1**). The failure probability is then calculated by [1]:

$$P_f = \Pr\left[G(x, y) \le 0\right] = \int_{G(x,y)\le 0} f_Y(y)\,\mathrm{d}y_1 \cdots \mathrm{d}y_n \tag{1}$$

where P_f is the failure probability, $f_Y(y)$ is the joint density function of the random variables Y and Pr [6] is the probability operator. The evaluation of integral (1) is not easy, because it represents a very small quantity and all the necessary information for the joint density function are not available. For these reasons, the *First and the Second Order Reliability Methods FORM/SORM* have been developed. They are based on the reliability index

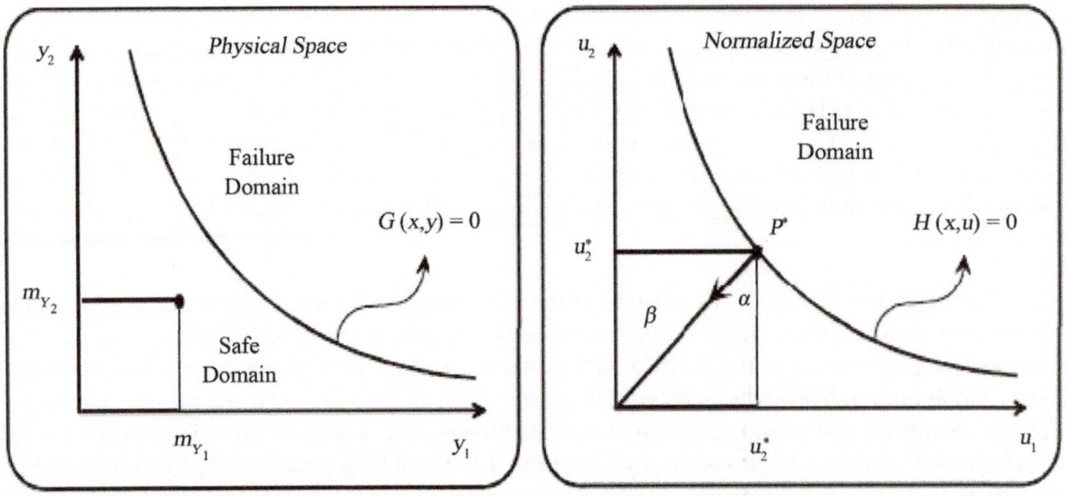

Figure 1. Physical and normalized spaces.

concept, followed by an estimation of the failure probability. The invariant reliability index β was introduced by working in the space of standard independent Gaussian variables instead of the space of physical variables. The transformation from the physical variables y to the normalized variables u is given by:

$$u = T(x,y) \quad \text{and} \quad y = T^{-1}(x,u) \tag{2}$$

This transformation $T(.)$ is called the *probabilistic transformation.* In this standard space, the limit state function takes the form:

$$H(x,u) \equiv G(x,y) = 0 \tag{3}$$

For practical engineering, Equation (3) gives sufficiently accurate estimation of the failure probability.

2.1. Approximate Reliability Methods (FORM & SORM)

The First Order Reliability Method (FORM) uses the closest point on the limit state function $H(u) = 0$ to the origin in the standard normal space as a measure of the reliability. This point is called design point u^* and its absolute beta value $\beta = \|u^*\|$ specifies the reliability or complementary the failure probability P_f, $P_f \approx \Phi(-\beta) = \Phi(-\|u^*\|)$, where $\Phi(.)$ denotes the standard normal cumulative distribution function. FORM leads to exact results in case a single linear limit state function $H(u) = 0$ is considered.

Second Order Reliability Methods (SORM) approximate the limit state function by an incomplete second order polynomial, which assumes in a rotated space the simple form

$$H(u) \approx \tilde{H}(u) = \beta - u_1 + \frac{1}{2}\sum_{i=2}^{n} k_i u_i^2 \tag{4}$$

The form is incomplete since the terms $\{(\beta - u_1)u_i\}_{i=2}^{n}$, leading to a complete second order approximation around the design point, are missing. An exact result in form of a one dimensional integral has been derived in [7] for the incomplete representation and an asymptotic result, sufficiently accurate for large β value, has been developed [8].

2.2. Importance Sampling

Importance Sampling has been one of the most prevalent approaches in the context of Simulation based methods for the estimation of structural reliability [9]. The underlying concept is to draw samples of the vector of random parameters y from a distribution $f(y)$ which is concentrated in the "important region" of the random parameter space.

2.3. Line Sampling

An alternative, quite suitable for large dimension n and efficient in context with FEA, is the following approach denoted as "line sampling". It requires an important direction a as starting point, defining the direction along the limit state $G(x, y)$ will be determined. Each point (x, y) in the standard point is decomposed into the one dimensional space [10].

2.4. Some Factors of Comparison

Procedures for estimating probabilities of failure developed over the last twenty years like FORM/SORM, Importance Sampling, and all its variants built on it, lack robustness or computational efficiency as the number of random variables (dimension n) tends to infinity. However, the robust and simple straight forward Line sampling procedures is able to overcome most difficulties encountered in traditional procedures. Summarizing the arguments, the following can be concluded:

- FORM provides a point estimate, subject to linearization errors, without confidence. Moreover, it requires the evaluation of the design point, which becomes difficult in high dimension for nonlinear limit state functions in the standard normal space. The efforts to compute the design point grows proportional with the dimension n.
- SORM requires in addition to the design point $n > 1$ main curvature with cannot be obtained in a feasible manner for high dimensions n. The procedure implies that the domain close to the design point is the important domain which is not the case for high dimensions.
- Importance Sampling is more robust and accurate as FORM or SORM, but not competitive to Line Sampling, because it is generally impossible to sample according to the optimal sampling density. The approach has difficulties to deal with multiple failure domains if they are not well separated.

Line Sampling is capable to take advantage of simple flat limit states in standard normal space and samples in the most important domain, without assuming a linear or quadratic limit state surface or requiring a design point computation.

3. Finite Element Method and TDF

3.1. Finite Element Method

The Finite Element Method is the standard tool for certain classes of partial differential equations arising in various fields of engineering and in particular for those arising in solid mechanics. For linear systems enforcing global static or dynamic equilibrium the FE Analysis leads to a system of linear equations, respectively

$$KU = F \tag{5}$$

$$M\ddot{U}(t) + C\dot{U}(t) + KU(t) = F(t) \tag{6}$$

where the matrices K and M are the global stiffness and mass matrices, respectively, obtained by adding the contributions of all element matrices

$$K = \sum_e K^e = \sum_e \int_{\Omega^e} D^e B^e \mathrm{d}\Omega^e \tag{7}$$

$$M = \sum_e M^e = \sum_e \int_{\Omega^e} \rho^e H^{e\mathrm{T}} H^e \mathrm{d}\Omega^e \tag{8}$$

where B^e is the matrix relating element displacements and strains, D^e is the elasticity matrix relating stresses and strains, ρ^e is the mass density, H^e is the shape functions and e is the spatial domain of the element. The global damping matrix C is typically formulated in terms of M and K.

3.2. Transforming Density Function TDF

The Transforming Density Function is based on the following theorem

Theorem: Suppose that X is a continuous random variable with PDF, $f(x)$ and $A \subset \Re$ is the one dimensional space where $f(X) \succ 0$, is differentiable and monotonic. Consider the random variable $Y = u(X)$, where $y = u(x)$ defines a one-to-one transformation that maps the set A onto a set $B \subset \Re$ so that the equation $y = u(x)$ can be uniquely solved for x in terms of y, say $x = u^{-1}(y)$. Then, the PDF of Y is

$$f_Y(Y) = f_X\left[u^{-1}(y)\right] * |J| \tag{9}$$

where $J = \frac{dx}{dy} = \frac{du^{-1}(y)}{dy}$ is the transformation Jacobean, which must be continuous for all points $y \in B$. The TDF is based on one-to-one mapping between the random output(s) and input(s) where the transformation Jacobean J can be computed. The PDF of the output(s) is then computed through the known joint PDF of the inputs multiplied by the determinant of transformation Jacobean matrix. The idea of TDF is based on the following formula [11]

$$f_u(u) = |J| \cdot f_z(z) = \left|\frac{dz}{du}\right| \cdot f_z(z) \tag{10}$$

$f_u(u)$ is the probability density function of the variable *u*. $f_z(z)$ is the probability density function of the variable *z*.

The general steps in the application of the TDF program: 1) the random variable input is generating and the stochastic equation of equilibrium is solved first using FEM software. 2) This solution is used to compute numerically the function between the input and the output by the Cubic Spline Interpolation, which is then inverted for the calculation of the determinant of the transformation Jacobean. Finally, the PDF of the response at any point in the domain can be deduced by using the formula (7). This approach has the advantage of giving a closed-form of the density function of the response, which is very helpful for reliability analysis of mechanical systems [3] [5]. To overcome the drawback of TDF, in analysis of complicated structures, we propose in this paper an interface between the FEM software and the TDF program.

4. Finite Element Method Coupled with Transforming Density Function: FEMCTDF

Before you begin to format your paper, first write and save the content as a separate text file. Keep your text and graphic files separate until after the text has been formatted and styled. Do not use hard tabs, and limit use of hard returns to only one return at the end of a paragraph. Do not add any kind of pagination anywhere in the paper. Do not number text heads—the template will do that for you.

Finally, complete content and organizational editing before formatting. Please take note of the following items when proofreading spelling and grammar:

The FEM software is used to perform the structural analysis to obtain the structure weight, maximal displacement, and maximal stress, corresponding to a set of given design variables. These analysis results are sent to the TDF program to conduct the Probabilistic Density Function PDF, and the probability of failure and generate new random variables. The newly generated variables are then used to update the input file. The FEM software is then invoked again to perform the structural analysis with the new input parameters. This process is repeated until satisfactory results are obtained.

4.1. Finite Element Analysis

Finite element codes are less complicated than many of the word processing and spreadsheet packages found on modern microcomputers. Nevertheless, they are complex enough that most users do not find it effective to program their own code. A number of prewritten commercial codes are available, representing a broad price range and compatible with machines from microcomputers to supercomputers 1. However, users with specialized needs should not necessarily shy away from code development, and may find the code sources available in such texts as that by Zienkiewicz 2 to be a useful starting point. Most finite element software is written in Fortran, but some newer codes such as felt are in C or other more modern programming languages. In practice, a finite element analysis usually consists of three principal steps:

1) Preprocessing: The user constructs a model of the part to be analyzed in which the geometry is divided into a number of discrete sub regions, or "elements," connected at discrete points called "nodes." Certain of these nodes will have fixed displacements, and others will have prescribed loads. These models can be extremely time consuming to prepare, and commercial codes vie with one another to have the most user-friendly

graphical "preprocessor" to assist in this rather tedious chore. Some of these preprocessors can overlay a mesh on a preexisting CAD file, so that finite element analysis can be done conveniently as part of the computerized drafting-and-design process.

2) Analysis: The dataset prepared by the preprocessor is used as input to the finite element code itself, which constructs and solves a system of linear or nonlinear algebraic equations $K_{ij}u_j = f_i$ where u and f are the displacements and externally applied forces at the nodal points. The formation of the K matrix is dependent on the type of problem being attacked, and this module will outline the approach for truss and linear elastic stress analyses. Commercial codes may have very large element libraries, with elements appropriate to a wide range of problem types. One of FEA's principal advantages is that many problem types can be addressed with the same code, merely by specifying the appropriate element types from the library.
3) Postprocessing: In the earlier days of finite element analysis, the user would pore through reams of numbers generated by the code, listing displacements and stresses at discrete positions within the model. It is easy to miss important trends and hot spots this way, and modern codes use graphical displays to assist in visualizing the results. A typical postprocessor display overlays colored contours representing stress levels on the model, showing a full-field picture.

4.2. Algorithm of Method FEMCTDF

We will implement our method in Matlab. The outline of the proposed technique FEMCTDF is as follows:

1) Generate the input random variables;
2) Calculate the value of output variables by FE software, for each value of input the correspondent value of output is estimating using FEM and it is stocking in solution file;
3) Approximate the function between input and output variables using Cubic Spline Interpolation;
4) Calculate the determinant of Jacobean of input and output variables (Equation (10));
5) Apply the basic relation of TDF (Equation (10));
6) Evaluate the graphic of PDF of output variable in function of this output variable (in our case the PDF of displacement in function of displacement);
7) Approximate the Probability of failure P_f.

We start with the creation of file variables. Out that serves to stock the random variables generating by TDF program in Matlab in order to use them by the executable file under FEM software. Once the FEM software receives these variables it introduces them in the modelling of the considered structure, and by the FEM and Static analysis done by the FEM software we get the results (nodal displacement, strains, and so on). These results are stocked in the file response. out. This later file is communicated to the TDF program in order to calculate the PDF of the output variable of the considered structure (in our case the vertical displacement u_y) and the probability of failure of this structure.

5. Application

Three case studies presented in this section all have implicit function of output variable in terms of the input random variable. The proposed technique use the FEM for modelling the structure in every iteration and calculate the value of displacement (output variable) corresponding at each input variable value, and in other hand it would be necessary to use the TDF program for estimating the value of P_f (probability of failure) of the considered structure. In order to validate the FEMCTDF technique and to compare the results, the Monte Carlo Simulation is also employed as reference method for comparison.

5.1. Three Bar Truss

In this example, the three-bar truss (**Figure 2** and **Figure 3**) is analysed using our proposed technique. For that purpose, statistic models must be defined for each random variable involved. Hence, uniform distribution in the range [5] [10] is assumed for the Young's modulus E and the other parameters are considered as deterministic variables. These parameters (Geometrical and material proprieties) are: the section of each beam $A = 1$ m^2, the length $l = 3$ m, the load $W = 1.2$ N, the mass density $\rho = 7800$ kg/m^3.

From **Figure 4**, the PDF's of the normalized vertical displacement u_y of the 3bar structure calculated by our proposed technique FEMCTDF are independent and uniformly distributed in the range [1.2505; 2.5008].

Figure 2. Real application of three bar truss.

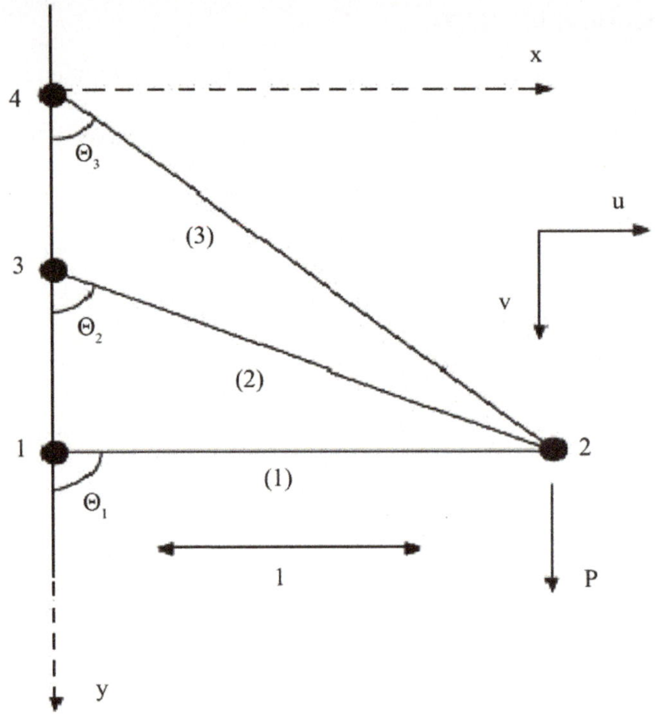

Figure 3. Three bar truss.

Also in this case the obtained results are accurate as shown by favourable comparison with classical Monte Carlo simulations. For Reliability Analysis, let us suppose that the limit displacement is $u_{\text{limit}} = 2\text{ mm}$, It is required to find the failure probability $P_f = P(u \le u_{\text{limit}})$.

The numerical values of probabilistic characteristics of the displacement of this 3bar structure are listed in this **Table 1**.

5.2. Spatial Truss

This application treats structural analysis on one hand of the arrow of a crane of construction that one assimilates to a spatial truss. It is constituted by identical beams (**Figure 5** and **Figure 6**). To the extremity of this truss is applied a load $M = 5$ t. The bars forming the structure are in steel of which Young's modulus E is uniformly distributed in the range [100 GPa, 300 GPa] and the Poisson coefficient $\sigma = 0.29$. The bars have circular sections of which the big one is worth $A1 = 22\text{ mm}^2$ and the small one $A2 = 11\text{ mm}^2$. The weight of each of the bars of

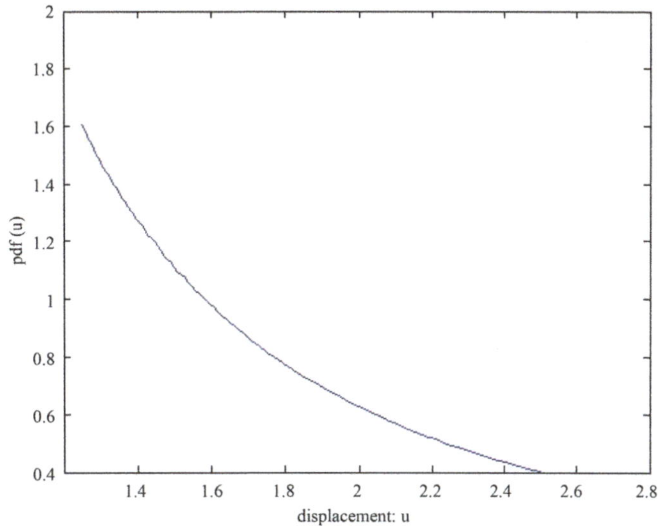

Figure 4. PDF (u) when E is uniformly distributed.

Figure 5. Real application of spatial truss.

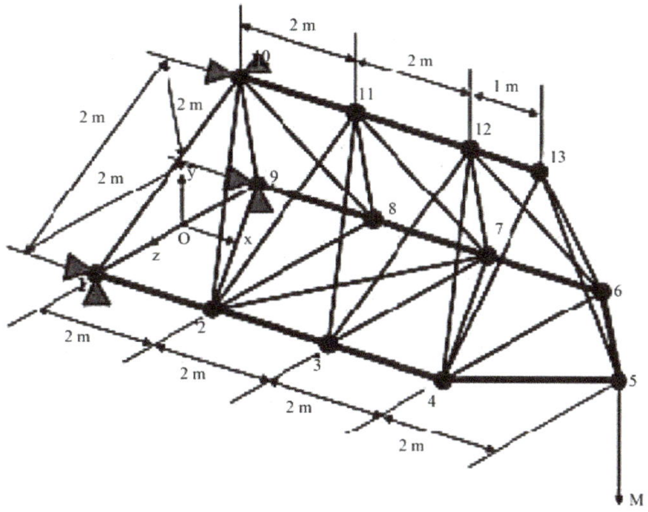

Figure 6. The spatial truss.

the truss is not negligible in front of the load M. The goal of this analysis is to determine the maximum displacement generated by the applied load to his extremity.

From **Figure 7**, the PDF's of the normalized vertical displacement u_y of the spatial truss calculated by our proposed technique FEMCTDF are independent and uniformly distributed in the range [0.00704, 0.02027]. Also the obtained results are accurate as shown by favourable comparison with classical Monte Carlo simulations. For Reliability Analysis, let us suppose that the limit displacement is u_{limit} =0.015 mm, It is required to find the failure probability $P_f = P(u \leq u_{\text{limit}})$. The numerical values of probabilistic characteristics of the displacement of this spatial truss are listed in this **Table 2**.

Table 1. FEMCTDF versus Monte Carlo simulation.

	FEMCTDF	Monte Carlo simulation
U_{max}	2.5009	2.5008
U_{min}	1.2505	1.2505
U_{mean}	1.7349	1.7334
Stdev	0.3556	0.3496
P_f	0.2536	0.2504

Table 2. FEMCTDF versus Monte Carlo simulation.

	FEMCTDF	Monte Carlo simulation
U_{max}	2.027e−2	2.11e−2
U_{min}	7.04e−3	7.039e−3
U_{mean}	1.120e−2	1.156e−2
Stdev	3.6e−3	3.75e−3
P_f	0.2031	0.203092

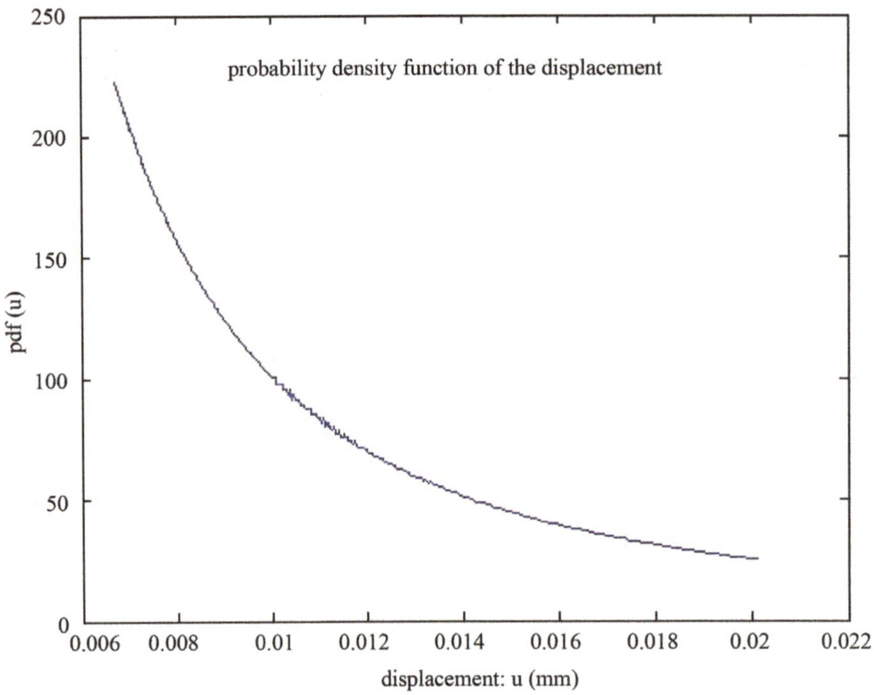

Figure 7. PDF (u) when E is uniformly distributed.

5.3. Pylon of a Line of Transportation of Electricity

We are going to analyze the reliability of a more complex structure, the pylon of a line of transportation of electricity. Two identical loads F of 1.8 KN are applied to the two superior extremities of the following pylon an angle of $\theta = 15$. The bars forming the pylon are in steel of which Young's modulus is uniformly distributed E = [100 GPa, 300 GPa] and the Poisson coefficient $\sigma = 0.29$. The section of every bar is worth A = 27.90 cm^2. The hypothesis for this problem is that the weight of each bar of the pylon is negligible in front of the applied efforts (see **Figure 5**). Using the proposed technique FEMCTDF, we obtain the following graph (see **Figure 8** and **Figure 9**):

The PDFs of the normalized vertical displacement u_y are plotted in **Figure 10** assuming that the variable u_y is independent and uniformly distributed in the range [2.9489e04, 8.8466e04]. Also in this case the results are accurate as shown by favorable comparison with classical Monte Carlo simulation. Let us suppose the limit displacement is u = 6.635e04 mm. It is required to find the failure probability $P_f = P(u \le u_{\text{limit}})$. Mean Squared Error (MSE), and Absolute Percentage Error (APE) are calculated by using the following equation:

Figure 8. Real application of pylon.

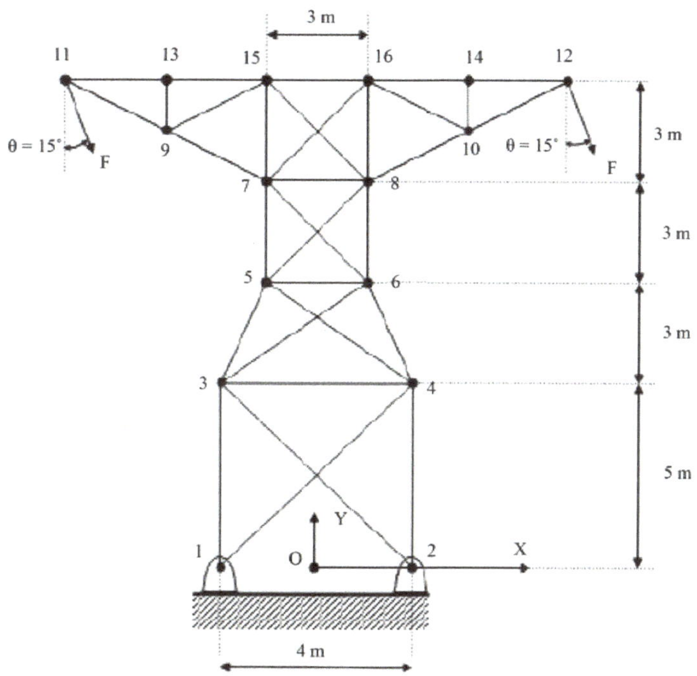

Figure 9. The pylon.

$$\text{MSE} = (u_1 - u_{mc})^2, \quad \text{APE} = \left|\frac{(u_1 - u_{mc})}{u_{mc}} \times 100\right|$$

where u_l is displacement obtained with FEMCTDF and u_{mc} the Monte Carlo simulation one.

Table 3 reports the results ($U_{\max}$, $U_{\min}$, U_{mean} are in mm) obtained by our technique and the Monte Carlo simulation (10,000 iterations). This table also illustrates the efficiency of the FEMCTDF, since a number of 800 (less than 1000) iterations suffice to obtain results close to those obtained by Monte Carlo simulation. To compare the results, the MSE and APE are calculated. The values of MSE and APE are very small, which shows the accuracy and efficiency of FEMCTDF.

6. Conclusion

In this paper, an efficient, robust technique is proposed to solve the reliability problems that required the evaluation

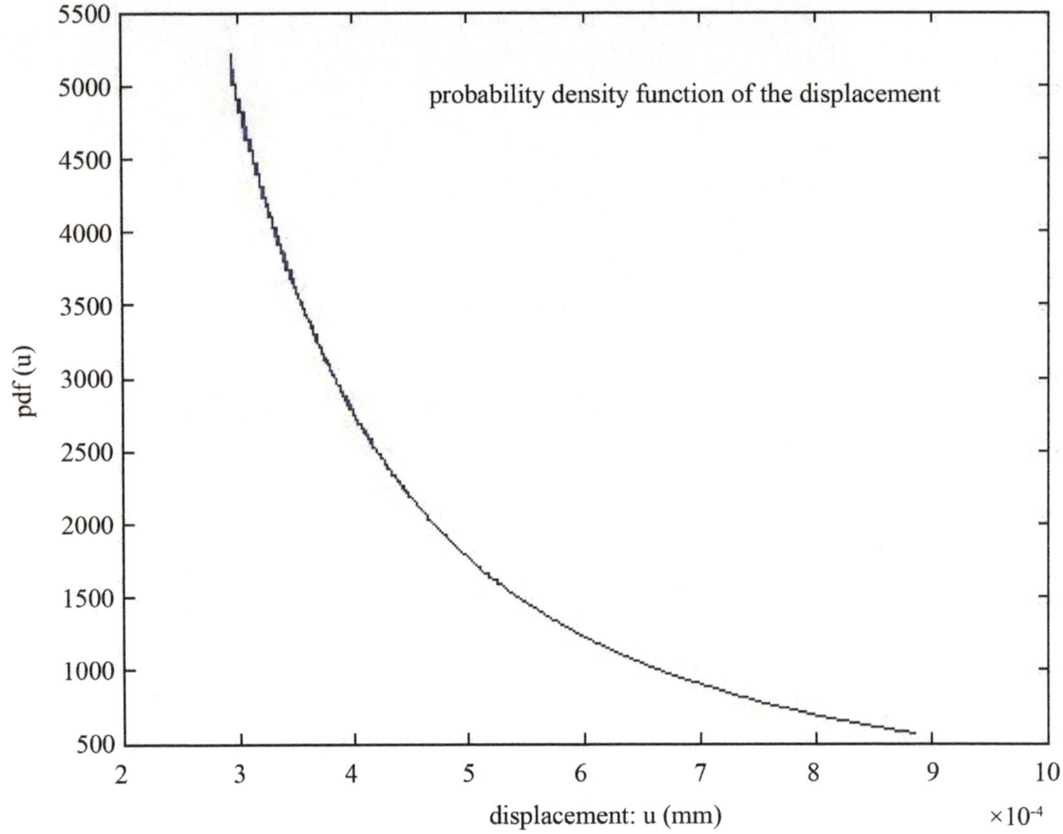

Figure 10. PDF (u) when E is uniformly distributed.

Table 3. FEMCTDF versus Monte Carlo simulation.

	FEMCTDF	M-C	MSE	APE (%)
$u_{\min}$	2.94e−04	2.94e−04	1.64e−16	0.0044
$u_{\max}$	8.84e−04	8.846e−04	2.53e−15	0.0057
u_{mean}	4.86e−04	4.85e−04	1.03e−14	0.0209
Stdev	1.57e−04	1.57e−04	9.10e−14	0.1919
P_f	1.88e−01	1.66e−01	4.78e−04	13.1181

of implicit function between the input variable of the structure and his response variable. This technique has been developed with the motivation to facilitate the application of structural reliability analysis methods to numerical models of mechanical structures. It integrates the treatment by the FEM software and the TDF program. In fact, the FEM is used to estimate the structural response function, once the implicit response function is numerically found that the TDF program can easily be applied to solve the structural reliability problem.

References

[1] Kadry, S. (2007) A Proposed Method for Reliability Analysis in Higher Dimension. *Engineering Letters*, **15**, 7-12.

[2] Liu, W.-K., Belytschko, T. and Mani, A. (1986) Random Field Finite Elements. *International Journal for Numerical Methods in Engineering*, **23**, 1831-1845. http://dx.doi.org/10.1002/nme.1620231004

[3] Deodatis, G. (1991) The Weighted Integral Method, I: Stochastic Stiffness Matrix. *Journal of Engineering Mechanics*, **117**, 1851-1864.

[4] Ditlevsen, O. and Madsen, H. (1996) Structural Reliability Methods. Wiley, Chichester.

[5] Ghanem, R.-G. and Spanos, P.-D. (1991) Stochastic Finite Elements—A Spectral Approach. Springer, Berlin. http://dx.doi.org/10.1007/978-1-4612-3094-6

[6] Kadry, S., Chateauneuf, A. and El-Tawil, K. (2006) One Dimensional Transformation Methods in Reliability Analysis. CST, Madrid.

[7] Tvedt, L. (1990) Distribution of Quadratic Forms in Normal Space-Application to Structural Reliability. *Journal of Engineering Mechanics*, **116**, 1183-1197. http://dx.doi.org/10.1061/(ASCE)0733-9399(1990)116:6(1183)

[8] Breitung, K. (1994) Asymtotic Approximations for Probability Integrals. Lecture Notes in Mathematics, 1592. Springer, Berlin.

[9] Schueller, G. (1998) Structural Reliability-Recent Advances-Freudenthal Lecture. In: Shiraishi, N., Shinozuka, M. and Wen, Y., Eds., *Proceedings of the 7th International Conference on Structural Safety and Reliability* (*ICOSSAR*'97), A.A. Balkema Publications, Rotterdam, 3-35.

[10] Koutsourelakis, P., Pradlwarter, H. and Schuëller, G. (2003) Reliability of Structures in High Dimensions. *Proceedings in Applied Mathematics and Mechanics*, **3**, 495-496. http://dx.doi.org/10.1002/pamm.200310517

[11] Papoulis, A. and Pillai, S.U. (2002) Probability, Random Variables and Stochastic Processes. 4th Edition, McGraw-Hill, Boston.

8

Curve Veering in Torsional Systems with Stepped Shafts

Mehdi Eshaghi*, **Rama Bhat**

Department of Mechanical and Industrial Engineering, Concordia University, Montreal, Canada
Email: *mehdi.eshaghi@gmail.com

Abstract

In this study, the influence of geometrical parameters on the curve veering phenomenon in a torsional system with stepped shaft is investigated. Three approximate solutions including finite element, Rayleigh-Ritz and discretization methods, along with an exact solution are employed to obtain the natural frequencies of the structure. The study reveals that, under specific circumstances, the results obtained by approximate methods are very close to the exact solution. The curve veering behavior is manifested irrespective of the method employed. It is concluded that for the structure studied the curve veering behavior is not because of the approximate techniques used to compute the natural frequencies, and is an inherent behavior of the structure.

Keywords

Curve Veering, Torsional System, Stepped Shaft

1. Introduction

Curve veering is defined as an abrupt veering of the natural frequency plots, when plotted against some system parameters [1]. This phenomenon was reported by Warburton [2], for the first time. Curve veering was observed, when variation of natural frequencies of the rectangular plate against side ratio was plotted. Leissa [3] observed curve veering in the vibration of square plates. When the variation of natural frequencies against aspect ratio was plotted, it was observed that the curves change smoothly everywhere except in some regions, where they show sudden changes. Leissa called these regions "transition zones" and showed that this behavior is attributed to approximate solution employed for finding the natural frequencies of the structure. The curve veering disappeared when an exact solution was employed. Schajer [4] reported an interesting feature of curve veering in vibration analysis of the vibrating string with a spring support. The study showed that curve veering is not limited

*Corresponding author.

to approximate solutions, and it may be an inherent behavior of some vibrating systems. Curve veering as an inherent property of the structure may be seen in rotating disks and plates [5], clamped beams on intermediate elastic supports [6] and vibration of disordered systems [7]. The significance of curve veering derives from the fact that, a small variation of frequency in the transition zone may yield a sudden change in the vibrational mode. If an external force excites the ith natural frequency of the structure in the transition zone, a small change of excitation frequency causes the $(i+1)$th frequency of the structure to be excited. As a result, a small change in excitation frequency yields a sudden change in the normal mode of the structure, and the satisfactory performance of the structure may be severely affected. The effect of frequency curve veering in instability of mechanical structures has been widely addressed. For example, mode localization reported in shallow arch [8], engineering structures [9] and cantilever beam [10] may be regarded as a result of curve veering in these structures. Moreover, estimation and veering analysis of imperfect structures such as cracked plate [11], nonlinear beam with geometry imperfection [12] and system with gyroscopic coupling [13] have been reported in the literature. It is worth noting that, frequency curve veering may cause localized buckling [14] or wrinkling in specific structures [15].

In high speed rotating machinery, a considerable number of studies have been carried out on the natural frequencies and mode shapes. Most often, in view of the complex geometry of the rotor systems, they are treated as lumped rotors mounted on shafts. In many practical situations, the shafts may have different cross sections and may have stepped configuration. Accurate determination of the natural frequencies is imperative in order to ensure that the system does not operate near resonant frequencies and particularly in the vicinity of curve veering ranges. Exact solutions are possible only in the case of well-defined uniform shaft geometries, and for practical rotors with many cross sectional changes, approximate techniques such as the discretization method, the Rayleigh Ritz method, finite element method are used. The first general theory for free vibration analysis of torsional systems was reported by Beddoe [16]. A one dimensional wave equation was employed to derive equation of motion of the structure. Maltbeak [17] [18] and Rao [19] studied free torsional vibration of uniform shafts with discrete inertias. Maltbeak assumed a sinusoidal angular displacement along the shaft. Wilson [20] utilized effective inertia method to analyze torsional vibration of a complicated system. To this end, the main structure was divided into some simple sub models, and the sub models were analyzed individually. Finally, Wilson found frequency characteristics of the main structure using a combination of the results obtained from the sub models. Leissa and So [21] applied three dimensional solution for estimation of natural frequencies of the shaft structure.

In the present study, a stepped shaft supporting a rotating disk at the tip is analyzed for its curve veering behavior by computing the natural frequencies by different methods. The effect of geometric parameters of the stepped shaft disk system on the curve veering phenomenon is investigated. Although approximate solutions exhibit curve veering in the structure, an exact method is also employed to confirm this phenomenon as an inherent property of the structure.

2. Mathematical Formulation

An isotropic, homogeneous torsional system composed of a stepped shaft with a lumped disk at the tip, as depicted in **Figure 1**, is used in the study. The length and diameter of upper and lower shafts are L_1, d_1 and L_2, d_2, respectively, and total length of the stepped shaft is L. Moreover, M and d denote the mass and diameter of the lumped disk.

2.1. Exact Solution

The equation of motion of the shaft is given by:

$$\frac{\partial^2\theta}{\partial t^2}=\left(\frac{G}{\rho}\right)\frac{\partial^2\theta}{\partial x^2} \tag{1}$$

where ρ, G and θ are density, shear modulus and twist angle of the shaft, respectively. The solution of Equation (1) may be found as follows [22]:

$$\theta(x,t)=\left[A\sin\left(\omega\sqrt{\rho/G}\right)x+B\cos\left(\omega\sqrt{\rho/G}\right)x\right]\left[C\sin\omega t+D\cos\omega t\right]=\Theta(x)f(t) \tag{2}$$

where ω is the frequency of vibration and A, B, C and D are unknown coefficients. The characteristic equation

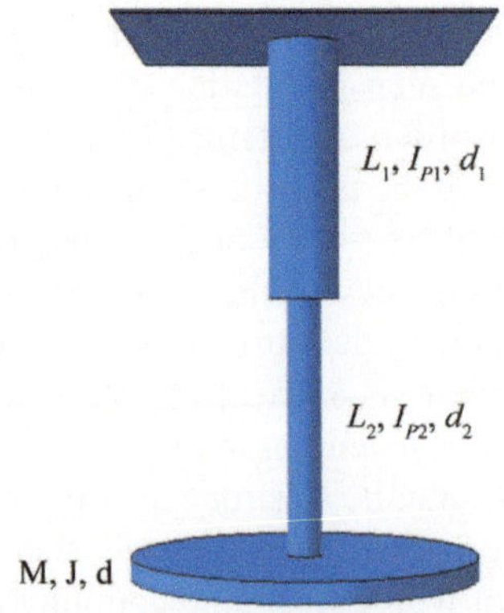

Figure 1. Stepped shaft connected to disk.

is obtained by substituting the boundary conditions as:

$$\begin{bmatrix} \sin(\lambda L_1) & -\cos(\lambda L_1) & -\sin(\lambda L_1) \\ I_{p1}\cos(\lambda L_1) & I_{p2}\sin(\lambda L_1) & -I_{p2}\cos(\lambda L_1) \\ 0 & J\omega^2\cos(\lambda L)+GI_{p2}\lambda\sin(\lambda L) & J\omega^2\sin(\lambda L)-GI_{p2}\lambda\cos(\lambda L) \end{bmatrix} \begin{Bmatrix} B_1 \\ A_2 \\ B_2 \end{Bmatrix} = \begin{Bmatrix} 0 \\ 0 \\ 0 \end{Bmatrix} \tag{3}$$

where $\lambda = \omega\sqrt{\rho/G}$ and I_{p1}, I_{p2} and J are polar moment of inertia of upper shaft, polar moment of inertia of lower shaft and polar mass moment of inertia of the disk, respectively, and A_2, B_1 and B_2 are unknown coefficients. For non-trivial solutions, the determinant of the matrix should be set to zero which will yield the natural frequencies.

2.2. The Rayleigh-Ritz Method

In the Rayleigh-Ritz solution, displacement field of a structure is defined as linear combination of admissible functions. In this study, the deflection shape is considered as:

$$\theta(x) = c_1\theta_1(x) + c_2\theta_2(x) + \cdots \tag{4}$$

where $\theta_i(x)$ are admissible functions satisfying at least the geometrical boundary conditions and c_i are arbitrary coefficients. For a shaft with distributed mass and elasticity, the kinetic and potential energy expressions are given by:

$$U_{\max} = \frac{1}{2} GI_p \int_0^L \left(\frac{\mathrm{d}\theta}{\mathrm{d}x}\right)^2 \mathrm{d}x \tag{5}$$

$$T_{\max} = \frac{1}{2} \rho I_p \omega^2 \int_0^L \theta^2(x)\mathrm{d}x \tag{6}$$

where ω is the frequency of vibration. For harmonic vibrations, we have:

$$U_{\max} = T_{\max} \tag{7}$$

The conditions for the stationary of the natural frequencies with respect to the arbitrary coefficients in the assumed deflection expression formulate the eigenvalue problem of the structure. It is well-known that the natural frequencies obtained by the Rayleigh-Ritz method are the upper bound. In this study, the following formulation is employed to obtain orthogonal admissible functions [23].

$$\theta_1(x) = x;\ \theta_2(x) = (x - B_2)\theta_1(x);\ \theta_i(x) = (x - B_i)\theta_{i-1}(x) - C_i\theta_{i-2}(x) \tag{8}$$

where

$$B_i = \frac{\int_0^L x(\theta_{i-1}(x))^2\,\mathrm{d}x}{\int_0^L (\theta_{i-1}(x))^2\,\mathrm{d}x} \tag{9}$$

$$C_i = \frac{\int_0^L x\theta_{i-1}(x)\theta_{i-2}(x)\,\mathrm{d}x}{\int_0^L (\theta_{i-2}(x))^2\,\mathrm{d}x} \quad (10)$$

It should be noted that, increasing the number of admissible functions improves the convergence of the results.

2.3. Discretization Method

Discretization technique may be regarded as the simplest, and the least accurate method that is used to find the fundamental frequency of the structure quickly. In this solution, the stiffness constants of the upper and lower shafts are found, individually. Total stiffness of the stepped shaft is obtained as a series combination of these two shaft segments. It should be noted that, solving the problem using this approach necessitates assuming linear torsional deflection through the stepped shaft, while the exact solution reveals trigonometric functions for the shaft deflection. The stiffness constants of the upper and lower shafts are

$$K_i = \frac{GI_{pi}}{L_i}, \quad i = 1,2 \quad (11)$$

where G, I_{pi} and L_i are shear modulus, polar moment of inertia and length of each shaft, respectively. Total stiffness of the stepped shaft is given by:

$$K_{eq} = K_1K_2/(K_1+K_2) \quad (12)$$

Finally, fundamental frequency of the structure is obtained as:

$$\omega = \sqrt{K_{eq}/J} \quad (13)$$

where J is polar mass moment of inertia of the disk.

3. Results and Discussion

In this study, the following base line values are assumed in the analysis: shear modulus of the structure is 79.3 Gpa, density is 7800 kg/m^3, L_1, L_2, d_1, d_2, d and M are chosen to be 1 m, 1 m, 0.1 m, 0.05 m, 0.5 m and 100 kg, respectively. It should be noted that, all natural frequencies are given in rad/s.

3.1. Comparison of the Results

The exact solution for the fundamental frequency of the structure is obtained as 121.022 rad/s. The results obtained by the Rayleigh-Ritz solution reveal that, using one term of admissible function gives fundamental frequency of the structure as 256.88 rad/s. It should be noted that, the admissible functions satisfy the geometric boundary condition, along with continuity of the angular displacement and torque at the point of step change in the shaft cross section. **Table 1** shows convergence of the fundamental frequency when the admissible functions satisfy only the geometric boundary conditions. The results given in **Table 1** indicate that, they converge to the exact value, although the rate of convergence is poor.

Table 1. Variation of fundamental frequency versus the number of admissible functions in Rayleigh-Ritz method, when d_1 = 0.1 m, d_2 = 0.05 m, d = 0.5 m and M = 100 kg.

Number of admissible functions	First frequency
1	256.868
2	165.885
3	137.485
4	136.992
5	130.562
6	130.562
7	127.832
8	127.790

Just using two elements in the finite element model yields the same result as obtained by the exact solution. It is interesting to note that the discretization technique gives fundamental frequency of the structure equal to 121.056 rad/s. The discretization technique considered the displacement field of the structure as a linear function, while the exact solution uses trigonometric functions to describe the displacement field. This is attributed to the magnitude of λ in the exact solution, which has a very small value. In fact, in this order of λ, $\sin(\lambda)$ may be assumed to be equal to λ; as a result, the frequency obtained by linear deflection assumption is the same as that of the exact solution. In order to clarify this behavior, another example is presented. In this example, all parameter values of the structure remain unchanged except the density, which is assumed to be $\rho = 78000000\ \text{kg/m}^3$. In this case, λ does not have a small value and $\sin(\lambda) \neq \lambda$. The fundamental frequency obtained by the exact solution is 40.8408 rad/s, while the frequency obtained using discretization technique is equal to 121.056 rad/s. Moreover, the Rayleigh-Ritz method—using one admissible functions satisfying geometric boundary condition and continuity of torque and displacement—yields $\omega = 46.7483\ \text{rad/s}$. The difference between these results derives from the fact that, for higher values of λ, the assumption of linear deflection through the shaft is incorrect. In this case, more terms in the Rayleigh-Ritz method is required to obtain accurate results.

3.2. Curve Veering

When the variation of natural frequencies against length ratio (L_1/L) is plotted, the curves change smoothly everywhere except in some regions, where they show sudden changes. As mentioned earlier, Leissa [2] called these regions "transition zones". This behavior was reported as a result of approximate solution method used to solve the problem, but in the present study curve veering is present whether the results are obtained by the exact method or the approximate methods. Hence, the curve veering may be regarded as an inherent behavior of the system. **Figure 2** shows the variation of the first five natural frequencies against length ratio, when $d_1 = 0.1\ \text{m}$, $d_2 = 0.01\ \text{m}$, $d = 0.5\ \text{m}$ and $M = 100$ kg. It is observed that, the variation of the fundamental frequency of the structure against length ratio is not significant. In fact, the curve veering does not occur, when the structure vibrates in the first vibrational mode. The curve veering may be observed in other modes of vibration. The graph

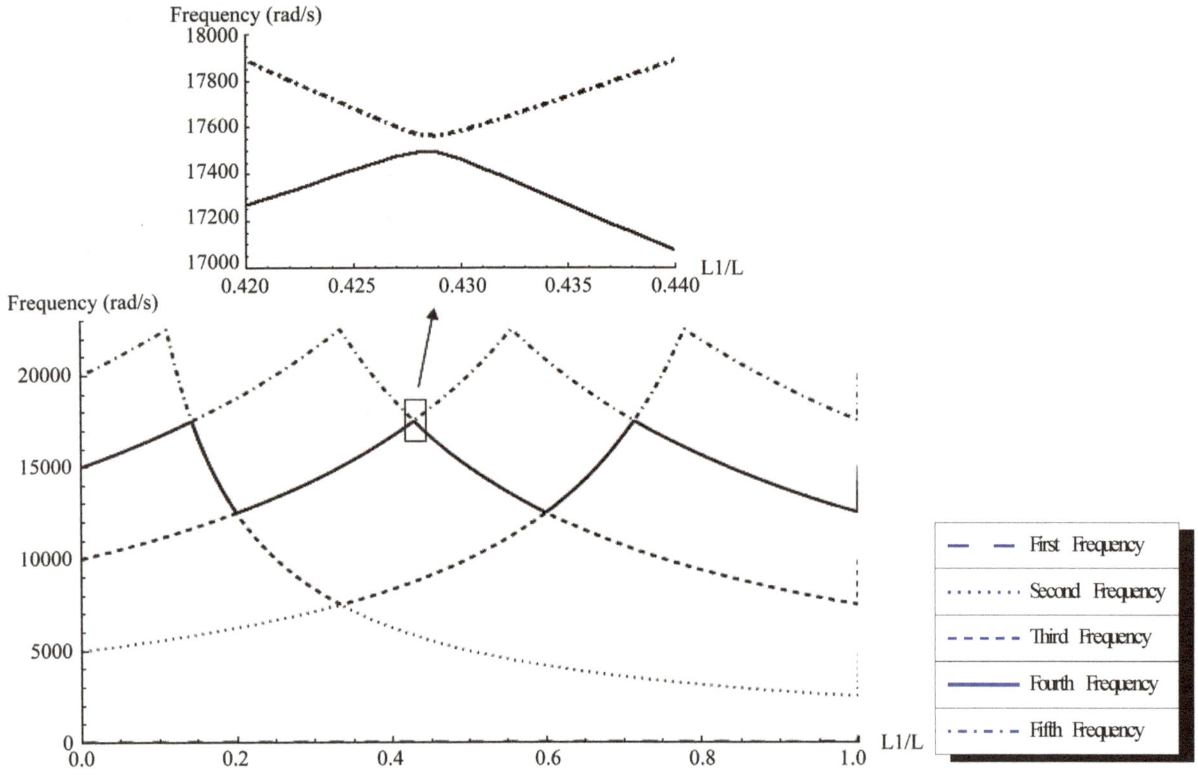

Figure 2. Variation of first five frequencies of the structure versus length ratio when $d_1 = 0.1$ m, $d_2 = 0.01$ m, $d = 0.5$ m and $M = 100$ kg.

shows that the number of transition zones increases when mode number increases. For instance, there is just one transition zone in the second mode of vibration, while three, five and seven transition zones may be observed in the third, fourth and fifth vibrational modes, respectively. In order to understand the curve veering phenomenon, transition zone corresponding to the fourth and fifth modes has been magnified. It can be seen that, natural frequencies approach each other and veer away in this region. This behavior is of great significance for the designers.

Table 2 shows the variation of the first five resonant frequencies of the structure, when $d_1 = 0.1$ m, $d_2 = 0.01$ m, $d = 0.5$ m and $M = 100$ kg and $0.05 < L_1/L < 0.25$. For the range of L_1/L considered, the shaft behaves, approximately, as a slender bar of $d_2 = 0.01$ m; as a result, the fundamental natural frequency of the structure becomes very low, in the range of 3 - 4 rad/s. The bold frequencies indicate the curve veering point in the transition zones. In **Figure 3** the variation of the first five natural frequencies versus length ratio of the structure is shown when $d_1 = 0.1$ m, $d_2 = 0.05$ m, $d = 0.5$ m and $M = 100$ kg. In this case, the curve veering is observed. **Figure 4** shows an analogous study on the structure, when $d_2 = 0.095$ m. It may be seen that variation of length ratio does not show a drastic change in the frequencies. It is attributed to the diameter of upper and lower shafts, which have almost the same magnitudes.

Table 2. Variation of first five frequency of the structure versus length ratio, when $d_1 = 0.1$ m, $d_2 = 0.01$ m, $d = 0.5$ m and $M = 100$ kg.

L_1/L	First	Second	Third	Fourth	Fifth
0.06	3.640	5329	10657	15985	21313
0.07	3.660	5386	10771	16157	21542
0.08	3.679	5444	10888	16332	21776
0.09	3.700	5504	11008	16512	22015
0.10	3.720	5565	11130	16695	22260
0.11	3.741	5628	11255	16883	22501
0.12	3.762	5692	11383	17074	20869
0.13	3.783	5757	11514	17270	19265
0.14	3.806	5824	11648	**17467**	**17680**
0.15	3.828	5893	11785	16694	17889
0.16	3.851	5963	11925	15652	17893
0.17	3.874	6035	12069	14732	18104
0.18	3.897	6108	12215	13914	18325
0.19	3.897	6184	12365	13183	18551
0.20	3.945	6261	**12482**	**12562**	18873
0.21	3.970	6340	11924	12682	19020
0.22	3.996	6421	11383	12844	19264
0.23	4.022	6505	10888	13010	19514
0.24	4.048	6590	10435	13181	19771

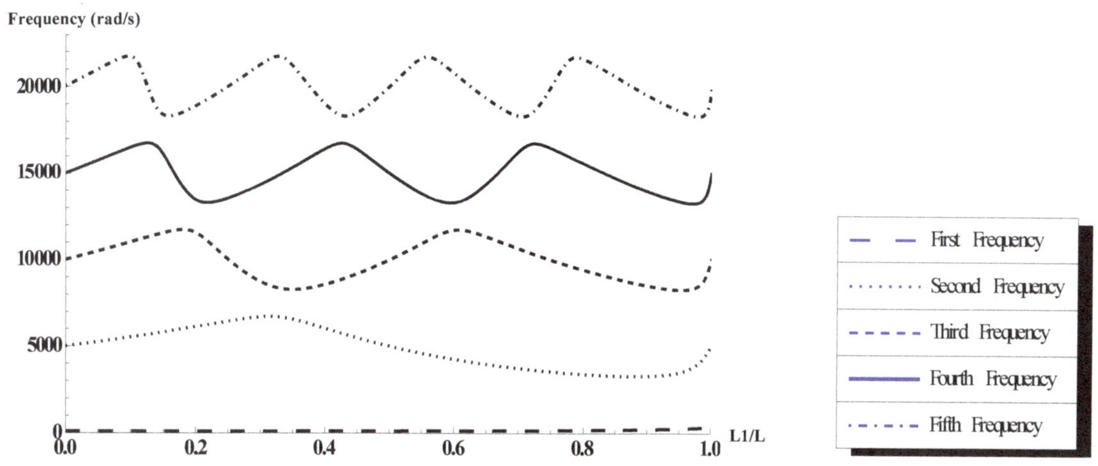

Figure 3. Variation of first five frequencies of the structure versus length ratio when $d_1 = 0.1$ m, $d_2 = 0.05$ m, $d = 0.5$ m and $M = 100$ kg.

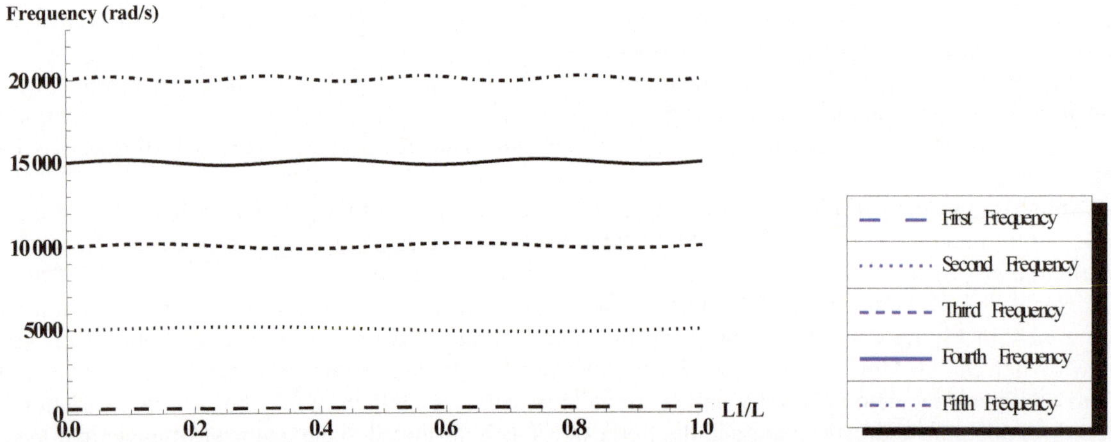

Figure 4. Variation of first five frequencies of the structure versus length ratio when $d_1 = 0.1$ m, $d_2 = 0.095$ m, $d = 0.5$ m and $M = 100$ kg.

4. Conclusion

This study deals with an analysis on the curve veering phenomenon in a torsional structure, which consists of a stepped shaft and a rotating disk. Different approximate techniques including the Rayleigh-Ritz, finite element and discretization methods, along with the exact solution were employed to extract natural frequencies of the structure. The results reveal that curve veering in this structure is not due to application of approximate solution, and it appears even if an exact solution is employed. As a result, the curve veering may be regarded as an inherent behavior of the structure. The geometric parameters affect the curve veering, noticeably. Moreover, a comparison of the results obtained by approximate solutions and those of the exact one was carried out. It was realized that, under some specific geometries and material properties, the frequencies obtained from approximate solutions are as accurate as the exact solution. Under such conditions, the trigonometric functions which describe angular displacement field can be replaced by a linear function.

References

[1] Bhat, R.B. (1954) Curve Veering Behavior of Some Vibrating Systems. *Shock and Vibration*, **7**, 241-249. http://dx.doi.org/10.1155/2000/841538

[2] Warburton, G.B. (1974) The Vibration of Rectangular Plates. *Shock and Vibration*, **8**, 371-384.

[3] Leissa, A.W. (1974) On a Curve Veering Aberration. *Journal of Applied Mathematics and Physics* (*ZAMP*), **25**, 99-110.

[4] Schajer, G.S. (1984) The Vibration of a Rotating Circular String Subject to Fixed Elastic Restraint. *Journal of Sound and Vibration*, **92**, 11-19. http://dx.doi.org/10.1016/0022-460X(84)90369-9

[5] Wang, J.T.S., Shaw, D. and Mahrenholtz, O. (1987) Vibration of Rotating Rectangular Plates. *Journal of Sound and Vibration*, **112**, 455-468. http://dx.doi.org/10.1016/S0022-460X(87)80111-6

[6] Rao, C.K. (1989) Frequency Analysis of Clamped-Clamped Uniform Beam with Intermediate Elastic Supports. *Journal of Sound and Vibration*, **133**, 502-509. http://dx.doi.org/10.1016/0022-460X(89)90615-9

[7] Pierre, C. (1988) Mode Localization and Eigenvalue Loci Veering Phenomena in Disordered Structures. *Journal of Sound and Vibration*, **126**, 485-502. http://dx.doi.org/10.1016/0022-460X(88)90226-X

[8] Breslavsky, I., Avramov, K.V., Mikhlin, Y. and Kochurov, R. (2008) Nonlinear Modes of Snap-Through Motions of a Shallow Arch. *Journal of Sound and Vibration*, **311**, 297-313. http://dx.doi.org/10.1016/j.jsv.2007.09.015

[9] Chan, H.C. and Liu, J.K. (2000) Mode Localization and Frequency Loci Veering in Disordered Engineering Structures. *Chaos, Solitons and Fractals*, **11**, 1493-1504. http://dx.doi.org/10.1016/S0960-0779(99)00073-9

[10] Bois, J.L.D., Adhikari, S. and Lieven, N.A.J. (2011) On the Quantification of Eigenvalue Curve Veering: A Veering Index. *Journal of Applied Mechanics*, **78**, 041007-1-041007-8.

[11] Saito, A., Castanier, M.P. and Pierre, C. (2009) Estimation and Veering Analysis of Nonlinear Resonant Frequencies of Cracked Plates. *Journal of Sound and Vibration*, **326**, 725-739. http://dx.doi.org/10.1016/j.jsv.2009.05.009

[12] Lacarbonara, W., Arafat, H. and Nayfeh, A.H. (2005) Large Non-Linear Interactions in Imperfect Beams at Veering. *International Journal of Non-Linear Mechanics*, **40**, 987-1003. http://dx.doi.org/10.1016/j.ijnonlinmec.2004.10.006

[13] Vidoli, S. and Vestroni, F. (2005) Veering Phenomena in Systems with Gyroscopic Coupling. *Journal of Applied Mechanics*, **72**, 641-647. http://dx.doi.org/10.1115/1.1940666

[14] Chen, J.S. and Li, Y.-T. (2006) Effects of Elastic Foundation on the Snap-Through Buckling of a Shallow Arch under a Moving Point Load. *International Journal of Solids and Structures*, **43**, 4220-4237. http://dx.doi.org/10.1016/j.ijsolstr.2005.04.040

[15] Al-Qaisia, A.A. and Hamdan, M.N. (2013) On Nonlinear Frequency Veering and Mode Localization of a Beam with Geometric Imperfection Resting on Elastic Foundation. *Journal of Sound and Vibration*, **332**, 4641-4655. http://dx.doi.org/10.1016/j.jsv.2013.03.031

[16] Beddoe, B. (1965) Wave Theory of Free Torsional Vibration of Composite Systems of Shafts. *Journal of Mechanical Engineering Science*, **7**, 48-56. http://dx.doi.org/10.1243/JMES_JOUR_1965_007_010_02

[17] Maltbaek, J.C. (1962) Influence of a Discrete Inertia on the Free Torsional Vibrations of a Uniform Bar. *International Journal of Mechanical Sciences*, **4**, 24-34. http://dx.doi.org/10.1016/0020-7403(62)90004-8

[18] Maltbaek, J.C. (1967) Torsional Vibration of a Stepped Shaft. *Engineer*, **223**, 972-974.

[19] Rao, D.K. (1978) Torsional Frequencies of a Multi-Stepped Shaft with Rotors. *International Journal of Mechanical Sciences*, **20**, 415-422. http://dx.doi.org/10.1016/0020-7403(78)90031-0

[20] Wilson, W.K. (1969) Practical Solution of Torsional Vibration Problems. 3rd Edition, Chapman & Hall, London.

[21] Leissa, A.W. and So, J.Y. (1995) Comparisons of Vibration Frequencies for Rods and Beams from One-Dimensional and Three-Dimensional Analysis. *Journal of Acoustic Society*, **98**, 2122-2135. http://dx.doi.org/10.1121/1.414331

[22] Thomson, W.T. and Dahleh, M.D. (1997) Theory of Vibrations with Applications. 5th Edition, Prentice Hall, Upper Saddle River.

[23] Bhat, R.B. (1985) Natural Frequencies of Rectangular Plates Using Characteristic Orthogonal Polynomials in Rayleigh-Ritz Method. *Journal of Sound and Vibration*, **102**, 493-499. http://dx.doi.org/10.1016/S0022-460X(85)80109-7

9

Joule Heating and Thermal Radiation Effects on MHD Boundary Layer Flow of a Nanofluid over an Exponentially Stretching Sheet in a Porous Medium

Jakkula Anand Rao[1], Gandamalla Vasumathi[1], Jakkula Mounica[2]

[1]Department of Mathematics, Osmania University, Hyderabad, Telangana
[2]Department of Mathematics, NGRI, Hyderabad, Telangana
Email: anand.rao.jakkula.ou@gmail.com, gandamallavasumathi@gmail.com, mounica.jakkula.iitk@gmail.com

Abstract

A numerical study on boundary layer flow behaviour, heat and mass transfer characteristics of a nanofluid over an exponentially stretching sheet in a porous medium is presented in this paper. The sheet is assumed to be permeable. The governing partial differential equations are transformed into coupled nonlinear ordinary differential equations by using suitable similarity transformations. The transformed equations are then solved numerically using the well known explicit finite difference scheme known as the Keller Box method. A detailed parametric study is performed to access the influence of the physical parameters on longitudinal velocity, temperature and nanoparticle volume fraction profiles as well as the local skin-friction coefficient, local Nusselt number and the local Sherwood number and then, the results are presented in both graphical and tabular forms.

Keywords

Exponentially Stretching Sheet, MHD, Thermal Radiation, Chemical Reaction, Joule Heating, Heat and Mass Transfer, Porous Medium

1. Introduction

The industrial processes like hot rolling, wire drawing, spinning of filaments, metal extrusion, crystal growing, glass fibre production, paper production, cooling of a large metallic plate in a bath, which may be an electrolyte, etc. to require the study of flow and heat transfer over a stretching surface. In all these cases, the quality of final product depends on the surface heat transfer rate and the skin friction coefficient. So this study has gained considerable attention in the recent years. Choi [1] was the first to sort heat transfer enhancement upon the invention of "nanofluid". These fluids are engineered colloidal suspensions composed of nanoparticles in a base fluid. Eastman [2] observed thermal conductivity enhancement in nanofluids. Metals (Al, Cu), oxides (Al_2O_3, TiO_2 and CuO), carbides (SiC), nitrides (AlN, SiN), or nonmetals (graphite, carbon nanotubes) and conductive fluids, such as water or ethylene glycol, or oil, other lubricants, bio-fluids, polymer solutions as base fluids are used in the manufacturing of nanofluids. 5% volume fraction of nanoparticles in these fluids ensure effective heat transfer enhancements which help them to exhibit enhanced thermal conductivity and the convective heat transfer coefficient compared with the base fluid. Routbort [3] found that the typical thermal conductivity enhancements and heat transfer coefficient enhancements are in the range of 15% - 40% and up to 40% over the base fluids. But still there must be other mechanisms to attribute to higher thermal conductivity. Increase in thermal conductivity cannot be considered as the sole reason. Here a survey conducted by Buongiorno [4] on convective transport in nanofluids implied that energy transfer by dispersion of nanoparticles was negligible. His model could not explain the abnormal heat transfer coefficients. The boundary layer has different properties due to the effect of temperature and thermophoresis. There may be a decrease in viscosity in the boundary layer, which will lead to heat transfer enhancement. Many other literatures support Buongiorno's explanation. An excellent assessment of nanofluid physics and developments had been provided by Cheng [5] and Ali [6]. Buongiorno and Hu [7] observed that although convective heat transfer enhancement had been suggested to be due to the dispersion of the suspended nanoparticles, this effect was too small to explain the observed enhancement.

It is often assumed in the problems of boundary layer flow over a stretching surface that the velocity of the stretching surface is linearly proportional to distance from the fixed origin. However, Gupta [8] had argued against the linearity condition. In the real world problem, the stretching of plastic sheet may not necessarily be linear. For example an exponentially stretching sheet was considered by Sanjayanand [9]. Heat and mass transfer on its boundary layers are investigated. Suction and heat transfer characteristics were addressed by Youn [10]. Heat transfer in a viscoelastic boundary layer flow over a stretching sheet with viscous dissipation and non-uniform heat source was studied by Subhas [11]. Thermal radiation effect on the flow was examined by Sohail Nadeem [12]. Anuar Ishak [13] studied the MHD boundary layer flow due to an exponentially stretching surface having the effect of radiation. Effect of magnetic field on boundary layer flow and heat transfer of a dusty fluid with an exponential temperature distribution on the exponentially stretching sheet was considered by Al-odat [14].

Hitesh Kumar [15] worked on the heat transfer MHD boundary-layer flow through a porous medium. Gopi Chand [16] considered an unsteady stretching surface in a porous medium and explained the viscous dissipation and Radiation effects on MHD flow over it. Flow through a porous medium bounded by a vertical surface in presence of hall current was explained by Sudhakar [17]. joules DISSIPATION over a nonlinear vertical stretching porous sheet was introduced by Subhas Abel [18]. M.M. Hamza [19] provided the study of Oscillatory Flow through a porous medium. P.R. Sharma [20] worked on a porous stretching sheet.

This paper provides the solution to the problem of flow and heat transfer of a nanofluid over an exponentially stretching porous sheet by considering the effect of chemical reaction, joule heating and thermal radiation parameters along with the suction parameter by adopting the Keller Box method.

2. Formulation of the Problem

The present problem is based on a steady two-dimensional incompressible viscous laminar flow of an electrically conducting nanofluid over a permeable exponentially stretching sheet in a porous medium as shown in **Figure 1**. The X-axis is taken along the stretching surface in the direction of motion and the Y-axis is perpendicular to it. The flow is confined to $y > 0$. Two equal and opposite forces are applied along the X-axis. Keeping the origin fixed, the sheet is then stretched with a velocity $U_w(x)$, varying exponentially with the distance from the slit. It is assumed that the surface temperature T_w and nanoparticle fraction C_w are constants and they differ from the ambient values of temperature and nanoparticle fraction T_∞ and C_∞, respectively. The usual boundary layer approximations are still applicable. The governing equations of continuity, momentum, energy and concentration in Cartesian coordinates x and y are:

$$\frac{\partial u}{\partial x}+\frac{\partial v}{\partial y}=0 \tag{1}$$

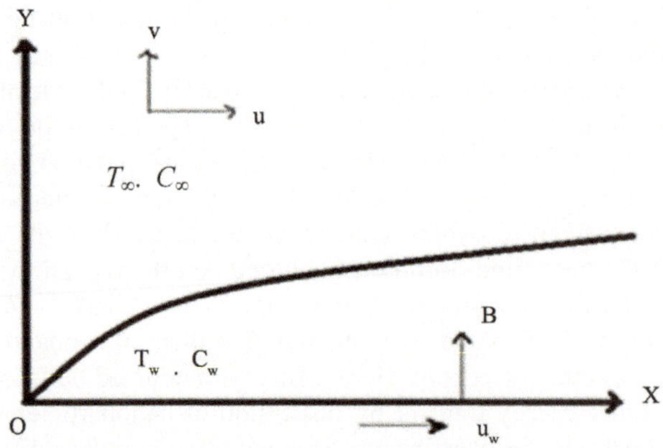

Figure 1. Physical modal and coordinate system.

$$u\frac{\partial u}{\partial x}+v\frac{\partial u}{\partial y}=\upsilon\frac{\partial^2 u}{\partial y^2}-\frac{\sigma B_0^2 u}{\rho}-\frac{\upsilon}{K_1}u \tag{2}$$

$$u\frac{\partial T}{\partial x}+v\frac{\partial T}{\partial y}=\alpha_m\frac{\partial^2 T}{\partial y^2}-\frac{1}{(\rho c)_f}\frac{\partial q_r}{\partial y}+\tau\left[D_B\frac{\partial C}{\partial y}\frac{\partial T}{\partial y}+\frac{D_T}{T_\infty}\left(\frac{\partial T}{\partial y}\right)^2\right]+\frac{\sigma B_0^2 u^2}{\rho c_p} \tag{3}$$

$$u\frac{\partial C}{\partial x}+v\frac{\partial C}{\partial y}=D_B\frac{\partial^2 C}{\partial y^2}+\frac{D_T}{T_\infty}\frac{\partial^2 T}{\partial y^2}-k_1(C-C_\infty) \tag{4}$$

where, u & v are the velocity components in X and Y-directions respectively, ρ-density of the nanofluid, B_0-induced magnetic field, σ-electrical conductivity, υ-kinematic viscosity, K_1-thermal conductivity, α_m-thermal diffusivity parameter, $(\rho c)_f$-the heat capacitance of the base fluid, τ-ratio between the effective heat capacity of the nanoparticle material and heat capacity of the nanofluid, D_B-the Brownian diffusion coefficient, D_T-the thermophoresis coefficient, q_γ-radiative heat flux, k_1-chemical reaction parameter, c_p-specific heat capacity of the nanoparticle, γ-chemical reaction parameter, α-the chemical reaction coefficient, T-temperature of the nanofluid, C-concentration of the nanofluid, T_w and C_w-the temperature and concentration along the stretching sheet, T_∞ and C_∞-the ambient temperature and concentration.

The boundary conditions are:

$$\begin{aligned} &u=U_w(x), v=-V_w(x), T=T_w, C=C_w, \text{ at } y=0 \\ &u=0, T\to T_\infty, C\to C_\infty \text{ as } y\to\infty \end{aligned} \tag{5}$$

By using the Rosseland approximation for radiative heat flux is defined as:

$$q_r=-\frac{4\sigma^*}{3K^*}\frac{\partial T^4}{\partial y} \tag{6}$$

where K^* is the mean absorption coefficient and σ^* is the Stefan-Boltzmann constant. Since the temperature differences within the flow field are assumed to be small, and then we linearize and expand T^4 into the Taylor series about T_∞, which after neglecting higher order forms takes the form.

$$T^4=4T_\infty^3 T-3T_\infty^4 \tag{7}$$

To examine the flow, the following transformations are used:

$$\begin{aligned} &u=U_0 e^{\frac{x}{L}}f'(\eta), \eta=\sqrt{\frac{U_0}{2\upsilon L}}e^{\frac{x}{2L}}y, v=-\sqrt{\frac{\upsilon U_0}{2L}}e^{\frac{x}{2L}}\left[f(\eta)+\eta f'(\eta)\right] \\ &\theta(\eta)=\frac{T-T_\infty}{T_W-T_\infty}, \phi(\eta)=\frac{C-C_\infty}{C_W-C_\infty} \end{aligned} \tag{8}$$

To determine the velocity, temperature distribution and rate of heat and mass transfer in the above boundary layer (5), we solve the equations related to the stretching sheet problem to obtain the following similarity equations using (8). In deriving these equations, the external electric field is assumed to be zero and the electric field due to polarization of charges in negligible.

$$f''' + ff'' - 2(f')^2 - (M+G)f' = 0 \tag{9}$$

$$\left(1+\frac{4}{3}R\right)\frac{1}{Pr}\theta'' + Nb\phi'\theta' + Nt\theta'^2 + Hf''^2 + f\theta' = 0 \tag{10}$$

$$\phi'' + Lef\phi' + \frac{Nt}{Nb}\theta'' - \gamma\phi = 0 \tag{11}$$

The transformed boundary conditions take the following forms:

$$f(0) = S,\ f'(0) = 1,\ f'(\infty) = 0,\ \theta(0) = 1,\ \theta(\infty) = 0,\ \phi(0) = 1,\ \phi(\infty) = 0 \tag{12}$$

where,

$$\begin{gathered} M = \frac{\sigma B_0^{\ 2}}{\rho c}, G = \frac{\upsilon}{cK}, c = \frac{U_0 e^{\frac{x}{L}}}{2L}, H = M\cdot Ec, Ec = \frac{U_0 e^{\frac{x}{L}}}{c_p}, Le = \frac{\upsilon}{D_B}, \gamma = \frac{\alpha}{C_w - C_\infty} \\ Nt = \frac{\tau D_T (T_w - T_\infty)}{\upsilon T_\infty}, Nb = \frac{\tau D_B (C_w - C_\infty)}{\upsilon}, R = \frac{4T^* T_\infty^{\ 3}}{k^* k}, \Pr = \frac{\upsilon}{\alpha_m}, S = \frac{V_w}{\sqrt{\upsilon c}} \end{gathered} \tag{13}$$

For the type of boundary layer flow, the skin-friction coefficient, heat transfer coefficient and mass transfer coefficients are important physical parameters.

They defined as:

$$\begin{gathered} C_f = \frac{\tau_w}{\frac{\rho U_w^{\ 2}}{2}} = \frac{\mu\left(\frac{\partial u}{\partial y}\right)_{y=0}}{\frac{\rho U_w^{\ 2}}{2}}, Nu_x = \frac{Lq_w}{k(T_w - T_\infty)} = -\frac{L\left(\frac{\partial T}{\partial y}\right)_{y=0}}{(T_w - T_\infty)} \\ \text{and } Sh_x = \frac{Lq_m}{k(C_w - C_\infty)} = -\frac{L\left(\frac{\partial C}{\partial y}\right)_{y=0}}{C_w - C_\infty} \end{gathered} \tag{14}$$

The dimensionless forms of these parameters are:

$$\sqrt{Re_x}Cf_x = \sqrt{2}f''(0),\ Nu_x = -\left(1+\frac{4}{3}R\right)\sqrt{\frac{Re_x}{2}}\theta'(0),\ \text{and}\ Sh_x = -\sqrt{\frac{Re_x}{2}}\phi'(0) \tag{15}$$

where the surface shear stress $\tau_w = \mu\left(\frac{\partial u}{\partial y}\right)_{y=0}$, the surface heat flux $q_w = k\left(\frac{\partial T}{\partial y} + \frac{\partial q_r}{\partial y}\right)_{y=0}$, the surface mass flux $q_m = k\left(\frac{\partial C}{\partial y}\right)_{y=0}$ and $Re_x = \frac{U_w L}{\upsilon}$ is the Reynolds number with μ and k being the dynamic viscosity and the thermal conductivity, respectively. The numerical values of $f''(0)$, $\theta'(0)$ and $\phi'(0)$ are proportional to the local skin-friction coefficient, local Nusselt number and local Sherwood number respectively and these are presented by **Table 1** for the values of the physical parameters.

3. Numerical Solution

Equations (9)-(11) subjected to the boundary conditions (12) are solved numerically using implicit finite difference method that is known as Keller Box in combination with the Newton's techniques as described by Cebeci and Bradshaw [21]. This method is unconditionally stable and has second order accuracy.

Table 1. Comparison of results for $-\theta'(0)$ with previous published works.

Pr	N.G. Rudraswamy [22]	Present work
0.72	0.6180	0.6152
1.0	0.7097	0.7093
1.5	0.7862	0.7862

The principal steps in using the Keller Box method are:

1) Reducing higher order ODEs (systems of ODES) in to systems of first order ODEs;
2) Writing the systems of first order ODEs into difference equations using central differencing scheme;
3) Linearizing the difference equations using Newton's method and wring it in vector form;
4) Solving the system of equations using block elimination method.

In order to solve the above differential equations numerically, we adopt Matlab software which is very efficient in using the well known Keller Box method. In accordance with the boundary layer analysis, the boundary condition (12) at $\eta = \infty$ is replaced by $\eta = 5$, and the step size $\Delta\eta = 0.05$ is used to obtain numerical solution with five decimal place accuracy as the criterion of convergence. Obtained coupled ordinary non-linear Equations (9)-(11) are solved by Keller Box method for boundary condition (12). Accuracy of this numerical-method shown in **Table 1** is being validated by direct comparison with the numerical results reported by N.G. Rudraswamy [22], Meraj Mustafa *et al.* [23]. The numerical comparisons of, $-\theta'(0)$, $-f''(0)$ and $-\phi'(0)$ for the values of Pr, Le, S, γ, R, Nt, Nb, M, G and H are shown in **Table 2**.

4. Results and Discussion

Referring from **Figure 2** larger Prandtl number has a relatively lower thermal diffusivity. This is because Pris defined as the ration of kinematic viscosity to thermal diffusivity. Thus an increase in Pr reduces thermal diffusivity; it decreases the thermal boundary layer thickness. Consequently, increases rate of heat transfer, and thereby increases the variation in the thermal characteristics. As expected, the variation in the temperature is more pronounced for smaller values of Pr than its larger values.

Figures 3-5 depict the effects of suction parameter S on velocity, temperature and concentration profiles, respectively at the boundary for exponentially stretching sheet. It is observed that velocity decreases significantly with increasing suction parameter where as fluid velocity is found to increase with blowing. It is observed that, when the wall suction ($S > 0$) is considered, this causes a decrease in the boundary layer thickness and the velocity field is reduced. Opposite behaviour is noted for blowing $S < 0$. In **Figure 4**, it is seen that temperature decreases with increasing suction parameter where as it increases due to blowing. Temperature overshoot is noted for blowing ($S < 0$) this feature prevails up to certain heights and then the process slowed down and at a far distance from the wall temperature vanishes. In **Figure 5**, it is observed that concentration decreases with increasing suction parameter where as it increases due to blowing.

Figure 6 shows the effect of Le on the dimensionless concentration for fixed values of other parameters. It is observed that for larger values of Le suppress the concentration profile *i.e.* inhibit nanoparticle species diffusion, as observed. There will be a much greater reduction in the concentration boundary layer thickness.

Figures 7-10 show that an abnormal increase in the concentration φ is found for a weaker Brownian motion Nb in fact an over shoot in the concentration function occurs Nt increases gradually. The effect is seen in the case of temperature profile. An appreciable increase in temperature profile is found for increasing values of Nt. In this case, an increase in temperature is found with increasing values of Nb.

Figure 11 and **Figure 12** exhibits the nature of velocity field for the variation of magnetic parameter M. With increasing M, velocity is found to decrease but the temperature increases. As the Lorentz force opposes the motion of the fluid, much fluid is not entering to the boundary layer. It enforces the thickening of thermal layer.

As shown in **Figure 13**, it is observed that the nanoparicle volume fraction decreases with increasing values of chemical reaction parameter whereas the velocity and temperature profiles are not significant with the chemical reaction parameter.

As depicted in **Figure 14**, it is noticed that an increase in R yields an increase in the nanofluid's temperature,

Table 2. Showing results of $-\theta'(0)$, $-f''(0)$ and $-\phi'(0)$ for the values of Pr, M, S, Le, R, γ, G, H, Nt, and Nb.

Pr	Le	S	γ	R	Nt	Nb	M_1	G	H	$\theta'(0)$	$f''(0)$	$\phi'(0)$
0.72	10	1	0.01	0.01	0.45	0.45	0.3	0.5	0.06	−0.5992	−2.1252	−10.3194
1										−0.6919	−2.1252	−10.3194
1.5										−0.7677	−2.1252	−10.2214
	10									−0.5992	−2.1252	−10.3194
	15									−0.5934	−2.1252	−15.3972
	20									−0.5903	−2.1252	−20.4433
		−0.2								−0.1744	−1.4741	−1.0406
		0.2								−0.3017	−1.6648	−3.5434
		0.4								−0.3494	−1.7701	−5.1324
			0.1							−0.5991	−2.1252	−10.3317
			1							−0.5986	−2.1252	−10.4507
				0.1						−0.5648	−2.1252	−10.3431
				0.2						−0.5302	−2.1252	−10.3674
				0.3						−0.4992	−2.1252	−10.3895
					0.4					−0.6102	−2.1252	−10.2551
					0.8					−0.531	−2.1252	−10.5394
					1.2					−0.4669	−2.1252	−10.6288
						0.4				−0.6182	−2.1252	−10.3511
						0.8				−0.4838	−2.1252	−10.1787
						1.2				−0.3829	−2.1252	−10.1445
							0.5			−0.5965	−2.1865	−10.3149
							1.5			−0.5856	−2.4636	−10.2956
							5			−0.5626	−3.2155	−10.2502
								0.5		−0.5992	−2.1252	−10.3194
								1.5		−0.5875	−2.4115	−10.2991
								5		−0.5636	−3.1783	−10.2522
									0.06	−0.5992	−2.1252	−10.3194
									0.9	−0.4896	−2.1252	−10.4189
									1	−0.4765	−2.1252	−10.4308

which leads to decrease in the heat transfer rate. Thus, the radiation should be at its minimum in order to facilitate the cooling process. All these physical behaviour are due to the combined effects of the strength of the Brownian motion and thermophoresis particle deposition.

Figure 15 and **Figure 16**, show effect of porosity parameter G on the temperature and velocity profiles, respectively. It is observed that the presence of the porous medium. The temperature profile whereas is reduces the velocity profile. This is because the porous medium inhibits the fluid not to move freely through the boundary layer. This leads the flow to increase thermal boundary layer thickness.

As shown in **Figure 17** and **Figure 18**, it is observed that an increase in joule heating parameter H the

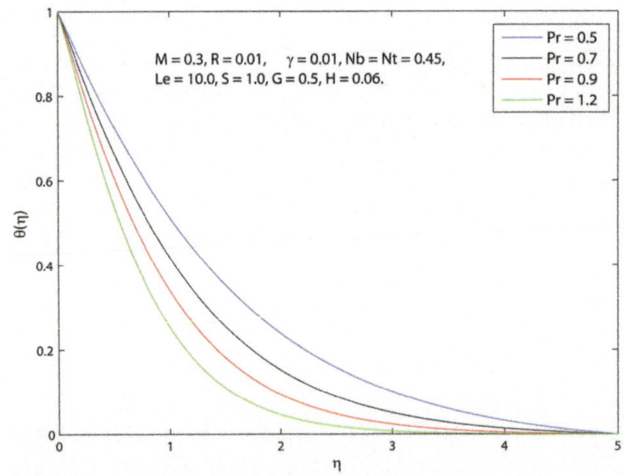

Figure 2. Variation of temperature $\theta(\eta)$ with η for several values of Prandtl number Pr.

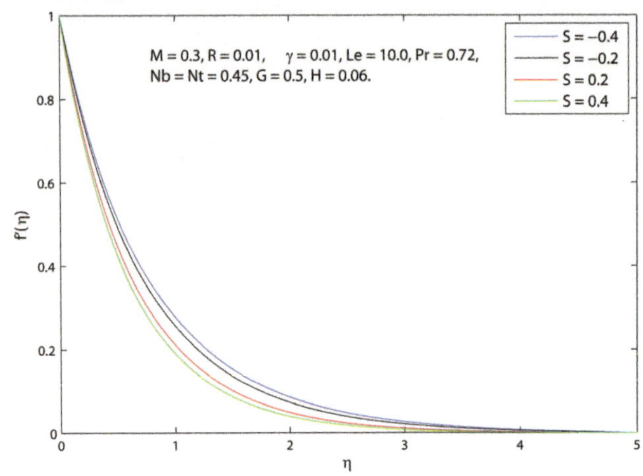

Figure 3. Variation of velocity $f'(\eta)$ with η for several values of suction parameter S.

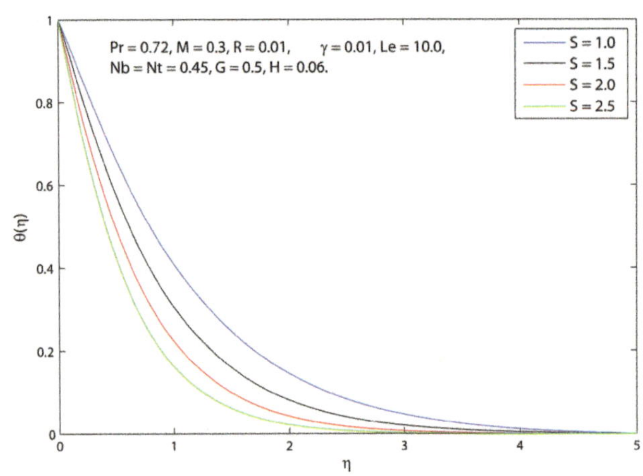

Figure 4. Variation of $\theta(\eta)$ with η for several values of suction parameter S.

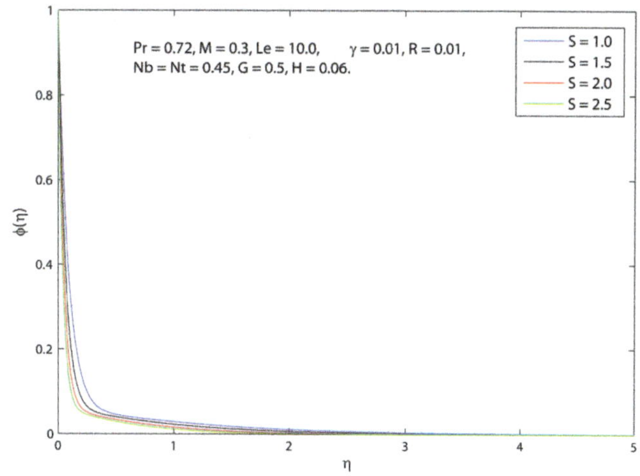

Figure 5. Variation of concentration $\phi(\eta)$ with η for several values of suction of S.

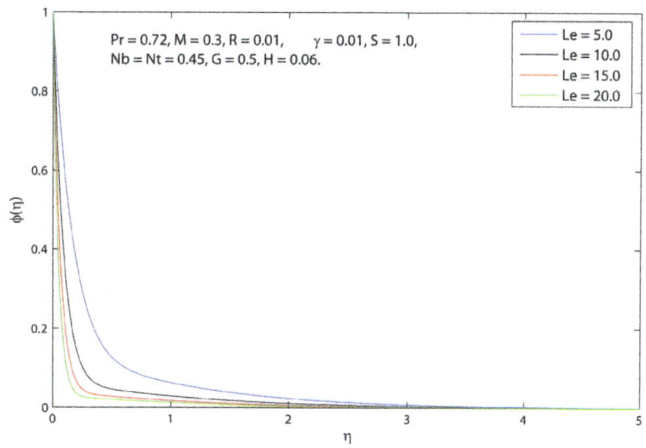

Figure 6. Variation of concentration $\phi(\eta)$ for with η for several values of Lewis number Le.

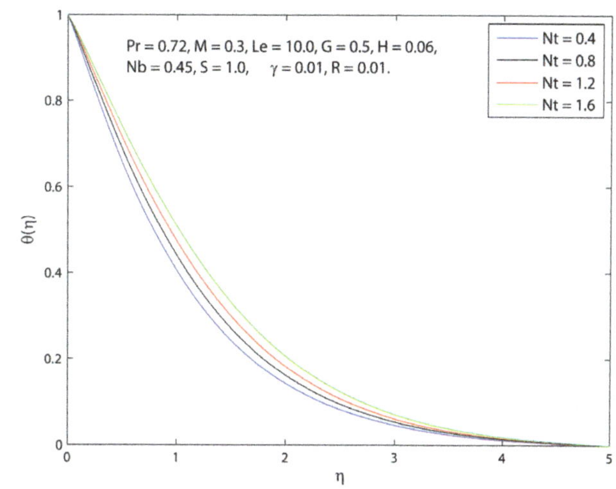

Figure 7. Variation of temperature $\theta(\eta)$ with ηseveral values of thermophoresis parameter Nt.

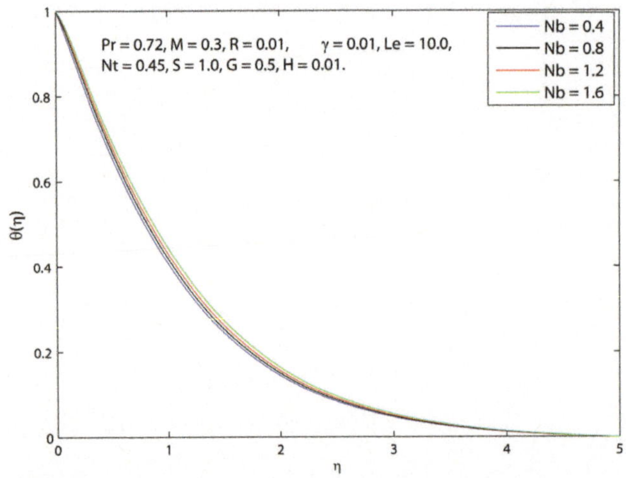

Figure 8. Variation of temperature $\theta(\eta)$ with η for several values of Brownian motion Nb.

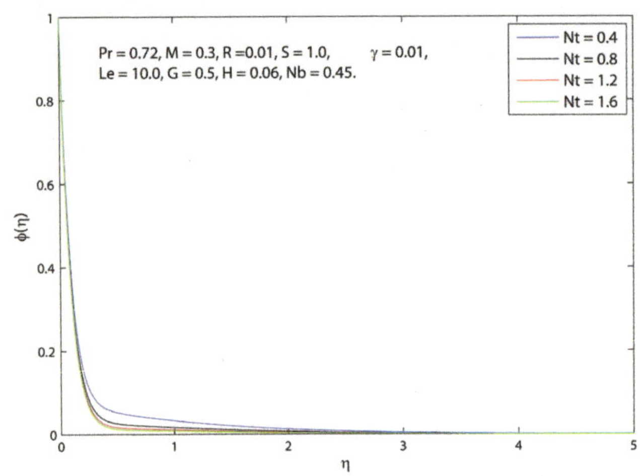

Figure 9. Variation of concentration of $\phi(\eta)$ with η for several values of thermophoresis parameter Nt.

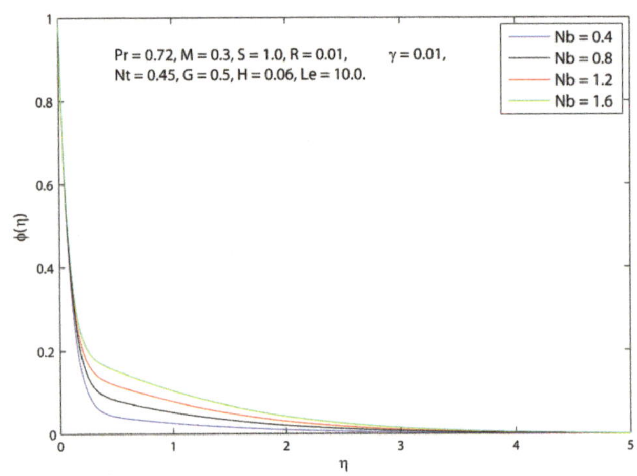

Figure 10. Variation of concentration of $\phi(\eta)$ with η for several values of Brownian motion parameter Nb.

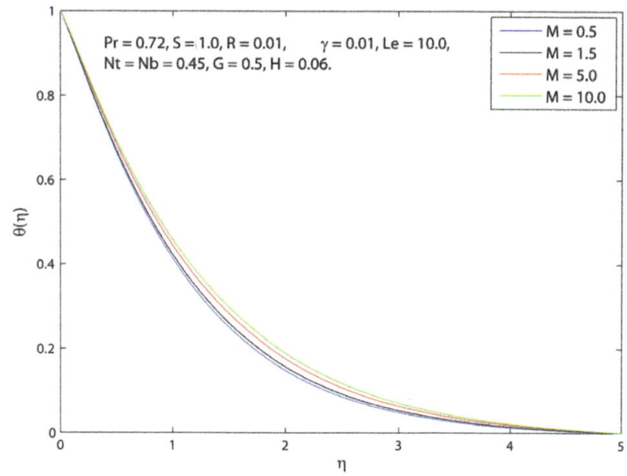

Figure 11. Variation of temperature $\theta(\eta)$ with η for several values of magnetic parameter M.

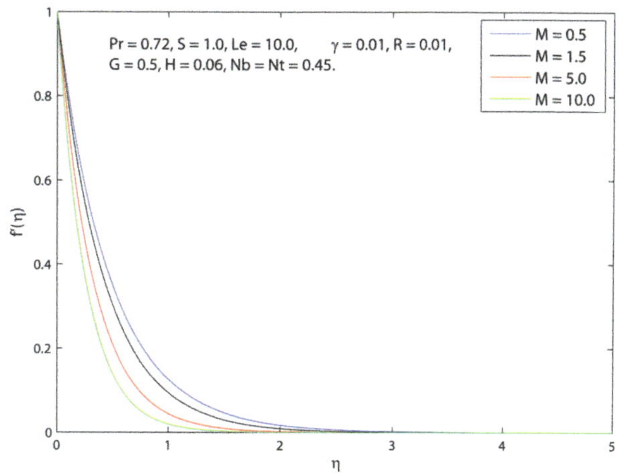

Figure 12. Variation of velocity $f'(\eta)$ with η with several values of Magnetic parameter M.

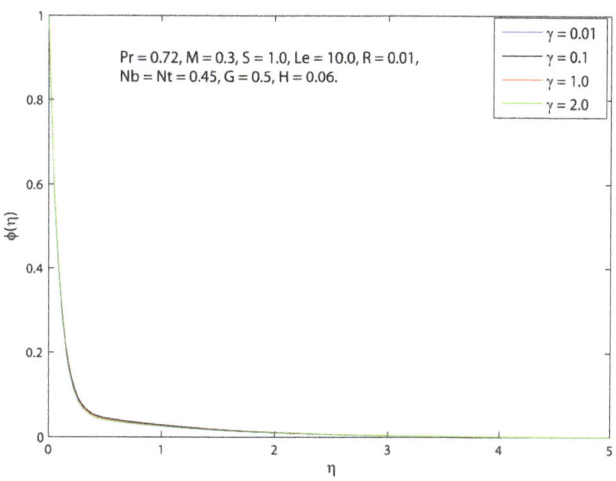

Figure 13. Variation of concentration $\phi(\eta)$ with η for several values of chemical reaction parameter γ.

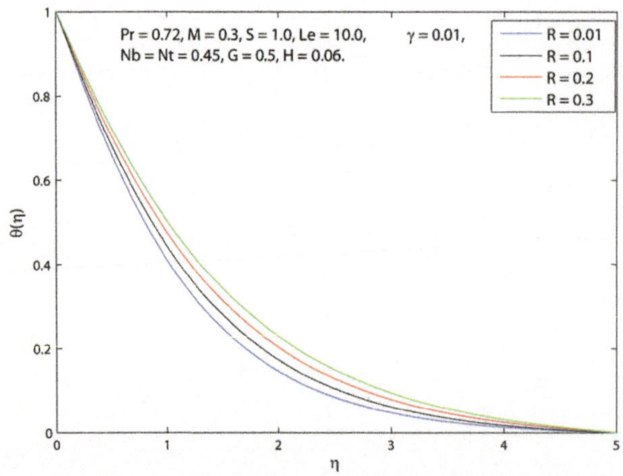

Figure 14. Variation of temperature $\theta(\eta)$ with η for several values of thermal radiation parameter R.

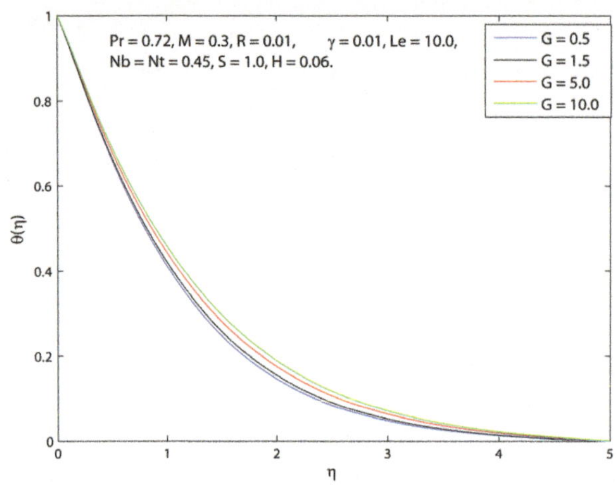

Figure 15. Variation of temperature $\theta(\eta)$ with η for several values of porosity parameter G.

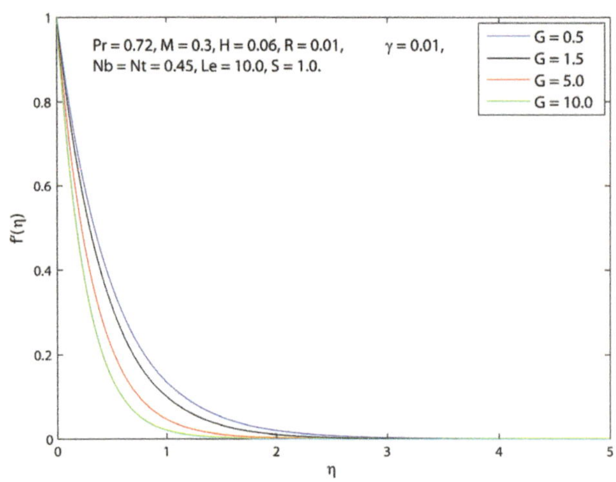

Figure 16. Variation of velocity $f'(\eta)$ with η for several values of porosity parameter G.

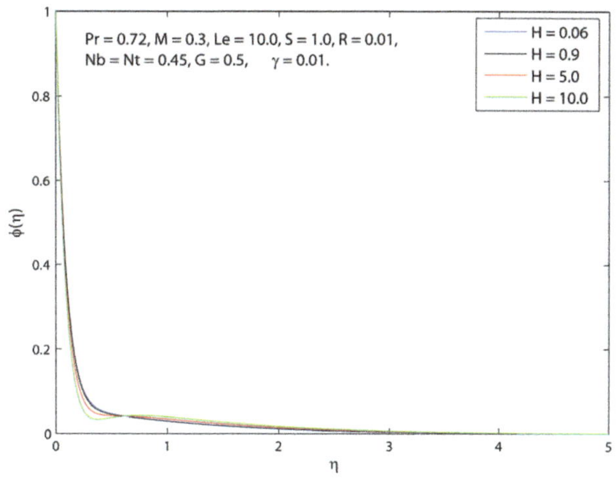

Figure 17. Variation of concentration (η) with η for several values of joule heating effect H.

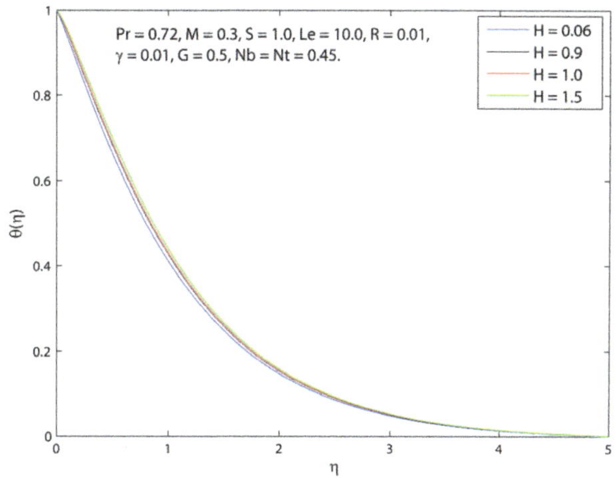

Figure 18. Variation of temperature $\theta(\eta)$ with η for several values of joule heating effect H.

profiles for temperature and concentration are decreasing.

5. Conclusions

A numerical study corresponding to the flow and heat transfer in a steady flow region of nanofluid over an exponential stretching surface and effects of chemical reaction, thermal radiation, magnetic, suction parameter, porosity parameter and joule heating parameters is examined and discussed in detail. The main observations of the present study are as follows.

An increase in suction parameter leads the velocity, temperature and concentration profiles to decrease. For larger values of Le suppress the concentration profile *i.e.* inhibit nanoparticle species diffusion, as observed. There will be a much greater reduction in the concentration boundary layer thickness. As Nt increases, temperature profile increases but the concentration profile decreases. With increasing values of Nb, both temperature and concentrations profiles increase. As increase in chemical reaction paramter γ leads the concentration profiles to decrease. The impact of porosity parameter shows that velocity profile is decreasing and temperature profile is increasing. This is because the porous medium inhibits the fluid not to move freely through the boundary layer. This leads the flow to increase thermal boundary layer thickness. Joule heating parameter reduces the temperature and concentration of nanofluid.

References

[1] Choi, S.U.S. (1995) Enhancing Thermal Conductivity of Fluids with Nanoparticles. In: Siginer, D.A. and Wang, H.P., Eds., *Developments and Applications of Non-Newtonian Flows*, ASME, New York, 99-105.

[2] Eastman, J.A. and Choi, S.U.S. (1997) Enhanced Thermal Conductivity through Development of Nanofluids. *MRS Proceedings*, **457**, 3-11. http://dx.doi.org/10.1557/proc-457-3

[3] Routbort, J.L., Yu, D.M. and Choi, S.U.S. (2008) Review and Comparison of Nanofluid Thermal Conductivity and Heat Transfer Enhancements. *Heat Transfer Engineering*, **29**, 432-460. http://dx.doi.org/10.1080/01457630701850851

[4] Buongiorno, J. (2006) Convective Transport in Nanofluids. *ASME Journal of Heat Transfer*, **128**, 240-250. http://dx.doi.org/10.1115/1.2150834

[5] Cheng, L. (2008) Nanofluid Two Phase Flow and Thermal Physics: A New Research Frontier of Nanotechnology and Its Challenges. *Journal of Nanoscience and Nanotechnology*, **8**, 3315-3332. http://dx.doi.org/10.1166/jnn.2008.413

[6] Rajabpour, A. and Akizi, F.Y. (2013) Molecular Dynamics Simulation of the Specific Heat Capacity of Water-Cu Nanofluids. *International Nano Letters*, **3**, 58. http://dx.doi.org/10.1186/2228-5326-3-58

[7] Hu, W. and Buongiorno, J. (2005) Nanofluid Coolants for Advanced Nuclear Power Plants. *Proceedings of ICAPP'*05, Seoul, May 2005, 15-19.

[8] Gupta, P.S. and Gupta, A.S. (1997) Heat and Mass Transfer on a Stretching Sheet with Suction or Blowing. *The Canadian Journal of Chemical Engineering*, **55**, 744-746. http://dx.doi.org/10.1002/cjce.5450550619

[9] Emmanuel, S. and Khan, S.K. (2006) On Heat and Mass Transfer in a Viscoelastic Boundary Layer Flow over an Exponentially Stretching Sheet. *International Journal of Thermal Sciences*, **45**, 819-828.

[10] Kim, Y.J. (2000) Unsteady MHD Convective Heat Transfer Past a Semi-Infinite Vertical Porous Moving Plate with Variable Suction. *International Journal of Engineering Science*, **38**, 833-845.

[11] Subhas Abel, M. and Siddheshwar, P.G. (2007) Heat Transfer in a Viscoelastic Boundary Layer Flow over a Stretching Sheet with Viscous Dissipation and Non-Uniform Heat Source. *International Journal of Heat and Mass Transfer*, **50**, 960-966. http://dx.doi.org/10.1016/j.ijheatmasstransfer.2006.08.010

[12] Nadeem, S., Zaheer, S. and Fang, T.G. (2011) Effects of Thermal Radiation on the Boundary Layer Flow of a Jeffrey Fluid over an Exponentially Stretching Surface. *Numerical Algorithms*, **57**, 187-205. http://dx.doi.org/10.1007/s11075-010-9423-8

[13] Ishak, A. (2011) MHD Boundary Layer Flow Due to an Exponentially Stretching Sheet with Radiation Effect. *Sains Malaysiana*, **40**, 391-395.

[14] Al-odat, M.Q., Damseh, R.A. and Al-azab, T.A. (2006) Thermal Boundary Layer on an Exponentially Stretching Continuous Surface in the Presence of Magnetic Field Effect. *International Journal of Applied Mechanics and Engineering*, **11**, 289-299.

[15] Kumar, H. (2013) Heat Transfer in MHD Boundary Layer Flow through a Porous Medium, Due to a Non-Isothermal Stretching Sheet, with Suction, Radiation and Heat Annihilation. *Chemical Engineering Communications*, **200**, 895-906. http://dx.doi.org/10.1080/00986445.2012.727509

[16] Chand, G. and Jat, R.N. (2014) Flow and Heat Transfer over an Unsteady Stretching Surface in a Porous Medium. *Thermal Energy and Power Engineering*, **3**, 266-272.

[17] Sudhakar, K., Srinivas Raju, R. and Rangamma, M. (2013) Hall Effect on Unsteady MHD Flow Past along a Porous Flat Plate with Thermal Diffusion, Diffusion Thermo and Chemical Reaction. *International Journal of Physical and Mathematical Sciences*, **4**, 370-395.

[18] Subhas Abel, M., Kumar, K.A. and Ravi kumara, R. (2011) MHD Flow, and Heat Transfer with Effects of Buoyancy, Viscous and Joules Dissipation over a Nonlinear Vertical Stretching Porous Sheet with Partial Slip. *Engineering*, **3**, 4.

[19] Hamza, M.M., Isah, B.Y. and Usman, H. (2011) Unsteady Heat Transfer to MHD Oscillatory Flow through a Porous Medium under Slip Condition. *International Journal of Computer Applications*, **33**, 266-272.

[20] Sharma, P.R. and Singh, G. (2010) Effects of Variable Thermal Conductivity, Viscous Dissipation on Steady MHD Natural Convection Flow of Low Prandtl Fluid on an Inclined Porous Plate with Ohmic Heating. *Meccanica*, **45**, 237-247.

[21] Cebeci, T. and Bradshaw, P. (1988) Physical and Computational Aspects of Convective Heat Transfer. Springer, New York. http://dx.doi.org/10.1007/978-1-4612-3918-5

[22] Rudraswamy, N.G. and Gireesha, B.J. (2014) Influence of Chemical Reaction and Thermal Radiation on MHD Boundary Layer Flow and Heat Transfer of a Nanofluid over an Exponentially Stretching Sheet. *Journal of Applied Mathematics and Physics*, **2**, 24-32. http://dx.doi.org/10.4236/jamp.2014.22004

[23] Mustafaa, M., Hayat, T. and Obaidat. S. (2013) Boundary Layer Flow of a Nanofluid over an Exponentially Stretching Sheet with Convective Boundary Conditions. *International Journal of Numerical Methods for Heat and Fluid Flow*, **23**, 945-959. http://dx.doi.org/10.1108/HFF-09-2011-0179

Thorn-Shape Waveform and Double-Strike Phenomenon Seen in the Impact Force of Soft Materials

Yukio Fujimoto, Chen Liu, Yoshikazu Tanaka, Eiji Shintaku, Toshiki Nakanishi

Department of Transportation and Environmental Engineering, Graduate School of Engineering, Hiroshima University, Hiroshima, Japan
Email: fujimoto@naoe.hiroshima-u.ac.jp

Abstract

In the impact tests of soft materials, we sometimes observe a thorn shape in the rising segment of the impact force waveform. However, the reason for the occurrence of the thorn shape has not been made clear. In this study, thorn-shape waveforms of several soft materials are measured using compact drop test equipment under the condition of a flat frontal impact. A flat frontal impact is the condition where a drop hammer with a flat bottom surface strikes a plate-like soft material in the normal direction. Synchronized impact forces are measured using two sensors installed on both the drop hammer side and the floor side. The examined soft materials are a sponge sheet, sponge rubber sheet, gel sheet, rubber sheet, flat oil clay, low-rebound urethane foam, cork sheet, sliced ham, pork ham steak, and pork. Based on the test results, the features of the thorn-shape waveforms are discussed from a bird's-eye view. Furthermore, the occurrence mechanism of the thorn-shape waveforms is discussed from the viewpoint of viscosity discontinuity and the double-strike phenomenon.

Keywords

Impact Force, Thorn-Shape Waveform, Soft Material, Drop Hammer, Viscosity Discontinuity, Strain Rate Dependence, Double-Strike Phenomenon

1. Introduction

In impact tests of soft materials such as rubber, clay, or soft plastic, we sometimes observe a thorn shape in the rising segment of the impulsive force waveform. Several thorn-shape waveforms can be seen in the impact test results of previous research studies [1]-[3]. However, the physical meaning of the thorn shape has not been discussed much.

Recently, we have developed an impact force sensor that can measure a high-speed impact force waveform accurately [4] [5], and we have performed impact tests on several soft materials using the sensor. While repeating the experiments, the authors realized that the thorn-shape waveform can be clearly measured under the condition of a flat frontal impact. Also, the authors pointed out the possibility that the thorn-shape waveform is related to the strain rate dependence of the material and the double-strike phenomenon [6].

In this study, free drop impact tests are performed on several soft materials using compact drop test equipment. The thorn-shape waveforms are observed from a bird's-eye view. Further, the occurrence mechanism of the thorn-shape waveform is discussed from the viewpoint of viscosity discontinuity and the double-strike phenomenon.

2. Thorn Shape and Double-Strike Phenomenon Seen in the Impact Force Waveform of Soft Materials

Figure 1(a) shows a soft material subjected to a flat frontal impact. The impact object is a free drop hammer. **Figure 1(b)** shows typical thorn-shape waveforms of the impact force. The features of the thorn-shape waveform are a steep slope on the rising segment (which is θ) and a thorn-shape waveform similar to the yield point of low-carbon steel. After that, at a certain point, it becomes a gentler slope (which is α) and generates a second mountain-shape waveform. The rising segment of the thorn is usually close to a straight line.

Figure 1(c) summarizes the occurrence mechanism of thorn-shape waveforms obtained by the following experimental observation. The impact force waveform of soft materials is the mixture of two types of waveforms. The first (thorn-shape) waveform shows a large viscosity resistance (red curve), and the second (mountain-shape) waveform shows a small viscosity resistance (dashed black curve). This resistance change occurs suddenly at the thorn peak; that is, a viscosity discontinuity (viscosity transient) is induced at the thorn peak. The compressive stress at the thorn peak is the trigger of the softening. The soft material surface is struck twice by the relative movement of collision objects. The reason why the slope α becomes smaller than the initial slope θ is because a viscosity discontinuity is induced by the first strike.

Figure 2 shows three types of thorn-shape waveforms measured in the following experiments:

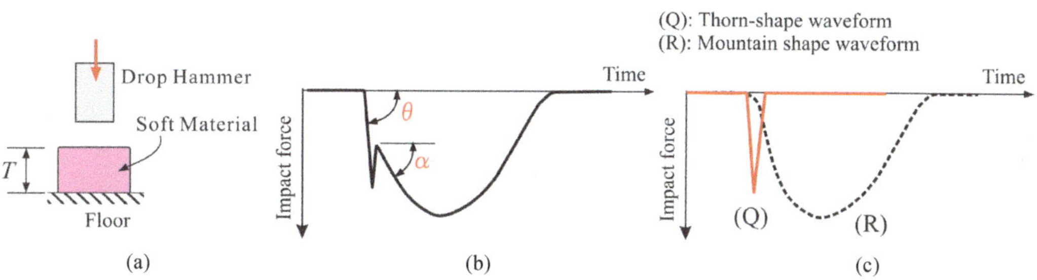

Figure 1. Impact force waveform of soft material. (a) Flat frontal impact by free drop hammer and plate-like soft material; (b) Typical impact force waveform of soft material including thorn; (c) Two types of waveforms contained in the impact force waveform of soft material.

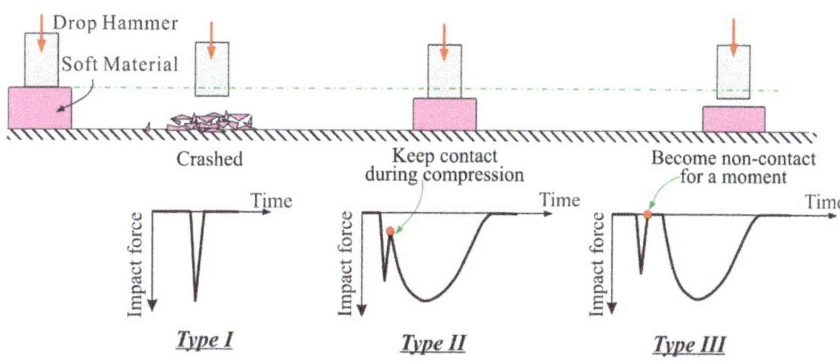

Figure 2. Three types of thorn-shape waveforms observed in the impact force of soft material.

Type I: In the case where the soft material is crushed into pieces by the first impact, only a thorn-shape waveform appears, and the second mountain-shape waveform does not appear.

Type II: This type of waveform appears in most of the soft materials. The slope θ and thorn peak are influenced by the material, impact velocity, material thickness, etc. In this waveform, a certain compressive force still remains after the first impact. This means that the hammer bottom and soft material surface remain in contact during unloading.

Type III: This type of waveform is observed in the thick soft material. The hammer bottom and the soft material surface have a non-contact condition for a moment after the first impact.

3. Condition of Flat Frontal Impact and Compact Drop Test Equipment

Through a number of experiments, we found that the thorn-shape waveform can be measured clearly under the condition of a flat frontal impact. In **Figure 3**, the impact object is a solid material with a flat contact surface. The soft material that receives the impact force is a plate-like material with a uniform thickness without surface irregularities. The impact object strikes the soft material in the normal direction. The thorn-shape waveform can be steadily measured in the above condition. When the bottom surface of the impact object is curved (**Figure 3(b)**), when the impact object is tilted (**Figure 3(c)**), or when the upper surface of the soft material is curved (**Figure 3(d)**), the thorn becomes unclear. Here, "unclear" does not mean that the thorn will not occur; it is unclear in that the contact area gradually expands from the local area, and the phenomenon is diluted.

Figure 4 shows the free drop test equipment used in this study. The device is braced by vertical columns on the left and right of the bottom plate. The top of the side column is connected with a metal plate. The drop hammer is made of an aluminum circular disk plate and a stainless cylinder (mass of 1.9 kg). The contact surface of the hammer bottom is a disk plate with a diameter of D = 60 mm. The upper sensor is an orbicular pad sensor and is attached near the drop hammer bottom. The lower sensor is a pad sensor of 100 mm square and is attached to the frame bottom. Both sensors are made from piezoelectric film. A high-speed voltage recorder with a sampling rate of 20 - 200 kHz is used for recording the sensor output. The output of the upper sensor is modified by 8% to account for the mass of the disk plate. Based on rigorous tests using a servo-controlled testing machine and a high-rigidity load cell, it was confirmed that both sensors can measure the impact force very accurately when the contact time is in the range between 0.1 msec and 2 sec. The drop height h is the height from

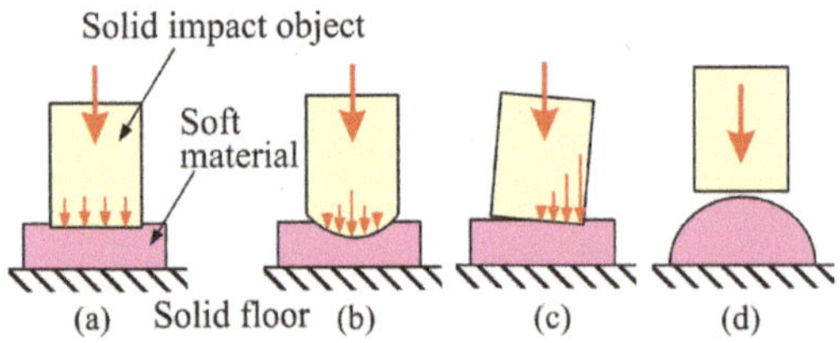

Figure 3. (a) Flat frontal impact condition that the thorn appears clearly. (b), (c), (d) are the conditions where the thorn becomes indistinct.

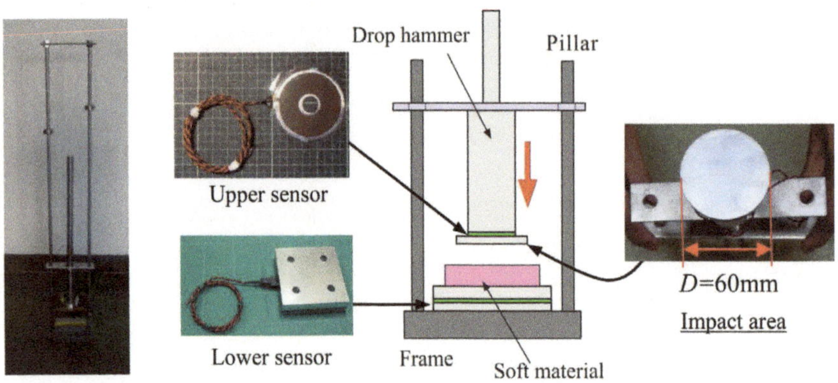

Figure 4. Drop test equipment and the upper and lower sensors used in the device.

the upper surface of the soft material to the bottom surface of the drop hammer. The impact velocity is $V_0 = 1.4$ m/sec when $h = 0.1$ m, $V_0 = 1.98$ m/sec when $h = 0.2$ m and $V_0 = 2.73$ m/sec when $h = 0.38$ m.

Figure 5 shows the importance of the flat frontal impact. **Figure 5(a)** shows woodblocks (triangular prism, semicircular column, and hemisphere) that are fixed to the bottom of the free drop hammer. The soft material is a nitrile rubber plate (NBR) with a 30 mm thickness. The drop height is $h = 380$ mm. **Figure 5(b)** shows the measured impact force waveforms. The waveform under the flat frontal impact condition (without a wood block) is also shown in the figure. The thorn-shape waveform appears clearly under the flat frontal impact. However, when the block corner or a curved surface locally strikes the rubber, the thorn-shape waveform is not obvious.

4. Experimental Results

4.1. Impact Force of Agar Jelly under Flat Frontal Impact (*Type I* Waveform)

Figure 6 shows two examples of impact force waveforms for an agar jelly with a 30 mm thickness. The height of the drop hammer is $h = 380$ mm. The initial strain rate at the collision is $\dot{\varepsilon}_0 = V_0/T = 91$ sec^{-1}, where V_0 is the hammer speed at the collision, and T is the soft material thickness. When the drop hammer collides, the agar jelly is crushed and scattered as juice. Therefore, the thorn-shape waveform (***Type I*** waveform) appears only in the upper sensor output. The thorn is close to an isosceles triangle, and its contact time is short at 0.5 msec. The impulse of the thorn part is quite small compared with the total momentum of the drop hammer (5.19 N·sec). The impact force seen in the lower sensor output is delayed. This may be caused by the crush effect. Because the jelly was crushed at the thorn peak, the impact force is reduced to zero. About 15 msec later, the hammer bottom strikes the lower sensor strongly without losing much of its falling velocity.

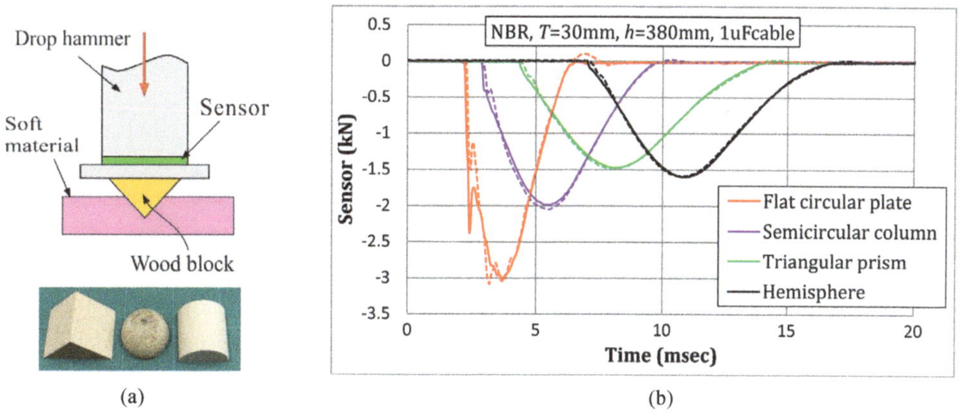

Figure 5. Influence of contact shape on impact force waveform. (a) Wood blocks attached to the drop hammer bottom; (b) Comparison of the waveform of the flat frontal impact and those of other impact conditions.

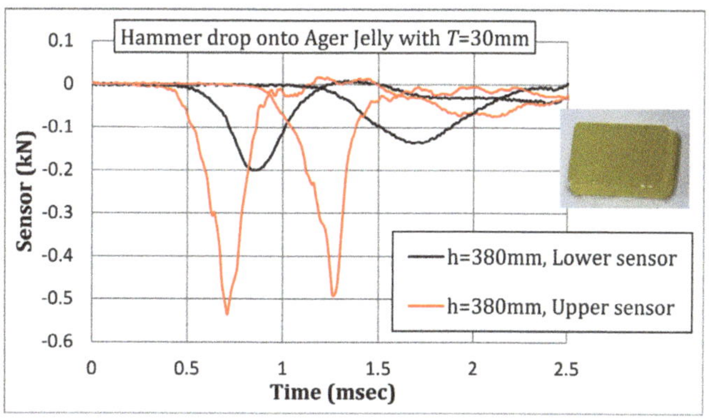

Figure 6. Impact force waveform induced by the hammer drop onto agar jelly.

4.2. Impact Force of Konjac under Flat Frontal Impact (*Type III* Waveform)

Figure 7 shows the impact force waveforms and an enlarged view of a konjac plate with a 40 mm thickness. The drop height is h = 100 mm ($\dot{\varepsilon}_0$ = V_0/T = 35 sec^{-1}) and 380 mm ($\dot{\varepsilon}_0$ = 68 sec^{-1}). The konjac plate is compressed greatly by the impact but not crushed. In the figure, the time interval between the thorn-shape waveform and the succeeding mountain-shape waveform is wide, and the double-strike phenomenon is obvious (***Type III*** waveform). In the magnified figure, the thorn is close to an isosceles triangle, and its contact time is short at 0.8 msec. Also, a small thorn can be seen in the lower sensor output. The impulse of the thorn part is quite small compared with the succeeding mountain-shape waveform. This means that most of the falling velocity is lost at the second mountain-shape waveform. Usually, the thorn shapes of the upper and lower sensor outputs are close when the material thickness is smaller.

4.3. Impact Force of Gel Sheet under Flat Frontal Impact (*Type II* Waveform)

Figure 8(a) shows the impact force waveforms of a gel sheet (product name: α gel) with a 20 mm thickness. **Figure 8(b)** is an enlarged view of the thorn part. The drop height is h=380 mm ($\dot{\varepsilon}_0$ = V_0/T = 136 sec^{-1}). The thorn-shape waveform can be seen clearly in both the upper and lower sensor outputs (***Type II*** waveform). In the enlarged view, we can observe that the slope θ of the rising segment is larger than the slope α of the segment past the thorn. The impulse of the thorn part is quite small compared with the succeeding mountain-shape waveform.

4.4. Impact Force of Nitrile Rubber (NBR) Plate under Flat Frontal Impact (*Type II* Waveform)

Figure 9(a) is the impact force of an NBR plate with T = 30 mm (rubber hardness of 50). The drop height is

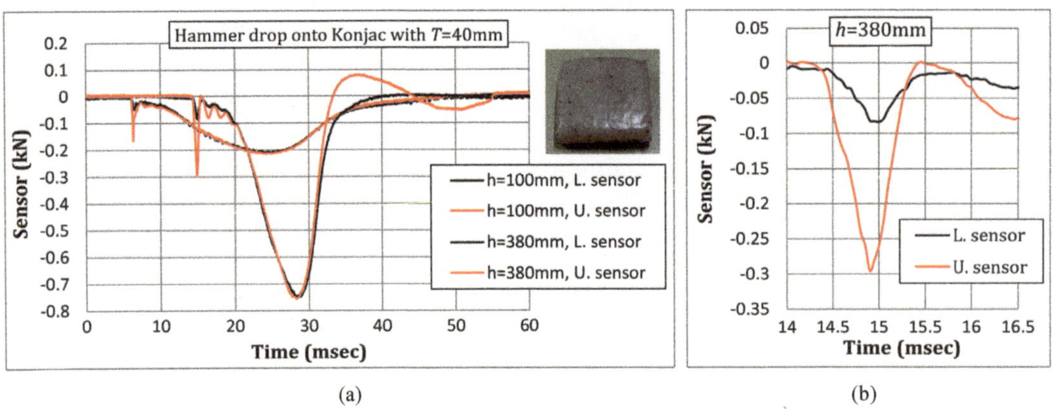

Figure 7. (a) Impact force waveforms induced by hammer drop onto konjac with a 40 mm thickness; (b) Enlarged figure of thorn part for the case of h = 380 mm.

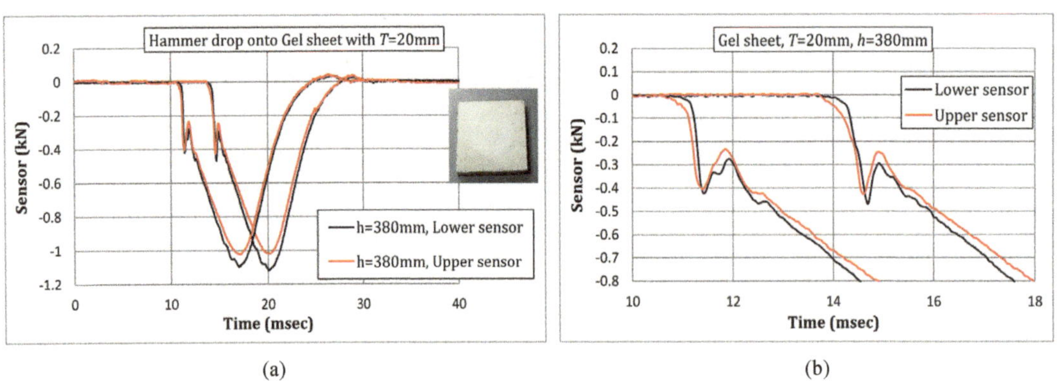

Figure 8. (a) Impact force waveform induced by the hammer drop onto a gel sheet with a 20 mm thickness; (b) Enlarged figure of the thorn part.

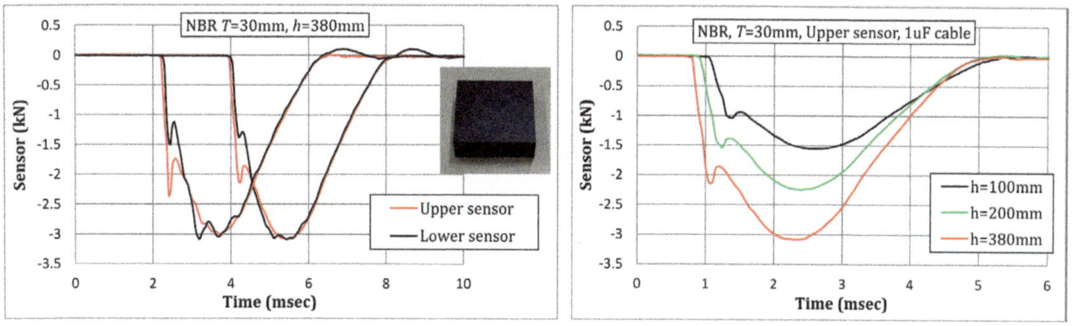

(a) (b)

Figure 9. Impulsive force induced by the hammer drop onto NBR with $T = 30$ mm.

$h = 380$ mm ($\dot{\varepsilon}_0 = 91$ sec^{-1}). The upper sensor output shows a relatively large thorn, and the lower sensor output shows a smaller thorn. The slopes θ of the upper and lower sensor outputs are almost the same. **Figure 9(b)** compares the impact force waveforms of the upper sensor for drop heights of $h = 100$, 200, and 380 mm. In the figure, when the impact velocity V_0 is faster, the slope θ of the rising segment and the height of the thorn peak are larger. This means that the slope θ and the height of the thorn peak are dependent on the impact velocity or strain rate.

4.5. Impact Force of Oil Clay Plate under Flat Frontal Impact (*Type II* Waveform)

Figure 10 shows the impact force waveform of an oil clay plate with a 50 mm thickness. The drop height is $h =$ 100 and 530 mm. A large thorn is seen by the upper sensor, and a small thorn is seen by the lower sensor. As the drop height increases (V_0 becomes faster), the slope θ becomes larger, and the thorn peak becomes taller. The thorns are more obvious than those of other materials. From the enlarged view, the thorn shape is close to an isosceles triangle.

4.6. Impact Force of Low-Rebound Urethane Foam under Flat Frontal Impact (*Type II* Waveform)

Figure 11 shows the impact force waveform of low-rebound urethane foam with a 40 mm thickness. The drop height is $h = 380$ mm. In this case, the bottom diameter of the drop hammer is 60 and 75 mm. In the figure, the steep slope of the rising segment is bent approximately 90 degrees at a certain force level. Also, a small thorn can be seen at the bent corner. Since low-rebound urethane foam is a high-viscosity material, the material shows a strong viscous resistance (hardening) until reaching a certain compressive stress.

4.7. Impact Force of Ham and Edible Meat under Flat Frontal Impact (*Type II* Waveform)

Figure 12 shows the impact force waveforms of sliced ham ($T = 21$ mm), pork ham steak ($T = 19$ mm), pork ($T = 25$ mm), and konjac ($T = 20$ mm). The drop heights are all $h = 380$ mm. The solid curve represents the upper sensor output, and the dashed curve represents the lower sensor output. We can see the steep slope θ in the ascending segment and the thorns in both the upper and lower sensor outputs.

Table 1 compares the falling momentum of the drop hammer and the impulse of the impact force waveform. The impulse is calculated by numerically integrating the area of the waveform. The impulse is about 10% larger than the falling momentum.

5. Measurement of Triple-Strike Waveform

Thin aluminum circular plates with a diameter of 25 mm and different thicknesses of $d = 0.5$, 1.0, 2.0, 3.0, and 5.0 mm are prepared. Each plate is attached to the bottom of the drop hammer with adhesive tape, as shown in **Figure 13(a)**. Then, the hammer is dropped from $h = 380$ mm onto the NBR plate with a 30 mm thickness. The hammer bottom collides in two stages. The contact area at the first impact is a circle with a 25 mm diameter, and the second is a ring-shape area with a diameter range from 25 to 60 mm.

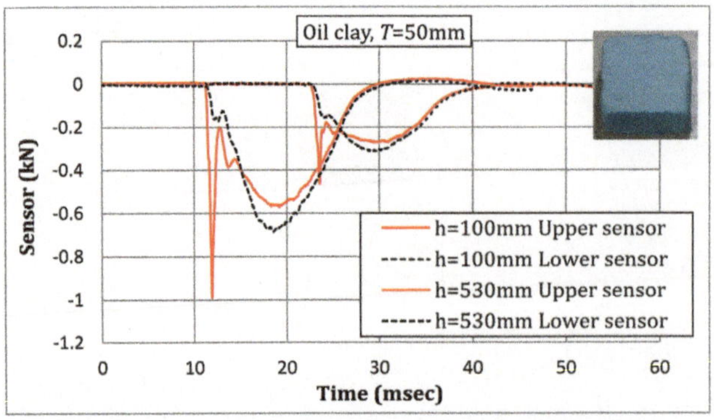

Figure 10. Impact force induced by the hammer drop onto oil clay with a 50 mm thickness.

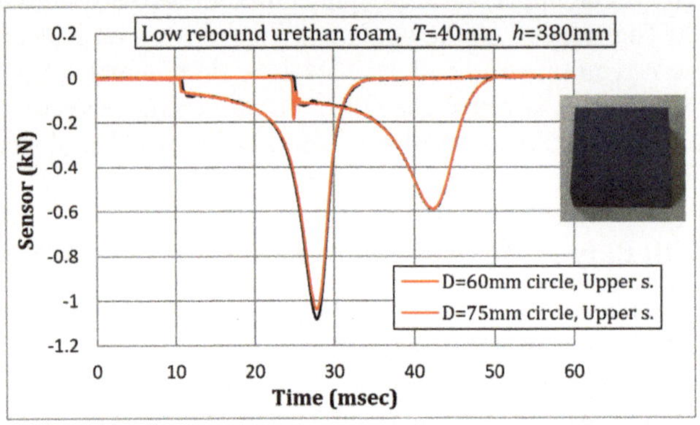

Figure 11. Impact force induced by the hammer drop onto low-rebound urethane foam.

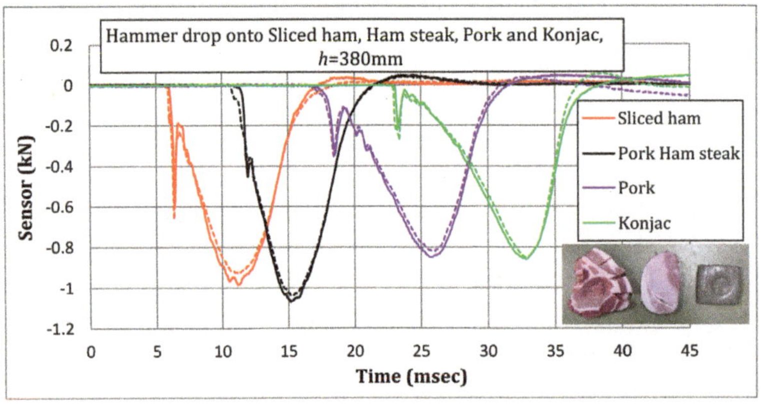

Figure 12. Impact force induced by the hammer drop onto sliced ham, pork ham steak, pork and konjac.

Table 1. Momentum and impulse of impact force induced by the hammer drop.

Mass of drop weight m (kg)	Falling height h (m)	Momentum mV (N·sec)	Material	Thickness T (mm)	Impulse (N·sec) Lower sensor
1.9	0.38	5.19	Sliced ham	21	5.92
			Pork ham steak	19	5.58
			Pork	25	5.58
			Konjac	20	5.28

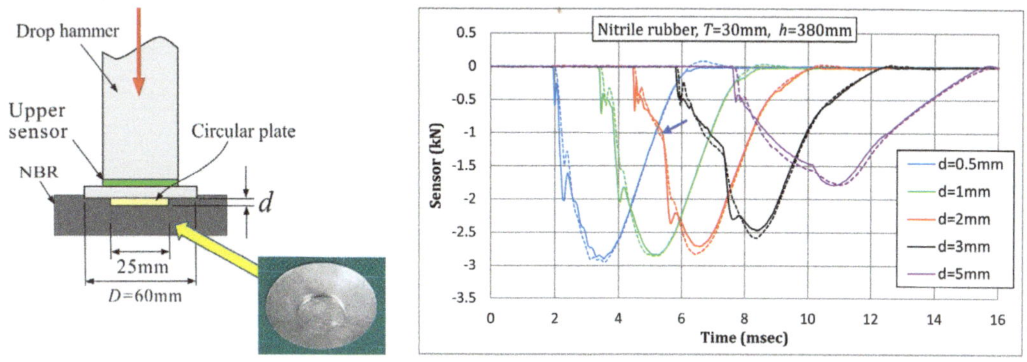

(a) (b)

Figure 13. Measurement of triple-strike waveform of impact force. (a) Aluminum circular plate is attached to the bottom of the drop hammer; (b) Impact force induced by the hammer drop onto NBR with a 30mm thickness.

Figure 13(b) shows the impact waveforms for all of the cases. The solid curves correspond to the upper sensor, and the dashed curves correspond to the lower sensor. We can see two thorn-shape waveforms and one mountain-shape waveform when d is 3 mm or less. The first thorn is due to the impact by the small circular plate. The second thorn is due to the impact by the ring-shape area. The bending point between the first and second thorns indicates the start of contact of the ring-shape area (shown by the blue arrow in the case of $d = 2$ mm).

In the typical double-strike phenomenon, the second strike makes a mountain-shape waveform. Therefore, the authors think that the second thorn such as that shown in **Figure 13(b)** will occur only if the drop hammer strikes the virgin surface of the soft material.

6. Mechanism of Thorn-Shape Waveform and Double-Strike Phenomenon

6.1. Contact Time of Thorn-Shape Waveform

Table 2 shows the contact time t_c of the thorn-shape waveform (approximated by an isosceles triangle) and $\dot{\varepsilon}_0$ for some soft materials. The value of t_c ranges between 0.4 msec and 1.0 msec. **Figure 14** shows the impact force waveform of a thick aluminum plate punched by a plastic hand hammer (PVC head). Its t_c is 0.74 msec, and this value is close to those of soft materials. The material hardness is the main factor that governs t_c. Therefore, this result means that the soft materials reached almost the same hardness as the plastic hammer head at the thorn.

6.2. Thickness Effect of Soft Material

Figure 15 shows the impact force waveforms of oil clay plates with $T = 10$, 20, and 50 mm. The solid curve is the upper sensor output, and the dashed curve is the lower sensor output. The drop heights are all $h = 380$ mm (impact velocity is $V_0 = 2.73$ m/sec). The strain rate at contact is $\dot{\varepsilon}_0 = V_0/T = 273$ sec^{-1}, 137 sec^{-1}, and 54.6 sec^{-1}, respectively. In the figure, it is clear that the slope θ and the peak force of the thorn are influenced by the strain rate. The second mountain-shape waveform is also influenced by the strain rate.

The authors also performed experiments on cork sheet, sponge sheet, sponge rubber, paper clay, tofu, relatively smooth areas of human skin, and some hollow balls. Among them, all of the materials except the cork sheet showed a thorn-shape waveform.

6.3. Mechanism of Thorn-Shape Waveform

Figure 16 shows the author's understanding of the occurrence mechanism for the thorn-shape waveform based on the above experimental observations.

1) In the rising segment P_1-P_2 of **Figure 16**, the viscosity resistance is excessively large, and the waveform shows a steep slope θ.

2) At the thorn peak P_2, the viscosity resistance changes suddenly (viscosity transient). The large viscous resistance is broken by the thorn peak force and is softened to a certain extent.

Table 2. Contact time of thorn-shape waveform for soft materials.

Soft Material	Contact Time t_c (msec)	$d\varepsilon_0/dt = V_0/T$ (sec^{-1})
Ager Jelly	0.5	91
Gel Sheet	0.8	136
Konjac	0.8	68
Oil Clay	0.4	273
Oil Clay	1.0	136
NBR	0.5	91
Low-Rebound Urethane Foam	0.7	68

h = 380 mm.

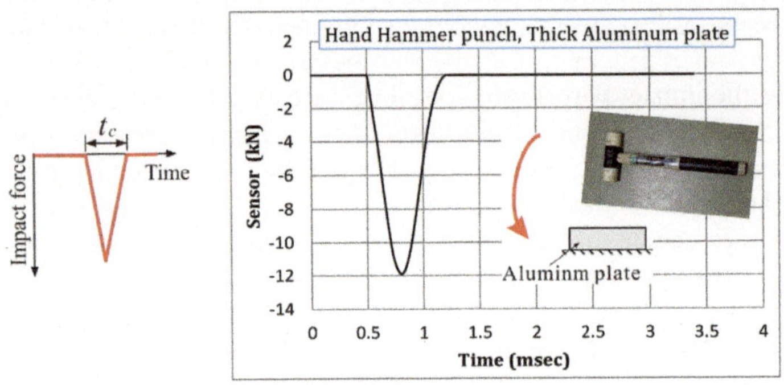

Figure 14. Impact force of aluminum plate punched by plastic hand hammer.

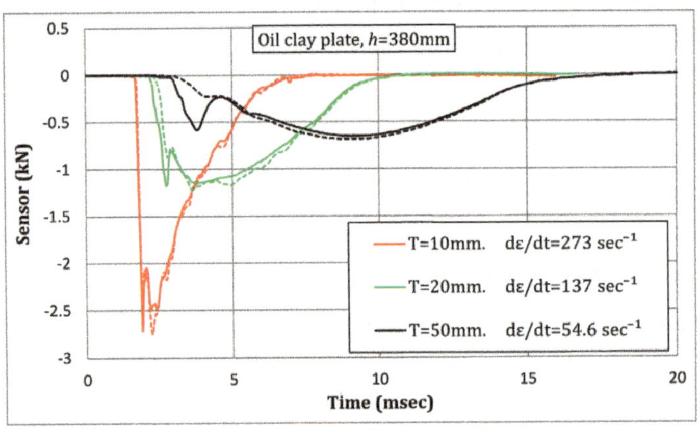

Figure 15. Thickness effect of oil clay plate on the impact force waveform. Drop heights are all h = 380 mm. Solid curve is upper sensor, and dashed curve is lower sensor.

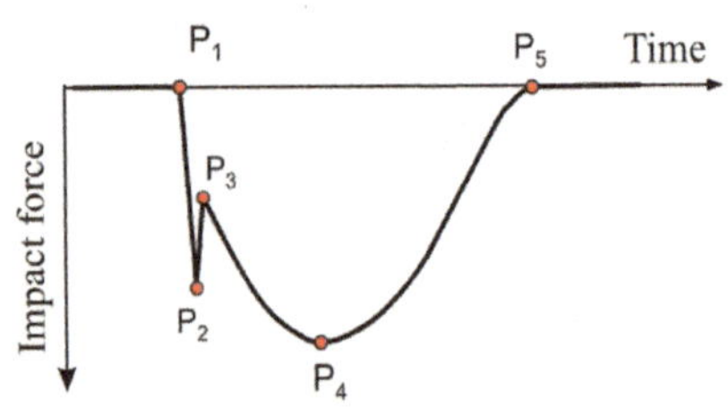

Figure 16. Mechanism of impact force waveform for soft materials based on the experimental observations.

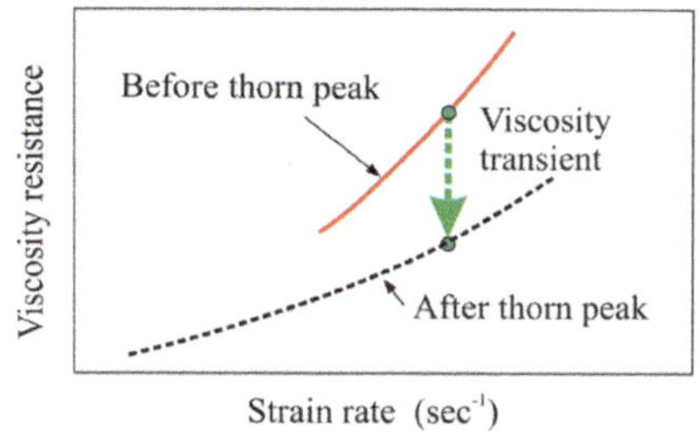

Figure 17. Viscosity transient vs. strain rate relationship.

3) When the viscosity resistance becomes small, the soft material is relaxed for a moment, and unloading occurs in the region P_2-P_3. When the softened material starts compressive deformation, the relative downward velocity between the hammer and the soft material surface decreases for a moment. This is the cause of unloading.

4) However, the hammer speed catches up to the soft material surface at the point P_3 and strikes it again. The second mountain-shape waveform shows a gentler slope α in the region P_3-P_4 because the viscosity resistance becomes small. The soft material shows a small viscosity resistance in the region P_3-P_5.

Based on the understanding of the above, the authors showed that the impact force waveform consisting of a thorn and mountain shape waveform can be simulated by using a simple dynamics model (SLS model) [7].

Figure 17 shows a schematic expression of the relationship between the viscosity resistance and the strain rate. The authors think that two types of viscosity resistance appear in the impact force of soft materials. One is the excessively large viscosity resistance (red curve) that appears in the first stage. The other is the small viscosity resistance (black dashed curve) that appears in the second stage. This viscosity transient (green dashed arrow) occurs at the thorn peak.

In some colloidal suspensions, the viscosity transient is known as "viscosity discontinuity" [8] [9]. Our experimental results indicate that the viscosity discontinuity may also occur in most soft materials.

It is known that the stress-strain curve of some soft materials under high strain rate is not a monotonically increasing function. It shows stress relaxation after the elastic peak [10]. This relaxation may be related to the viscosity discontinuity.

In running sports, it is known that a thorn-shape waveform called an "initial spike" can be observed [11]. When the heel strikes the ground first, it rebounds rapidly because the heel is hard. As a result, the heel generates a thorn-like waveform. Along with the movement of the center of gravity of the human body, the forefoot striking generates the second mountain-shape waveform. This is also understood as the double-strike phenomenon consisting of the first hard heel strike and the second soft forefoot strike.

7. Conclusions

Using compact drop test equipment, this study experimentally observed the impact force waveforms for several kinds of soft materials. The results obtained here are as follows:

1) Thorn-shape waveform of soft material appears clearly under the condition of a flat frontal impact. If either surface of the drop hammer or the soft material is curved, the thorn-shape waveform becomes unclear.

2) Most of the soft materials (except cork sheet) such as sponge, sponge rubber, nitrile rubber, oil clay, paper clay, low-rebound urethane foam, ham, pork, and konjac show a thorn-shape waveform under a flat frontal impact.

3) The thorn-shape waveform is close to an isosceles triangle. The rising slope θ and the height of the thorn peak are related to the strain rate dependence.

4) The thorn-shape waveform and succeeding mountain-shape waveform are induced by the double-strike phenomenon.

5) Viscosity discontinuity before and after the thorn waveform is thought to be the cause of the impact force waveform in soft materials.

References

[1] Zhao, H., Gary, G. and Klepaczko, J.P. (1997) On the Use of a Viscoelastic Split Hopkinson Pressure Bar. *International-*

al Journal of Impact Engineering*, **19**, 319-330. http://dx.doi.org/10.1016/S0734-743X(96)00038-3

[2] Song, B., Chen, W.W., Ge, Y. and Weerasooriya, T. (2007) Radial Inertia Effects in Kolsky Bar Testing of Extra-Soft Specimens. *Experimental Mechanics*, **67**, 659-670. http://dx.doi.org/10.1007/s11340-006-9017-5

[3] Kwon, J. and Subhash, G. (2010) Compressive Strain Rate Sensitivity of Ballistic Gelatin. *Journal of Biomechanics*, **43**, 420-425. http://dx.doi.org/10.1016/j.jbiomech.2009.10.008

[4] Fujimoto, Y., Shintaku, E., Tanaka, Y. and Fujiyoshi, J. (2012) Pad Type Compressive Force Sensor Suitable for High-Speed Impact Force Measurement. *Transactions of the Japan Society of Mechanical Engineers, Series C*, **78**, 2438-2449. (in Japanese) http://dx.doi.org/10.1299/kikaic.78.2438

[5] Fujimoto, Y., Liu, C., Uesugi, T., Tanaka, Y. and Shintaku, E. (2013) Pipe Surface Sensor for Impulsive Force Measurement. *Transactions of the Japan Society of Mechanical Engineers, Series C*, **79**, 113-123. (In Japanese) http://dx.doi.org/10.1299/kikaic.79.1949

[6] Fujimoto, Y., Liu, C., Tanaka, Y. and Shintaku, E. (2013) Measurement of Double-Strike Phenomenon of Impulsive Force Using Fluctuating Load Detection Plate. *The Japanese Society for Experimental Mechanics*, **13**, 112-120. (In Japanese)

[7] Liu, C., Tanaka, Y. and Fujimoto, Y. (2015) Viscosity Transient Phenomenon during Drop Impact Testing and Its Simple Dynamics Model. *World Journal of Mechanics*, **5**, 33-41. http://dx.doi.org/10.4236/wjm.2015.53004

[8] Hoffman, R.L. (1972) Discontinuous and Dilatant Viscosity Behavior in Concentrated Suspensions. I. Observation of a Flow Instability. *Transactions of the Society of Rheology*, **16**, 155-173. http://dx.doi.org/10.1122/1.549250

[9] Bender, J. and Wagner, N.J. (1996) Reversible Shear Thickening in Monodisperse and Bidisperse Colloidal Dispersions. *Journal of Rheology*, **40**, 889-915. http://dx.doi.org/10.1122/1.550767

[10] Richeton, J., Ahzi, S., Vecchio, K.S., Jiang, F.C. and Adharapurapu, R.R. (2006) Influence of Temperature and Strain Rate on the Mechanical Behavior of Three Amorphous Polymers: Characterization and Modeling of the Compressive Yield Stress. *International Journal of Solids and Structures*, **43**, 2318-2335. http://dx.doi.org/10.1016/j.ijsolstr.2005.06.040

[11] Lieberman, D.E., Venkadesan, M., Werbel, W.A., Daoud, A.I., D'Andrea, S., Davis, I.S., Mang'eni, R.O. and Pitsiladis, Y. (2010) Foot Strike Patterns and Collision Forces in Habitually Barefoot Versus Shod Runners. *Nature*, **463**, 531-535. http://dx.doi.org/10.1038/nature08723

11

A Short Vector Solution of the Foucault Pendulum Problem

I. A. Ciureanu[1], D. Condurache[2]

[1]Department of Medical Informatics and Biostatistics, University of Medicine and Pharmacy "Gr. T. Popa", Iasi, Romania
[2]Department of Theoretical Mechanics, Technical University of Iasi, Iasi, Romania
Email: adrian.ciureanu@umfiasi.ro, daniel.condurache@gmail.com

Abstract

The paper studies the motion of the Foucault Pendulum in a rotating non-inertial reference frame and provides a closed form vector solution determined by vector and matrix calculus. The solution is determined through vector and matrix calculus in both cases, for both forms of the law of motion (for the Foucault Pendulum Problem and its "Reduced Form"). A complex vector which transforms the motion equation in a first order differential equation with constant coefficients is used. Also, a novel kinematic interpretation of the Foucault Pendulum motion is given.

Keywords

Foucault Pendulum, Non-Inertial Reference Frame, Closed Form Vector Solution, Complex Vector

1. Introduction

Swinging with elegance across the meridian of Paris inside the grand hall of the observatory, the pendulum built by Bernard Léon Foucault (1819-1868) proved the rotation of the Earth for the first time by terrestrial methods. It was a true kick for both mathematicians and physicists because none of them could write the equations or imagine this simple experiment. As we now know, Cauchy never thought that is possible that a pendulum can change the oscillation plan and Poisson said in 1827 that a pendulum cannot move such way.

The "non-mathematician" Foucault, as the members of the French Academy named him, wrote the first equation which computes the period of the whole rotation of the oscillation plan depending of the latitude of the place of oscillation. The as-known "Foucault formula" or "The law of sinus" is $T = 24/\sin\theta$ (h), with θ being the latitude [1] [2].

The famous experience done by Léon Foucault in 1851 emphasized the movement of the Earth around the poles, without the need for astronomical observations. The problem is very important out of the theoretical point of view. Modeling this experiment involves the study of a harmonic oscillator with respect to a non-inertial frame of reference with uniform rotation.

But finding the equation of the movement of the pendulum proved to be for mathematicians a really "hard nut" due to the non-inertial character of the reference frame. Long time, the solution had been obtained after many approximations which had to simplify the differential equations.

The type of motion that will be named "Foucault Pendulum-like motion" is described by the non-linear initial value problem [3] [4]:

$$\begin{cases} \ddot{\boldsymbol{r}} + 2\boldsymbol{\omega}\times\dot{\boldsymbol{r}} + \boldsymbol{\omega}\times(\boldsymbol{\omega}\times\boldsymbol{r}) + f(r)\dfrac{\boldsymbol{r}}{r} = \boldsymbol{0}; \ \ \boldsymbol{\omega}\in V_3; \ \omega_0 \in \mathbb{R}_+^* \\ \boldsymbol{r}(0) = \boldsymbol{r}_0; \\ \dot{\boldsymbol{r}}(0) = \boldsymbol{v}_0. \end{cases} \tag{1.1}$$

where $f:\mathbb{R}_+ \to \mathbb{R}$ is a continuous real valued map, r denotes the magnitude of vector $\boldsymbol{r}$ and $\boldsymbol{\omega}$ is a differentiable vector value map ($\mathbb{R}$ denotes the set of real numbers). The above equation models the motion in a non-inertial reference frame with instantaneous angular velocity $\boldsymbol{\omega}$ in a central force field.

The motion which is described by the below linear initial value problem will be named "Foucault Pendulum motion":

$$\begin{cases} \ddot{\boldsymbol{r}} + 2\boldsymbol{\omega}\times\dot{\boldsymbol{r}} + \boldsymbol{\omega}\times(\boldsymbol{\omega}\times\boldsymbol{r}) + \omega_0^2\boldsymbol{r} = \boldsymbol{0}; \ \ \boldsymbol{\omega}\in V_3; \ \omega_0 \in \mathbb{R}_+^* \\ \boldsymbol{r}(0) = \boldsymbol{r}_0; \\ \dot{\boldsymbol{r}}(0) = \boldsymbol{v}_0. \end{cases} \tag{1.2}$$

In this case, the function f from Equation (1.1) has the particular expression of a constant real number and r is the position vector, ω is the angular velocity of the reference frame (an arbitrary differential vector map) and ω_0 is the pulsation of the pendulum which depends on its length and the gravitational acceleration at the experiment place. The relation (1.2) represents the initial value problem that describes a motion of a harmonic oscillator related to a rotating reference frame.

Many times is used the simplified form of (1.2) written below, when the inertial centripetal force is ignored (see for example [1]-[5]):

$$\begin{cases} \ddot{\boldsymbol{r}} + 2\boldsymbol{\omega}\times\dot{\boldsymbol{r}} + \omega_0^2\boldsymbol{r} = \boldsymbol{0}; \ \ \boldsymbol{\omega} = \overrightarrow{\text{const}}; \ \omega_0 \in \mathbb{R}_+^* \\ \boldsymbol{r}(0) = \boldsymbol{r}_0 \\ \dot{\boldsymbol{r}}(0) = \boldsymbol{v}_0 \end{cases} \tag{1.3}$$

where $\boldsymbol{\omega} = \boldsymbol{\omega}_{\text{Earth}}\sin\theta$ is considered to be constant, θ is the latitude of the place of the experiment and $\boldsymbol{\omega}_{\text{Earth}}$ represents the angular velocity of the Earth. In Equation (1.3), the inertial centripetal force $\boldsymbol{\omega}\times(\boldsymbol{\omega}\times\boldsymbol{r})$ is neglected and only the Coriolis forceis considered. As we know, there is no vector closed form explicit solution in any work. An approximate solution to (1.3) is given by Arnold using the isomorphism between plane vectors and complex numbers (see [5]). In this paper, Equation (1.3) will be named Reduced Foucault Pendulum Problem and will find its vector solution in Section 5.

The present paper presents a closed form vector solution which exploits the benefits of the dualism of vector calculus and matrix calculus with extension to tensors. It is structured in five sections described below.

In the second section, two theorems which put the basis of the correspondence between vector operations and their matrix representation are stated. Two symbolic representations are defined which creates the two ways of the cross-representations of equations in vector and matrix forms.

The third section presents the vector solution of the Foucault pendulum problem (1.2) using the two symbolic representations. Here a workaround is used through a complex vector which transforms (1.2) in a first order differential equation with constant coefficients.

Section 4 prepares the next one because it presents the tensor method of representation of vector functions which will be very useful when we will find the vector solution of the Reduced Foucault Pendulum Problem (1.3). Therefore, the transformation $\boldsymbol{F}_{\boldsymbol{\omega}}$ will be defined and its properties will be listed, so that, through them, the solution of (1.3) to be found.

Finally, in Section 5, we will compute the solution of (1.3) and we will be able to extract the surprising conclusion that the solution of the Reduced Foucault Pendulum Problem is less simple than the solution of the whole Cauchy problem (1.2).

Many times, the solution to the Cauchy problem (1.3) is given only for the planar case, using polar coordinates [6] or Cartesian coordinates [7].

2. Mathematical Preliminaries

Consider the vector space V_3 of free vectors from Euclidean space E_3 and $\boldsymbol{B} = \{\boldsymbol{e}_1, \boldsymbol{e}_2, \boldsymbol{e}_3\}$ an orthonormal basis of this space. A given vector $\boldsymbol{v}$ from V_3 can be uniquely written as:

$$\boldsymbol{v} = v_1\boldsymbol{e}_1 + v_2\boldsymbol{e}_2 + v_3\boldsymbol{e}_3 \in R, \quad k = \overline{1,3} \tag{2.1}$$

Consider V_3 the vector space of column matrix, with three rows of real numbers. An element of V_3 has the following shape:

$$v = [v_1, v_2, v_3]^{\mathrm{T}}; \quad v_k \in \mathbb{R}, \ k \in \{1,2,3\}. \tag{2.2}$$

A function

$$\psi : V_3 \to V_3; \quad \psi(\boldsymbol{v}) = v \tag{2.3}$$

is an isomorphism of vectors spaces.

If $\boldsymbol{\omega} \in V_3$, fixed, then the function:

$$\begin{aligned} &\alpha : V_3 \to V_3; \\ &\alpha(\boldsymbol{v}) = \boldsymbol{\omega} \times \boldsymbol{v}; \quad \forall \boldsymbol{v} \in V_3, \ \boldsymbol{\omega} \in V_3 \setminus \{\boldsymbol{0}\} \end{aligned} \tag{2.4}$$

is an endomorphism of V_3.

If $\tilde{\omega} \in m_{3\times3}(\boldsymbol{R})$ is a square matrix of order three, with real elements, fixed, non-zero, then the function:

$$\begin{aligned} &\beta : V_3 \to V_3 \\ &\beta(v) = \tilde{\boldsymbol{\omega}} \cdot v; \quad \forall v \in V_3, \ \tilde{\boldsymbol{\omega}} \in m_{3\times3}(\mathbb{R}); \ \tilde{\boldsymbol{\omega}} \neq 0, \text{fixed} \end{aligned} \tag{2.5}$$

is an endomorphism on V_3 [8].

We want to find the link between the vector $\boldsymbol{\omega}$ and $\tilde{\boldsymbol{\omega}}$ defined above as a square matrix of order three, with real elements, fixed, non-zero, for which the function (2.3) is a symbolic representation regarding the two endomorphism $\alpha(\boldsymbol{v})$ and $\beta(v)$.

So, if $\boldsymbol{B} = \{\boldsymbol{e}_1, \boldsymbol{e}_2, \boldsymbol{e}_3\}$ is a rectangle orthonormal basis and:

$$\boldsymbol{\omega} = \omega_1\boldsymbol{e}_1 + \omega_2\boldsymbol{e}_2 + \omega_3\boldsymbol{e}_3; \quad \omega_k \in \mathbb{R}, \ k = \overline{1,3} \tag{2.6}$$

then:

$$\boldsymbol{\omega} \times \boldsymbol{v} = (\omega_2 v_3 - \omega_3 v_2)\boldsymbol{e}_1 - (\omega_1 v_3 - \omega_3 v_1)\boldsymbol{e}_2 + (\omega_1 v_2 - \omega_2 v_1)\boldsymbol{e}_3. \tag{2.7}$$

Using (2.6) and (2.7) we have:

$$\psi(\boldsymbol{\omega} \times \boldsymbol{v}) = \begin{bmatrix} \omega_2 v_3 - \omega_3 v_2 \\ \omega_3 v_1 - \omega_1 v_3 \\ \omega_1 v_2 - \omega_2 v_1 \end{bmatrix} = \begin{bmatrix} 0 & -\omega_3 & \omega_2 \\ \omega_3 & 0 & -\omega_1 \\ -\omega_2 & \omega_1 & 0 \end{bmatrix} \begin{bmatrix} v_1 \\ v_2 \\ v_3 \end{bmatrix}. \tag{2.8}$$

Therefore, with the notation:

$$\boldsymbol{\omega} = \begin{bmatrix} 0 & -\omega_3 & \omega_2 \\ \omega_3 & 0 & -\omega_1 \\ -\omega_2 & \omega_1 & 0 \end{bmatrix} \tag{2.9}$$

from (2.8) results the relation:

$$\psi(\boldsymbol{\omega}\times\boldsymbol{v})=\tilde{\omega}\cdot\boldsymbol{v};\quad \forall\boldsymbol{v}\in V_3;\quad \boldsymbol{\omega}\in V_3. \tag{2.10}$$

with $\boldsymbol{\omega}$ and $\tilde{\omega}$ from (2.6) and (2.9), respectively, we have:

Theorem 2.1. The function $\psi:V_3\to V_3$; defined by $\psi(\boldsymbol{v})=\boldsymbol{v}$ is an isomorphism of vectors spaces with the property:

$$\psi(\boldsymbol{\omega}\times\boldsymbol{v})=\tilde{\omega}\psi(\boldsymbol{v});\quad \forall\boldsymbol{v}\in V_3;\quad \boldsymbol{\omega}\in V_3,\ \text{fixed}. \tag{2.11}$$

So, it is an exact symbolic representation, with respect to the endomorphism (2.4) of vector space V_3 over vector space V_3 with the operator $\tilde{\omega}$ (2.9).

Note: The matrix $\tilde{\omega}$ is a matrix representation of a skew-symmetric second order tensor associated with the vector $\boldsymbol{\omega}$ (2.6) in the basis $\left\{\boldsymbol{e}_p\otimes\boldsymbol{e}_q\right\}_{p,q=\overline{1,3}}$. We'll note $\boldsymbol{\omega}=\text{vect}\tilde{\omega}$.

The characteristic polynomial of the skew-symmetric matrix (2.9) is:

$$p(\lambda)=\det\left[\tilde{\omega}-\lambda\tilde{\omega}^0\right]=-\lambda^3-\omega^2\lambda. \tag{2.12}$$

Solutions (roots) of the equation $p(\lambda)=0$ are: $\lambda_1=0$, $\lambda_{2,3}=\pm j|\boldsymbol{\omega}|$, $j^2=-1$.

Using one of the known proceedings for determination of an exponential matrix, it follows that:

Theorem 2.2. If $\boldsymbol{\omega}\in V_3-\{\boldsymbol{0}\}$ and $\boldsymbol{\omega}=\text{vect}\tilde{\omega}$, then

$$\exp(-\tilde{\omega}t)=e^{-\tilde{\omega}t}=\varphi_0\tilde{\omega}^0+\varphi_1\tilde{\omega}+\varphi_2\tilde{\omega}^2;\ \forall t\in\mathbb{R} \tag{2.13}$$

where $\varphi_k:\mathbb{R}\to\mathbb{R}$, $k\in\{0,1,3\}$:

$$\begin{aligned}\varphi_0&=1, && \forall t\in\mathbb{R};\\ \varphi_1&=-\frac{\sin|\boldsymbol{\omega}|t}{|\boldsymbol{\omega}|}, && \forall t\in\mathbb{R};\\ \varphi_2&=\frac{1-\cos|\boldsymbol{\omega}|t}{\boldsymbol{\omega}^2}, && \forall t\in\mathbb{R}.\end{aligned} \tag{2.14}$$

Note: due to the Cayley-Hamilton theorem, any square matrix verifies her characteristic equation; consequently, from Equation (2.12), it follows that:

$$\tilde{\omega}^3=-\omega^2\tilde{\omega}. \tag{2.15}$$

If we denote by ψ^{-1} the inverse of the relation (2.11), we obtain, with (2.6) and (2.9), putting $\boldsymbol{v}=\psi^{-1}(\boldsymbol{v})$:

$$\tilde{\omega}\cdot\boldsymbol{v}\xrightarrow{\psi^{-1}}\boldsymbol{\omega}\times\boldsymbol{v}; \tag{2.16}$$

$$\tilde{\omega}^2\cdot\boldsymbol{v}\xrightarrow{\psi^{-1}}\boldsymbol{\omega}\times(\boldsymbol{\omega}\times\boldsymbol{v}); \tag{2.17}$$

$$\tilde{\omega}^3\cdot\boldsymbol{v}\xrightarrow{\psi^{-1}}-\omega^2(\boldsymbol{\omega}\times\boldsymbol{v}). \tag{2.18}$$

If we denote by V_3^* the set of vector functions of the real variable and by V_3^* the set of matrix functions, with three rows and a column, of the real variable, symbolic representation (2.11) induce:

$$\psi:V_3^*\to V_3^* \tag{2.19}$$

$$\psi(\boldsymbol{\omega}\times\boldsymbol{v})=\tilde{\omega}\cdot\psi(\boldsymbol{v});\quad \forall\boldsymbol{v}\in V_3^*\setminus\{0\}\ \text{fixed}. \tag{2.20}$$

3. Short Solution of Foucault Pendulum Problem

The mathematical model of this experiment is given by the Cauchy problem:

$$\begin{cases}\ddot{\boldsymbol{r}}+2\boldsymbol{\omega}\times\dot{\boldsymbol{r}}+\boldsymbol{\omega}\times(\boldsymbol{\omega}\times\boldsymbol{r})+\omega_0^2\boldsymbol{r}=\boldsymbol{0};\quad \omega_0>0\\ \dot{\boldsymbol{r}}(0)=\dot{\boldsymbol{r}}_0\\ \dot{\boldsymbol{r}}(0)=\boldsymbol{v}_0.\end{cases} \tag{3.1}$$

In Equation (3.1), $\boldsymbol{r}$ is the position vector of the particle which corresponds to the relative equilibrium position $\boldsymbol{r}=\boldsymbol{0}$, $\boldsymbol{\omega}$ angular velocity of the Earth, ω_0 the feature pulsation of the oscillator [9].

Using the symbolic representation, we will find a vector exact solution for the problem (3.1). Applying to the problem (3.1) the correspondence ψ (2.11), we will obtain the matrix form:

$$\begin{cases}\ddot{\boldsymbol{r}}+2\boldsymbol{\omega}\cdot\dot{\boldsymbol{r}}+\tilde{\boldsymbol{\omega}}^2\cdot\boldsymbol{r}+\omega_0^2\boldsymbol{r}=\boldsymbol{0}\\ \boldsymbol{r}(0)=\boldsymbol{r}_0\\ \dot{\boldsymbol{r}}(0)=\boldsymbol{v}_0\end{cases}. \tag{3.2}$$

We will consider now the column matrix with complex functions elements given by:

$$\xi=\boldsymbol{r}-j\frac{\dot{\boldsymbol{r}}+\tilde{\boldsymbol{\omega}}\cdot\boldsymbol{r}}{\omega_0};\quad j^2=-1 \tag{3.3}$$

First, we will differentiate this column matrix:

$$\dot{\xi}=\dot{\boldsymbol{r}}-j\frac{\ddot{\boldsymbol{r}}+\tilde{\boldsymbol{\omega}}\cdot\dot{\boldsymbol{r}}}{\omega_0};\quad j^2=-1 \tag{3.4}$$

Replacing $\ddot{\boldsymbol{r}}$ in (3.4) from (3.2) and adding and subtracting $\tilde{\boldsymbol{\omega}}\cdot\boldsymbol{r}$, we will obtain:

$$\dot{\xi}=\dot{\boldsymbol{r}}+\tilde{\boldsymbol{\omega}}\cdot\boldsymbol{r}+j\frac{\tilde{\boldsymbol{\omega}}\cdot\dot{\boldsymbol{r}}+\tilde{\boldsymbol{\omega}}^2\cdot\boldsymbol{r}}{\omega_0}-\tilde{\boldsymbol{\omega}}\cdot\boldsymbol{r}+j\omega_0\boldsymbol{r};\quad j^2=-1 \tag{3.5}$$

After developing, (3.5) becomes

$$\dot{\xi}=\dot{\boldsymbol{r}}+j\frac{\tilde{\boldsymbol{\omega}}\cdot\dot{\boldsymbol{r}}}{\omega_0}+\frac{\omega_0\tilde{\boldsymbol{\omega}}\cdot\boldsymbol{r}}{\omega_0}+j\frac{\tilde{\boldsymbol{\omega}}^2\cdot\boldsymbol{r}}{\omega_0}-\tilde{\boldsymbol{\omega}}r+j\omega_0 r;\quad j^2=-1 \tag{3.6}$$

Grouping the terms, (3.6) becomes:

$$\dot{\xi}=-j\frac{\dot{\boldsymbol{r}}}{\omega_0}\left(-\tilde{\boldsymbol{\omega}}+j\omega_0\boldsymbol{I}_3\right)-j\frac{\tilde{\boldsymbol{\omega}}\cdot\boldsymbol{r}}{\omega_0}\left(-\tilde{\boldsymbol{\omega}}+j\omega_0\boldsymbol{I}_3\right)+\boldsymbol{r}\left(-\tilde{\boldsymbol{\omega}}+j\omega_0\boldsymbol{I}_3\right);\quad j^2=-1 \tag{3.7}$$

and this means that

$$\dot{\xi}=\left(-\tilde{\boldsymbol{\omega}}+j\omega_0\boldsymbol{I}_3\right)\left(\boldsymbol{r}-j\frac{\dot{\boldsymbol{r}}-\tilde{\boldsymbol{\omega}}\cdot\boldsymbol{r}}{\omega_0}\right);\quad j^2=-1 \tag{3.8}$$

We will note $-\tilde{\boldsymbol{\omega}}+j\omega_0\boldsymbol{I}_3$ with $\tilde{\boldsymbol{\omega}}^*$ and knowing that $\xi=\boldsymbol{r}-j\dfrac{\dot{\boldsymbol{r}}+\tilde{\boldsymbol{\omega}}\cdot\boldsymbol{r}}{\omega_0}$; $j^2=-1$, (3.8) becomes:

$$\dot{\xi}=\tilde{\boldsymbol{\omega}}^*\xi;\quad \tilde{\boldsymbol{\omega}}^*=-\tilde{\boldsymbol{\omega}}+j\omega_0\boldsymbol{I}_3. \tag{3.9}$$

It results that the function ξ is the solution of the following Cauchy problem:

$$\begin{cases}\dot{\xi}=\tilde{\boldsymbol{\omega}}^*\xi\\ \xi(\boldsymbol{0})=\boldsymbol{r}_0-j\dfrac{\boldsymbol{v}_0+\tilde{\boldsymbol{\omega}}\boldsymbol{r}_0}{\omega_0}\end{cases}. \tag{3.10}$$

and the solution of the problem (3.10) is:

$$\xi=\mathrm{e}^{\tilde{\boldsymbol{\omega}}^* t}\xi(\boldsymbol{0}). \tag{3.11}$$

Using Equation (3.3), and knowing that $\mathrm{Re}\{z_1z_2\}=\mathrm{Re}\{z_1\}\mathrm{Re}\{z_2\}-\mathrm{Im}\{z_1\}\mathrm{Im}\{z_2\}$ with z_1 and z_2 complex numbers, it follows that the solution of the problem (3.2) is the real part of Equation (3.11):

$$\boldsymbol{r}=\mathrm{Re}\xi=\mathrm{Re}\left(\mathrm{e}^{\tilde{\boldsymbol{\omega}}^* t}\right)\cdot\mathrm{Re}\xi(\boldsymbol{0})-\mathrm{Im}\left(\mathrm{e}^{\tilde{\boldsymbol{\omega}}^* t}\right)\cdot\mathrm{Im}\xi(\boldsymbol{0}). \tag{3.12}$$

Using Theorem 2.2 and the definition of a matrix exponential, it follows:

$$\mathrm{e}^{\tilde{\boldsymbol{\omega}}^* t}=\mathrm{e}^{-\tilde{\boldsymbol{\omega}}\cdot t+j\omega_0 t\boldsymbol{I}_3}=\mathrm{e}^{-\tilde{\boldsymbol{\omega}}\cdot t}\mathrm{e}^{j\omega_0 t\boldsymbol{I}_3}=\mathrm{e}^{-\tilde{\boldsymbol{\omega}}\cdot t}\mathrm{e}^{j\omega_0 t}\boldsymbol{I}_3=\left(\cos\omega_0 t+j\sin\omega_o t\right)\mathrm{e}^{-\tilde{\boldsymbol{\omega}}\cdot t}. \tag{3.13}$$

with (3.3) and (3.13), Equation (3.12) becomes:

$$\boldsymbol{r} = \cos\omega_0 t \mathrm{e}^{-\tilde{\boldsymbol{\omega}}\cdot t}\cdot\boldsymbol{r}_0 + \sin\omega_0 t \mathrm{e}^{-\tilde{\boldsymbol{\omega}}\cdot t}\cdot\frac{\left(\boldsymbol{v}_0+\tilde{\boldsymbol{\omega}}\cdot\boldsymbol{r}_0\right)}{\omega_0}. \tag{3.14}$$

After restructuring, Equation (3.14) looks like:

$$\boldsymbol{r} = \mathrm{e}^{-\tilde{\boldsymbol{\omega}}\cdot t}\left[\boldsymbol{r}_0\cos\omega_0 t + \frac{\left(\boldsymbol{v}_0+\tilde{\boldsymbol{\omega}}\cdot\boldsymbol{r}_0\right)}{\omega_0}\sin\omega_0 t\right]. \tag{3.15}$$

with the notation:

$$\boldsymbol{\psi}_0 = \boldsymbol{r}_0\cos\omega_0 t + \frac{\left(\boldsymbol{v}_0+\tilde{\boldsymbol{\omega}}\cdot\boldsymbol{r}_0\right)}{\omega_0}\sin\omega_0 t;\quad t\in\mathbf{0} \tag{3.16}$$

Equation (3.15) becomes:

$$\boldsymbol{r} = \mathrm{e}^{-\tilde{\boldsymbol{\omega}}\cdot t}\boldsymbol{\psi}_0. \tag{3.17}$$

Applying to Equation (3.17), the correspondence ψ^{-1}, using the results of Theorem 2.2, we will obtain the vector solution of the problem (3.1):

$$\boldsymbol{r} = \boldsymbol{\psi}_0 - \sin\omega t\frac{\boldsymbol{\omega}\times\boldsymbol{\psi}_0}{\omega} + \left(1-\cos\omega t\right)\frac{\boldsymbol{\omega}\times\left(\boldsymbol{\omega}\times\boldsymbol{\psi}_0\right)}{\bar{\omega}^2} \tag{3.18}$$

where:

$$\boldsymbol{\psi}_0 = \boldsymbol{r}_0\cos\omega_0 t + \frac{\boldsymbol{v}_0+\boldsymbol{\omega}\times\boldsymbol{r}_0}{\omega_0}\sin\omega_0 t \tag{3.19}$$

After elementary transformation, the solution (3.18) of the Cauchy problem (3.1) it will be written:

$$\boldsymbol{r} = \frac{\boldsymbol{\omega}\cdot\boldsymbol{\psi}_0}{\boldsymbol{\omega}^2}\boldsymbol{\omega} - \sin\omega t\frac{\boldsymbol{\omega}\times\boldsymbol{\psi}_0}{\omega} - \cos\omega t\frac{\boldsymbol{\omega}\times\left(\boldsymbol{\omega}\times\boldsymbol{\psi}_0\right)}{\boldsymbol{\omega}^2} \tag{3.20}$$

Note:

1) The function (3.19) is the solution of the Cauchy problem:

$$\begin{cases}\ddot{\boldsymbol{r}} + \omega_0^2\cdot\boldsymbol{r} = \mathbf{0};\\ \boldsymbol{r}(0) = \boldsymbol{r}_0;\\ \dot{\boldsymbol{r}} = \boldsymbol{v}_0 + \boldsymbol{\omega}\times\boldsymbol{r}_0.\end{cases} \tag{3.21}$$

The differential equation of the problem (3.21) can be found from the differential equation of the Cauchy problem (3.1) for $\boldsymbol{\omega} = \mathbf{0}$.

2) The solution (3.20) is the vector form of an equation matrix (3.17). This has special significance.

Let $\boldsymbol{Q}$ be the matrix function given by:

$$\boldsymbol{Q}(t) = \mathrm{e}^{-\tilde{\boldsymbol{\omega}}\cdot t};\quad \forall t\in\mathbb{R}. \tag{3.22}$$

The function $\boldsymbol{Q}(t)$ from (3.22) is the orthogonal matrix for any $t\in\mathbb{R}$. Indeed, because the matrix $\tilde{\boldsymbol{\omega}}$ is skew-symmetric, $\left(\tilde{\boldsymbol{\omega}} = -\tilde{\boldsymbol{\omega}}^{\mathrm{T}}\right)$ we will have:

$$\boldsymbol{Q}\cdot\boldsymbol{Q}^{\mathrm{T}} = \mathrm{e}^{-\tilde{\boldsymbol{\omega}}\cdot t}\cdot\mathrm{e}^{-\tilde{\boldsymbol{\omega}}^{\mathrm{T}}\cdot t} = \mathrm{e}^{-\tilde{\boldsymbol{\omega}}\cdot t}\cdot\mathrm{e}^{\tilde{\boldsymbol{\omega}}\cdot t} = \boldsymbol{I}_3. \tag{3.23}$$

Because from Equation (3.22), we have $\det\boldsymbol{Q}(t) = \mathrm{e}^{-\mathrm{trace}\tilde{\boldsymbol{\omega}}t} = \mathrm{e}^0 = 1,\ \forall t\in\mathbb{R}$, the transformation given by Equation (3.17) is an own rotation for $\forall t\in\mathbb{R}$.

The angular velocity corresponding to this rotation is:

$$\boldsymbol{\omega}^* = \mathrm{vect}\dot{\boldsymbol{Q}}\cdot\boldsymbol{Q}^{\mathrm{T}} = \mathrm{vect}\left(-\tilde{\boldsymbol{\omega}}\cdot\mathrm{e}^{-\tilde{\boldsymbol{\omega}}\cdot t}\cdot\mathrm{e}^{\tilde{\boldsymbol{\omega}}\cdot t}\right) = -\mathrm{vect}\tilde{\boldsymbol{\omega}} = -\boldsymbol{\omega} \tag{3.24}$$

The transformation (3.17) is therefore an own rotation with angular velocity $-\boldsymbol{\omega}$. We will note with $\boldsymbol{F}_{-\boldsymbol{\omega}}$ the tensor operator that matrix transcription is given by Equation (3.17). The solution for the Cauchy problem (3.1) is now written:

$$\boldsymbol{r} = \boldsymbol{F}_{-\boldsymbol{\omega}}(\boldsymbol{\psi}_0). \tag{3.25}$$

with the above observation, we can obtain the next theorem:

Theorem 3.1. The solution of the Cauchy problem

$$\begin{cases} \ddot{\boldsymbol{r}} + 2\boldsymbol{\omega}\times\dot{\boldsymbol{r}} + \boldsymbol{\omega}\times(\boldsymbol{\omega}\times\boldsymbol{r}) + \omega_0^2\boldsymbol{r} = \boldsymbol{0}; \quad \boldsymbol{\omega}\in V_3;\ \omega_0 \in \mathbb{R}_+^* \\ \boldsymbol{r}(0) = \boldsymbol{r}_0 \\ \dot{\boldsymbol{r}}(0) = \boldsymbol{v}_0 \end{cases} \tag{3.26}$$

will be obtained applying the tensor of the rotation operator with the angular velocity $-\boldsymbol{\omega}$:

$$\begin{aligned} &\boldsymbol{F}_{-\boldsymbol{\omega}} : V_3^* \to V_3^* \\ &\boldsymbol{F}_{-\boldsymbol{\omega}}(\cdot) = \frac{\boldsymbol{\omega}\cdot(\cdot)}{\boldsymbol{\omega}^2} - \sin\omega t\frac{\boldsymbol{\omega}\times(\cdot)}{\omega} - \cos\omega t\frac{\boldsymbol{\omega}\times[\boldsymbol{\omega}\times(\cdot)]}{\boldsymbol{\omega}^2} \end{aligned} \tag{3.27}$$

to the solution of the next Cauchy problem:

$$\begin{cases} \ddot{\boldsymbol{r}} + \omega_0^2\boldsymbol{r} = \boldsymbol{0}; \\ \boldsymbol{r}(0) = \boldsymbol{r}_0; \\ \dot{\boldsymbol{r}}(0) = \boldsymbol{v}_0 + \boldsymbol{\omega}\times\boldsymbol{r}_0. \end{cases} \tag{3.28}$$

Note: The hodograph of the solution of the problem (3.21) $\psi_0 = \boldsymbol{r}_0\cos\omega_o t + \frac{\boldsymbol{v}_0 + \boldsymbol{\omega}\times\boldsymbol{r}_0}{\omega_0}\sin\omega_o t$ is an ellipse, possibly degenerate, having the conjugated diameters with directions given by the vectors $\boldsymbol{r}_0$ and $\boldsymbol{v}_0 + \boldsymbol{\omega}\times\boldsymbol{r}_0$.

The solution of the problem (3.1) can be viewed by the rotation of the plane of the ellipse, with the angular velocity $-\boldsymbol{\omega}$. As a rule, the hodograph of the vector function—solution of a Cauchy problem (3.1)—is a space curve.

The tensor relation (3.25) suggests a direction to approach the symbolic representation of a vector function of real variable which will be developed in the next paragraph.

4. The Symbolic Tensor Representation of a Vector Functions

This section describes the tensor method of representation the vector functions which will be used in the next chapter when we will give the solution to the Reduced Foucault Pendulum Problem.

We will denote by $\boldsymbol{SO}_3$ the orthogonal group of second order tensors (rotation tensors), by $\boldsymbol{SO}_3^{\mathbb{R}}$ the set of the maps defined on $\mathbb{R}$ with values in $\boldsymbol{SO}_3$, by $\boldsymbol{SO}_3$ the group of skew-symmetric second order tensors and by $\boldsymbol{SO}_3^{\mathbb{R}}$ the set of maps defined on $\mathbb{R}$ with values in $\boldsymbol{SO}_3$.

Let $V_3^* = \{f : \boldsymbol{I}\subseteq\mathbb{R}\to V_3\}$ be the set of vector function of real variable and let $\boldsymbol{\omega}\in V_3^*$ be a constant.

Let $\tilde{\boldsymbol{\omega}}$ be a second order skew-symmetric tensor corresponding to $\boldsymbol{\omega}$ with $\tilde{\boldsymbol{\omega}}\in\boldsymbol{SO}_3^{\mathbb{R}}$. We consider, in the set of second order tensors, a first order Cauchy problem:

$$\begin{cases} \dot{\boldsymbol{Q}} = \boldsymbol{Q}\cdot\tilde{\boldsymbol{\omega}}; \quad \tilde{\boldsymbol{\omega}}^{\mathrm{T}} = -\tilde{\boldsymbol{\omega}} \\ \boldsymbol{Q}(0) = \boldsymbol{I}_3. \end{cases} \tag{4.1}$$

The problem (4.1) has a unique solution $\boldsymbol{Q}\in\boldsymbol{SO}_3^{\mathbb{R}}$, if $\tilde{\boldsymbol{\omega}} = \tilde{\boldsymbol{\omega}}(t)$, $t\in\mathbb{R}$ is a continual skew-symmetric tensor function, the solution tensor being orthogonal, proper.

Indeed, be a tensor function of a real variable:

$$\boldsymbol{A} = \boldsymbol{Q}\boldsymbol{Q}^{\mathrm{T}} \tag{4.2}$$

Using Equation (4.1), it follows that $\boldsymbol{A}$ verify the Cauchy problem:

$$\begin{cases} \dot{\boldsymbol{A}} = \tilde{\boldsymbol{\omega}}\boldsymbol{A} - \boldsymbol{A}\tilde{\boldsymbol{\omega}}; \\ \boldsymbol{A}(0) = \boldsymbol{I}_3. \end{cases} \tag{4.3}$$

The solution of the problem (4.3) is unique and because the identity tensor $\boldsymbol{A} = \boldsymbol{I}_3$ verifies the Equation

(4.3), it follows that $A = I_3$, $\forall t \in \mathbb{R}$, therefore the solution of the problem (4.1) has the property $QQ^{\mathrm{T}} = I_3$, and Q is an orthogonal tensor. Knowing that $\det Q \in \{-1,1\}$ and $\det Q(0) = 1$, it follows that $\det Q = 1$, so Q is a proper orthogonal tensor.

Let

$$\boldsymbol{\omega} = \operatorname{vect}\tilde{\boldsymbol{\omega}}, \quad \boldsymbol{\omega} \in V_3^* . \tag{4.4}$$

The unique solution of Equation (4.1) will be further named as "the rotation tensor corresponding to the angular velocity $\boldsymbol{\omega}$".

If $\boldsymbol{\omega} \in V_3^*$ is a differentiable vector function on the range $I \subseteq \mathbb{R}$, we can define the transformation $F_{-\omega}$ as below:

$$\begin{cases} F_\omega : V_3^* \to V_3^*; \\ F_\omega(\boldsymbol{r}) \overset{\text{def}}{=} Q\boldsymbol{r}; \quad \boldsymbol{r} \in V_3^* . \end{cases} \tag{4.5}$$

where Q is the tensor function of the unique solution of Equation (4.1).

If the vector function $\boldsymbol{\omega} \in V_3^*$ has fixed direction:

$$\boldsymbol{\omega} = \omega(t)\boldsymbol{u} \tag{4.6}$$

where $\boldsymbol{u}$ is a constant vector and $\omega : I \subseteq \mathbb{R} \to \mathbb{R}$ is a real function of a real variable, the corresponding tensor function $\tilde{\boldsymbol{\omega}}$ has the property of auto-commutativity:

$$\tilde{\boldsymbol{\omega}}(t_1)\cdot\tilde{\boldsymbol{\omega}}(t_2) = \tilde{\boldsymbol{\omega}}(t_1)\cdot\tilde{\boldsymbol{\omega}}(t_2), \quad \forall t_1,\ t_2 \in \mathbb{R} . \tag{4.7}$$

In this condition, the solution of the Cauchy problem (4.1) will be written in explicit form:

$$Q = \mathrm{e}^{\int_0^t \tilde{\omega}(\tau)\mathrm{d}\tau} \tag{4.8}$$

Using one of the known procedures to determine the exponential matrix, we will have:

$$Q = \tilde{\boldsymbol{u}}^0 + \sin\alpha(t)\cdot\tilde{\boldsymbol{u}} + \left[1-\cos\alpha(t)\right]\tilde{\boldsymbol{u}}^2 \tag{4.9}$$

where:

$$\alpha(t) = \int_0^t \omega(\tau)\mathrm{d}\tau; \quad t \in \mathbb{R}_+ \tag{4.10}$$

and $\tilde{\boldsymbol{u}}$ is the skew-symmetric matrix corresponding to the vector $\boldsymbol{u}$ from Equation (4.6). The transformation (4.5) can be written:

$$F_\omega(\boldsymbol{r}) = \boldsymbol{r} + \boldsymbol{u}\times\boldsymbol{r}\sin\alpha(t) + \left(1-\cos\alpha(t)\right)\boldsymbol{u}\times(\boldsymbol{u}\times\boldsymbol{r}) \tag{4.11}$$

or:

$$F_\omega(\boldsymbol{r}) = (\boldsymbol{u}\cdot\boldsymbol{r})\boldsymbol{u} + \sin\alpha(t)\boldsymbol{u}\times\boldsymbol{r} - \cos\alpha(t)\boldsymbol{u}\times(\boldsymbol{u}\times\boldsymbol{r}) . \tag{4.12}$$

Using the relation $\boldsymbol{u} = \dfrac{\boldsymbol{\omega}}{\omega}$, (4.12) becomes:

$$F_\omega(\boldsymbol{r}) = \frac{\boldsymbol{\omega}\cdot\boldsymbol{r}}{\omega^2}\boldsymbol{\omega} + \sin\alpha(t)\frac{\boldsymbol{\omega}\times\boldsymbol{r}}{\omega} - \cos\alpha(t)\frac{\boldsymbol{\omega}\times(\boldsymbol{\omega}\times\boldsymbol{r})}{\omega^2} \tag{4.13}$$

where $\alpha(t)$ is given by the relation (4.10).

If $\boldsymbol{\omega} = \text{const}$ in the Equation (4.10) $\alpha(t) = \omega t$ and the Equation (4.12) becomes:

$$F_\omega(\boldsymbol{r}) = \frac{\boldsymbol{\omega}\cdot\boldsymbol{r}}{\omega^2}\boldsymbol{\omega} + \sin\omega t\frac{\boldsymbol{\omega}\times\boldsymbol{r}}{\omega} - \cos\omega t\frac{\boldsymbol{\omega}\times(\boldsymbol{\omega}\times\boldsymbol{r})}{\omega^2} \tag{4.14}$$

Theorem 4.1.

If $\boldsymbol{\omega} \in V_3^*$ has fixed direction, then the transform (4.5) has the following properties:

1) F_ω is $\mathbb{R}$-linear;

2) $F_\omega(\omega\times r)=\omega\times F_\omega(r);\quad \forall r\in V_3^*$;

3) $F_\omega(\dot r+\omega\times r)=\dot F_\omega(r);\quad \forall r\in V_3^*$;

4) $F_\omega(\ddot r+2\omega\times\dot r+\omega\times(\omega\times r)+\dot\omega\times r)=\ddot F_\omega(r);\quad \forall r\in V_3^*$;

5) $|F_\omega(r)|=|r|;\quad \forall r\in V_3^*$;

6) $\begin{cases} F_\omega(r)\big|_{t=0}=r_0 \\ \dot F_\omega(r)\big|_{t=0}=v_0+\omega_0\times r_0 \end{cases}$ where: $r_0=r(0);\ v_0=\dot r(0);\ \omega_0=\omega(0)$;

7) F_ω is invertible and $(F_\omega)^{-1}=F_{-\omega}$.

Proof:

1) $\mathbb{R}$-linearity of the correspondence (4.5) is a trivial consequence of a definition.

2) F_ω is a tensor function, given for any $t\in\mathbb{R}$ by a proper orthogonal tensor. The vector ω is a proper vector of this transformation for any $t\in\mathbb{R}$, his direction, fixed, is the instantaneous axis of rotation:

$$F_\omega(\omega)=\omega;\quad \omega\in V_3^*. \tag{4.15}$$

Equation (4.15) can be obtained directly from Equation (4.14).

The proper tensor F_ω preserves, for any $t\in\mathbb{R}$, the cross products, because it preserves also the metrics and the orientation:

$$F_\omega(\omega\times r)=F_\omega(\omega)\times F_\omega(r);\quad \forall r\in V_3^*;\ \omega\in V_3^*. \tag{4.16}$$

From (4.16) using (4.15) it follows:

$$F_\omega(\omega\times r)=\omega\times F_\omega(r);\quad \forall r\in V_3^*. \tag{4.17}$$

The matrix form of the tensor relation (4.17) is:

$$Q\cdot\tilde\omega\cdot r=\tilde\omega\cdot\tilde\omega\cdot r,\quad \forall r\in V_3^* \tag{4.18}$$

From Equation (4.18) it follows that the matrix Q and $\tilde\omega$ are commutative for any $t\in\mathbb{R}$:

$$Q\cdot\tilde\omega=\tilde\omega\cdot Q,\quad \forall t\in\mathbb{R}. \tag{4.19}$$

3) We will prove the matrix form of Equation (3). Let $F=Qr$ be a solution of Equation (4.1), where Q is a matrix function corresponding to the proper orthogonal tensor, and r is the column matrix associated with the vector function r. The derivation of F with respect to time is:

$$\dot F=\dot Q\cdot r+Q\cdot\dot r. \tag{4.20}$$

Using Equation (4.1), Equation (4.20) will be written:

$$\dot F=\tilde\omega\cdot Q\cdot r+Q\cdot\dot r. \tag{4.21}$$

Using Equation (4.18) we will have:

$$\dot F=Q\cdot\tilde\omega\cdot r+Q\cdot\dot r=Q\cdot[\dot r+\tilde\omega\cdot r]. \tag{4.22}$$

The corresponding tensor of Equation (4.22) is:

$$\dot F_\omega(r)=F_\omega(\dot r+\omega\times r);\quad \forall r\in V_3^*. \tag{4.23}$$

4) We will apply twice Equation (4.23):

$$\begin{aligned} F_\omega\left[\ddot r+2\omega\times\dot r+\omega\times(\omega\times r)+\dot\omega\times r\right]&=F_\omega\left[\frac{\mathrm{d}}{\mathrm{d}t}(\dot r+\omega\times r)+\omega\times(\dot r+\omega\times r)\right]\\ &=\dot F_\omega(\dot r+\omega\times r)=\ddot F_\omega(r). \end{aligned} \tag{4.24}$$

5) The transformed (4.5) being a proper rotation is also an isometry, so:

$$\left|\boldsymbol{F}_{\omega}(\boldsymbol{r})\right|=|\boldsymbol{r}|; \quad \forall \boldsymbol{r}\in \boldsymbol{V}_3^* \tag{4.25}$$

6) In matrix notation we have $\boldsymbol{F}_{\omega}(\boldsymbol{r})=\boldsymbol{Q}\cdot\boldsymbol{r}$. For $t=0$ we will have:

$$\boldsymbol{F}_{\omega}(\boldsymbol{r})\big|_{t=0}=\boldsymbol{Q}(0)\cdot\boldsymbol{r}(0)=\boldsymbol{I}_3\cdot\boldsymbol{r}(0)=\boldsymbol{r}_0\,. \tag{4.26}$$

Equation (4.26) can be also written:

$$\boldsymbol{F}_{\omega}(\boldsymbol{r})\big|_{t=0}=\boldsymbol{r}_0; \quad \forall \boldsymbol{r}\in \boldsymbol{V}_3^*\,. \tag{4.27}$$

To compute $\dot{\boldsymbol{F}}_{\omega}(\boldsymbol{r})\big|_{t=0}$ using 3° and after Equation (4.27):

$$\dot{\boldsymbol{F}}_{\omega}(\boldsymbol{r})\big|_{t=0}=\boldsymbol{F}_{\omega}(\dot{\boldsymbol{r}}+\boldsymbol{\omega}\times\boldsymbol{r})\big|_{t=0}=\dot{\boldsymbol{r}}(0)+\boldsymbol{\omega}(0)\times\boldsymbol{r}(0)=\boldsymbol{v}_0+\boldsymbol{\omega}_0\times\boldsymbol{r}_0\,. \tag{4.28}$$

7) Being an orthogonal transformation, $\boldsymbol{F}_{\omega}$ is invertible. If the transformation tensor $\boldsymbol{F}_{\omega}$ is $\boldsymbol{Q}$, the transformation tensor $(\boldsymbol{F}_{\omega})^{-1}$ is $\boldsymbol{Q}^{\mathrm{T}}$. To demonstrate that $(\boldsymbol{F}_{\omega})^{-1}=\boldsymbol{F}_{-\omega}$ we will show that $\boldsymbol{Q}^{\mathrm{T}}$ verifies Equation (2.1) with $-\tilde{\boldsymbol{\omega}}$ instead of $\tilde{\boldsymbol{\omega}}$.

From (4.1), by transposition, it follows:

$$\dot{\boldsymbol{Q}}^{\mathrm{T}}=\boldsymbol{Q}^{\mathrm{T}}\tilde{\boldsymbol{\omega}}^{\mathrm{T}}\,. \tag{4.29}$$

Knowing that $\tilde{\boldsymbol{\omega}}$ is skew-symmetric, we have $\tilde{\boldsymbol{\omega}}^{\mathrm{T}}=-\tilde{\boldsymbol{\omega}}$. Equation (4.29) will be written:

$$\dot{\boldsymbol{Q}}^{\mathrm{T}}=-\boldsymbol{Q}^{\mathrm{T}}\tilde{\boldsymbol{\omega}}\,. \tag{4.30}$$

From Equation (4.19), by transposition and considering $\tilde{\boldsymbol{\omega}}^{\mathrm{T}}=-\tilde{\boldsymbol{\omega}}$, it follows:

$$\boldsymbol{Q}^{\mathrm{T}}\tilde{\boldsymbol{\omega}}=\tilde{\boldsymbol{\omega}}\boldsymbol{Q}^{\mathrm{T}}\,. \tag{4.31}$$

From Equations (4.30) and (4.31) we have:

$$\dot{\boldsymbol{Q}}^{\mathrm{T}}=-\tilde{\boldsymbol{\omega}}\cdot\boldsymbol{Q}^{\mathrm{T}}\,. \tag{4.32}$$

Therefore $\boldsymbol{F}_{-\omega}(\boldsymbol{r})=\boldsymbol{Q}^{\mathrm{T}}\cdot\overline{r}$, so $(\boldsymbol{F}_{\omega})^{-1}=\boldsymbol{F}_{-\omega}$.

Note: The transformation $\boldsymbol{F}_{\omega}:V_3^*\to V_3^*$, given by Equation (4.5), is a symbolic representation of a vector space V_3^* in itself. The third property of Theorem 3.1 shows that we can define like unary operation on V_3^* the relative derivation with respect to angular velocity $\boldsymbol{\omega}$, and like operator for symbolic representation the trivial derivation of vector functions of real variable with respect to the variable.

This transformation "gives an algebraic form" to a class of vector differential equations that model the motion of mechanical systems in non-inertial frames, whom are in the motion of non-uniform rotation, on fixed direction, also the motion with respect to the inertial frames in the fields of gyroscopic forces.

5. The Solution of the Reduced Form Problem

The motion of the Foucault Pendulum is described by the following non-linear initial value problem:

$$\begin{cases}\ddot{\boldsymbol{r}}+2\boldsymbol{\omega}\times\dot{\boldsymbol{r}}+\boldsymbol{\omega}\times(\boldsymbol{\omega}\times\boldsymbol{r})+f(r)\dfrac{\boldsymbol{r}}{r}=\boldsymbol{0}; \quad \boldsymbol{\omega}\in V_3\\ \boldsymbol{r}(0)=\boldsymbol{r}_0\\ \dot{\boldsymbol{r}}(0)=\boldsymbol{v}_0\end{cases} \tag{5.1}$$

If the force field is elastic, the type $\boldsymbol{F}=\omega_*^2\boldsymbol{r}$, $\omega_*\in\mathbb{R}_+^*$, we will have the mathematic model of Foucault pendulum on a non-inertial frame that is rotating non-uniform on fixed direction:

$$\begin{cases}\ddot{\boldsymbol{r}}+2\boldsymbol{\omega}\times\dot{\boldsymbol{r}}+\boldsymbol{\omega}\times(\boldsymbol{\omega}\times\boldsymbol{r})+\omega_*^2\boldsymbol{r}=\boldsymbol{0}; \quad \boldsymbol{\omega}\in V_3\\ \boldsymbol{r}(0)=\boldsymbol{r}_0\\ \dot{\boldsymbol{r}}(0)=\boldsymbol{v}_0\end{cases} \tag{5.2}$$

We will use the present method in order to resolve the reduced form of the problem (5.2):

$$\begin{cases} \ddot{\boldsymbol{r}} + 2\boldsymbol{\omega}\times\dot{\boldsymbol{r}} + \omega_0^2\boldsymbol{r} = \boldsymbol{0};\ \ \boldsymbol{\omega} = \overrightarrow{\text{const}};\ \omega_0 \in \mathbb{R}_+^* \\ \boldsymbol{r}(0) = \boldsymbol{r}_0 \\ \dot{\boldsymbol{r}}(0) = \boldsymbol{v}_0 \end{cases} \tag{5.3}$$

The mathematical model of the Foucault pendulum is presented of the type (5.3) in the theoretical mechanics [10] [11]. The problem (5.3) can be found from (5.2) neglecting the term $\boldsymbol{\omega}\times(\boldsymbol{\omega}\times\boldsymbol{r})$, which represents the contribution of centrifugal force of inertia in the hypothesis $\boldsymbol{\omega} = \overrightarrow{\text{const}}$. We will see that, paradoxically, the exact solution of the short form (5.3) is more difficult to obtain than for the complete problem (5.2).

1) If $\boldsymbol{r}_0\times\boldsymbol{\omega} = \boldsymbol{0}$ and $\boldsymbol{v}_0\times\boldsymbol{\omega} = \boldsymbol{0}$, we are looking for the solution of the problem (5.3) with the property $\boldsymbol{r}\times\boldsymbol{\omega} = \boldsymbol{0}$, $\dot{\boldsymbol{r}}\times\boldsymbol{\omega} = \boldsymbol{0}$. A such solution, if it exists, verifies the differential of the equation:

$$\ddot{\boldsymbol{r}} + \omega_0^2\boldsymbol{r} = \boldsymbol{0} \tag{5.4}$$

The solution of the problem (5.4):

$$\boldsymbol{r} = \boldsymbol{r}_0\cos\omega_0 t + \frac{\boldsymbol{v}_0}{\omega_0}\sin\omega_0 t \tag{5.5}$$

verifies the initial conditions of (5.3) and has the property: $\boldsymbol{r}\times\boldsymbol{\omega} = \boldsymbol{0}$, $\dot{\boldsymbol{r}}\times\boldsymbol{\omega} = \boldsymbol{0}$.

2) If $\boldsymbol{r}_0\cdot\boldsymbol{\omega} = 0$ and $\boldsymbol{v}_0\cdot\boldsymbol{\omega} = 0$, we are searching the solution of the problem (5.3) with the property $\boldsymbol{r}\cdot\boldsymbol{\omega} = 0$. Let be $\boldsymbol{\psi} = \boldsymbol{F}_{\boldsymbol{\omega}}(\boldsymbol{r})$:

$$\boldsymbol{F}_{\boldsymbol{\omega}}(\boldsymbol{r}) = \frac{\boldsymbol{\omega}\cdot\boldsymbol{r}}{\boldsymbol{\omega}^2}\boldsymbol{\omega} + \sin\omega t\frac{\boldsymbol{\omega}\times\boldsymbol{r}}{\omega} - \cos\omega t\frac{\boldsymbol{\omega}\times(\boldsymbol{\omega}\times\boldsymbol{r})}{\boldsymbol{\omega}^2} \tag{5.6}$$

Assuming $\boldsymbol{\omega}\cdot\boldsymbol{r} = 0$ from (5.6) it follows that $\boldsymbol{\omega}\cdot\boldsymbol{F}_{\boldsymbol{\omega}}(\boldsymbol{r}) = 0 \Leftrightarrow \boldsymbol{\psi}\cdot\boldsymbol{\omega} = 0$. Now we will have:

$$\begin{aligned} \ddot{\boldsymbol{\psi}} &= \boldsymbol{F}_{\boldsymbol{\omega}}\left[\ddot{\boldsymbol{r}} + 2\boldsymbol{\omega}\times\dot{\boldsymbol{r}} + \boldsymbol{\omega}\times(\boldsymbol{\omega}\times\boldsymbol{r})\right] = \boldsymbol{F}_{\boldsymbol{\omega}}\left[\ddot{\boldsymbol{r}} + 2\boldsymbol{\omega}\times\dot{\boldsymbol{r}}\right] + \boldsymbol{F}_{\boldsymbol{\omega}}\left[\boldsymbol{\omega}\times(\boldsymbol{\omega}\times\boldsymbol{r})\right] \\ &= \boldsymbol{F}_{\boldsymbol{\omega}}(\ddot{\boldsymbol{r}} + 2\boldsymbol{\omega}\times\dot{\boldsymbol{r}}) + \boldsymbol{\omega}\times\left[\boldsymbol{\omega}\times\boldsymbol{F}_{\boldsymbol{\omega}}(\bar{r})\right] = \boldsymbol{F}_{\boldsymbol{\omega}}(\ddot{\boldsymbol{r}} + 2\boldsymbol{\omega}\times\dot{\boldsymbol{r}}) + \boldsymbol{\omega}\times(\boldsymbol{\omega}\times\boldsymbol{\psi}), \end{aligned} \tag{5.7}$$

Therefore:

$$\boldsymbol{F}_{\boldsymbol{\omega}}(\ddot{\boldsymbol{r}} + 2\boldsymbol{\omega}\times\dot{\boldsymbol{r}}) = \ddot{\boldsymbol{\psi}} - \boldsymbol{\omega}(\boldsymbol{\omega}\times\boldsymbol{\psi}) \tag{5.8}$$

Applying the transformation $\boldsymbol{F}_{\boldsymbol{\omega}}$ on the Cauchy problem (5.3), considering (5.8), we will have the problem:

$$\begin{cases} \ddot{\boldsymbol{\psi}} - \boldsymbol{\omega}\times(\boldsymbol{\omega}\times\boldsymbol{\psi}) + \omega_0^2\boldsymbol{\psi} = \boldsymbol{0} \\ \boldsymbol{\psi}(0) = \boldsymbol{r}_0 \\ \dot{\boldsymbol{\psi}}(0) = \boldsymbol{v}_0 + \boldsymbol{\omega}\times\boldsymbol{r}_0 \end{cases} \tag{5.9}$$

The solutions of (5.9) with the property $\boldsymbol{\psi}\cdot\boldsymbol{\omega} = 0$ is searched. In this hypothesis, the problem (5.9) becomes:

$$\begin{cases} \ddot{\boldsymbol{\psi}} + (\omega^2 + \omega_0^2) = \boldsymbol{0} \\ \boldsymbol{\psi}(0) = \boldsymbol{r}_0 \\ \dot{\boldsymbol{\psi}}(0) = \boldsymbol{v}_0 + \boldsymbol{\omega}\times\boldsymbol{r}_0 \end{cases} \tag{5.10}$$

The solution of the problem (5.10) is:

$$\boldsymbol{\psi}_* = \boldsymbol{r}_0\cos\omega_* t + \frac{\boldsymbol{v}_0 + \boldsymbol{\omega}\times\boldsymbol{r}_0}{\omega_*}\sin\omega_* t;\quad \omega_* = \sqrt{\omega^2 + \omega_0^2} \tag{5.11}$$

In the hypothesis $\boldsymbol{r}_0\cdot\boldsymbol{\omega} = 0$ and $\boldsymbol{v}_0\cdot\boldsymbol{\omega} = 0$, it follows $\boldsymbol{\psi}_*\cdot\boldsymbol{\omega} = 0$, therefore, (5.11) is also the solution for the problem (5.9). On the same hypothesis the solution of the problem (5.3) is:

$$\boldsymbol{r} = \boldsymbol{F}_{-\boldsymbol{\omega}}(\boldsymbol{\psi}_*) = \frac{\boldsymbol{\omega}\cdot\boldsymbol{\psi}_*}{\boldsymbol{\omega}^2}\boldsymbol{\omega} - \sin\omega t\frac{\boldsymbol{\omega}\times\boldsymbol{\psi}_*}{\omega} - \cos\omega t\frac{\boldsymbol{\omega}\times(\boldsymbol{\omega}\times\boldsymbol{\psi}_*)}{\boldsymbol{\omega}^2} \tag{5.12}$$

Considering the fact that $\boldsymbol{\psi}_*\cdot\boldsymbol{\omega} = 0$, the solution (5.12) will be written:

$$\boldsymbol{r} = -\sin\omega t\frac{\boldsymbol{\omega}\times\boldsymbol{\psi}_*}{\omega} + \cos\omega t\boldsymbol{\psi}_* \tag{5.13}$$

Now, let be the Cauchy problems:

$$\begin{cases} \ddot{\boldsymbol{r}} + 2\boldsymbol{\omega}\times\dot{\boldsymbol{r}} + \omega_0^2\boldsymbol{r} = \boldsymbol{0}; \\ \boldsymbol{r}(0) = \dfrac{\boldsymbol{\omega}\cdot\boldsymbol{r}_0}{\boldsymbol{\omega}^2}\boldsymbol{\omega} = \boldsymbol{r}_{01}; \\ \dot{\boldsymbol{r}}(0) = \dfrac{\boldsymbol{\omega}\cdot\boldsymbol{v}_0}{\boldsymbol{\omega}^2} = \boldsymbol{v}_{01}. \end{cases} \tag{5.14}$$

$$\begin{cases} \ddot{\boldsymbol{r}} + 2\boldsymbol{\omega}\times\dot{\boldsymbol{r}} + \omega_0^2\boldsymbol{r} = \boldsymbol{0}; \\ \boldsymbol{r}(0) = \dfrac{\boldsymbol{\omega}\times(\boldsymbol{r}_0\times\boldsymbol{\omega})}{\boldsymbol{\omega}^2}\boldsymbol{\omega} = \boldsymbol{r}_{02}; \\ \dot{\boldsymbol{r}}(0) = \dfrac{\boldsymbol{\omega}\times(\boldsymbol{v}_0\times\boldsymbol{\omega})}{\boldsymbol{\omega}^2} = \boldsymbol{v}_{02}. \end{cases} \tag{5.15}$$

In accordance with those shown in the points a) and b) the solutions of the problems (5.14) and (5.15) are:

$$\boldsymbol{r}_1 = \boldsymbol{r}_{01}\cos\omega_0 t + \frac{\boldsymbol{v}_{01}}{\omega_0}\sin\omega_0 t = \frac{\boldsymbol{\omega}\cdot\boldsymbol{\psi}_0}{\boldsymbol{\omega}^2}\boldsymbol{\omega} \tag{5.16}$$

where:

$$\boldsymbol{\psi}_0 = \boldsymbol{r}_0\cos\omega_0 t + \frac{\boldsymbol{v}_0 + \boldsymbol{\omega}\times\boldsymbol{r}_0}{\omega_0}\sin\omega_0 t \tag{5.17}$$

respectively:

$$\boldsymbol{r}_2 = -\sin\omega t\frac{\boldsymbol{\omega}\times\boldsymbol{\psi}_2}{\omega} + \cos\omega t\boldsymbol{\psi}_2 \tag{5.18}$$

where:

$$\boldsymbol{\psi}_2 = \boldsymbol{r}_{02}\cos\omega_* t + \frac{\boldsymbol{v}_{02} + \boldsymbol{\omega}\times\boldsymbol{r}_{02}}{\omega_*}\sin\omega_* t = \frac{\boldsymbol{\omega}\times(\boldsymbol{\psi}_*\times\boldsymbol{\omega})}{\boldsymbol{\omega}^2} \tag{5.19}$$

with $\boldsymbol{\psi}_*$ given by Equation (5.11). From (5.19) and (5.20), after elementary calculus, it follows:

$$\boldsymbol{r}_2 = -\sin\omega t\frac{\boldsymbol{\omega}\times\boldsymbol{\psi}_*}{\omega} + \cos\omega t\frac{\boldsymbol{\omega}\times(\boldsymbol{\psi}_*\times\boldsymbol{\omega})}{\boldsymbol{\omega}^2} \tag{5.20}$$

The relations $\boldsymbol{r}_0 = \boldsymbol{r}_{01} + \boldsymbol{r}_{02}$, $\boldsymbol{v}_0 = \boldsymbol{v}_{01} + \boldsymbol{v}_{02}$ and of the linearity of the differential Equation (5.3) have the consequence that the solution of the Cauchy problem (5.3) is given by $\boldsymbol{r} = \boldsymbol{r}_1 + \boldsymbol{r}_2$ with $\boldsymbol{r}_1$ and $\boldsymbol{r}_2$ given by the relations (5.16) and (5.20). Finally, the results can be summarized as following:

Theorem 5.1:

The solution of Cauchy problem:

$$\begin{cases} \ddot{\boldsymbol{r}} + 2\boldsymbol{\omega}\times\dot{\boldsymbol{r}} + \omega_0^2\boldsymbol{r} = \boldsymbol{0}; \quad \boldsymbol{\omega} = \overrightarrow{\text{const}}; \quad \omega_0 \in \mathbb{R}_+^* \\ \boldsymbol{r}(0) = \boldsymbol{r}_0 \\ \dot{\boldsymbol{r}}(0) = \boldsymbol{v}_0 \end{cases} \tag{5.21}$$

is given by the vector function:

$$\boldsymbol{r} = \frac{\boldsymbol{\omega}\cdot\boldsymbol{\psi}_0}{\boldsymbol{\omega}^2}\boldsymbol{\omega} - \sin\omega t\frac{\boldsymbol{\omega}\times\boldsymbol{\psi}_*}{\omega} - \cos\omega t\frac{\boldsymbol{\omega}\times(\boldsymbol{\omega}\times\boldsymbol{\psi}_*)}{\boldsymbol{\omega}^2}; \quad t \geq 0 \tag{5.22}$$

where

$$\psi_0 = r_0 \cos\omega_0 t + \frac{v_0 + \omega \times r_0}{\omega_0} \sin\omega_0 t; \quad t \geq 0 \tag{5.23}$$

$$\psi_* = r_0 \cos\omega_* t + \frac{v_0 + \omega \times r_0}{\omega_*} \sin\omega_* t; \quad t \geq 0; \quad \omega_* = \sqrt{\omega^2 + \omega_0^2} \tag{5.24}$$

Note: Also, the problem (5.21) shapes the movement of the vibration for a class of gyroscopic instruments. Even in the case of planar motion, the literature shows only the approximate solutions assuming $\omega^* \gg \omega_0$ with particular initial conditions as in the case of harmonic oscillator in magnetic field. The presented solution (5.22) is accurate.

6. Conclusion

The work presents the closed form vector solution for the well-known Foucault pendulum problem. Both forms of the Foucault problems (the whole form and the as known "reduced form" when the centripetal force is neglected) are considered. The last one models the movement of the harmonic oscillator in uniform magnetic field, also. Therefore, a specific isomorphism between the free vectors map and the column matrix map is used. The short solution of the Foucault pendulum problem is obtained using vectors as column matrix of complex numbers adequate defined. With this method, the second order Cauchy vector problem which describes the spatial movement of the Foucault pendulum becomes a first order differential matrix equation with constant coefficients. The closed form vector solution obtained in this way allows a suggestive kinematic representation of the spatial movement of the Foucault pendulum. The closed form vector solution for Foucault pendulum problem is obtained by means of a time dependent tensor operator which reduces this problem to only two classic problems very easy to be solved. The tensor operator as introduced can extend the study of all Foucault type movements in the case of non-inertial reference frame with time dependent angular velocity.

References

[1] Foucault, J.B.L. (1851) Physical Demonstration of the Rotation of the Earth by Means of the Pendulum. *Comptes Rendus de l'Académie des Sciences de Paris*, **51**, 350-353.

[2] Anonymous (1851) On Foucault's Pendulum Experiments. *Journal of the Franklin Institute*, **52**, 419-421.

[3] Condurache, D. and Martinusi, V. (2007) Relative Spacecraft Motion in a Central Force Field. *AIAA Journal of Guidance, Control and Dynamics*, **30**, 873-876. http://dx.doi.org/10.2514/1.26361

[4] Condurache, D. and Martinusi, V. (2008) Exact Solution to the Relative Orbital Motion in a Central Force Field. *The 2nd International Symposium on Systems and Control in Aeronautics and Astronautics*, Shenzhen.

[5] Arnold, V. (1989) Mathematical Methods of Classical Mechanics. New York. (Translated from the 1974 Russian Original by K. Vogtmann and A. Weinstein, Springer, Berlin.)

[6] Appell, P. (1926) Traité de Mécanique Rationelle. 5 Volumes, Gauthier-Villars, Paris.

[7] Levi-Civita, T. and Amaldi, U. (1922-1926) Lezioni di mecanica razionale. N. Zanichelli (Ed.).

[8] Lurie, A.I. (2002) Analytical Mechanics. Springer, Berlin, 864. http://dx.doi.org/10.1007/978-3-540-45677-3

[9] Condurache, D. and Matcovschi, M.-H. (1997) An Exact Solution to Foucault's Pendulum Problem. *Buletinul Institutului Politehnic Din Iasi*, **XLI (XLVII)**, 83-92.

[10] Goldstein, H., Poole, C.P. and Safko, J.L. (2002) Classical Mechanics. 3rd Edition, Addison-Wesley, Reading.

[11] Landau, L. and Lifschitz, E. (1981) Mécanique. Mir, Moscou.

12

MHD Mixed Convection Flow of a Casson Fluid over an Exponentially Stretching Surface with the Effects of Soret, Dufour, Thermal Radiation and Chemical Reaction

Kankanala Sharada, Bandari Shankar

Department of Mathematics, Osmania University, Hyderabad, India
Email: kankan.sharada@gmail.com

Abstract

The present study deals with MHD (magneto hydrodynamics) mixed convection flow of a Casson fluid over an exponentially stretching sheet with the effects of Soret and Dufour, thermal radiation, chemical reaction. The governing partial differential equations are converted into ordinary differential equations by using similarity transformations. These equations are then solved numerically by applying finite difference scheme known as the Keller Box method. The effects of various parameters on velocity, temperature and concentration profiles are presented graphically to interpret and the results are discussed.

Keywords

Casson Fluid, Magneto Hydrodynamics, Mixed Convection, Exponentially Stretching Surface, Soret and Dufour Effects, Thermal Radiation, Chemical Reaction

1. Introduction

The study of magneto hydrodynamics, boundary layer flow with heat and mass transfer from a stretching surface has many applications in industrial and engineering fields. Magnetic fields can be used to manage thermal energy in flowing electrically conducting polymers (Garnier [1]). After the pioneering studies of Sakiadas [2] and Crane [3], several researchers further investigate the problem of stretching sheet boundary layer flow to obtain thermal and kinematic behaviour by considering the various types of stretching velocity. Magyari and Keller [4] examine the similarity solutions of thermal boundary layer and flow over an exponentially stretching surface with an exponential temperature distribution. Numerical solution of boundary layer flow over an exponentially stretching sheet including thermal radiation is investigated by Bidin and Nazar [5], and analytical study is investigated by Sajid and Hayat [6].

Casson fluid is one type of fluid model for non-Newtonian fluid, which has significant applications in polymer processing industries and biomechanics. Casson fluid can be defined as a shear thinning liquid which is presumed to have an infinite viscosity at zero rate of shear and a yield stress under which no flow occurs and zero viscosity at an infinite rate of shear. Casson fluid model is used in many foodstuffs and biological materials, especially blood. It describes the steady shear stress, shear rate behaviour of blood. Merill *et al.* [7] and Mc Donald [8] investigate the behaviour of blood. Nadeem *et al.* [9] investigate the magneto hydrodynamic boundary layer flow of Casson fluid over exponentially permeable shrinking sheet. Casson fluid over an unsteady stretching surface is analyzed by Mukhopadhyay [10]. The application of the minimal energy phenomenon to a Casson fluid is studied by John L. McGregor [11]. A similarity solution technique for mixed convection heat transfer from an exponentially stretching surface with the effect of viscous dissipation is investigated by Partha *et al.* [12]. Recently Dual pal [13] studies mixed convection heat transfer in the boundary layers over an exponentially stretching surface including magnetic field. The effects on mixed free forced convective and mass transfer steady, laminar boundary layer flow over a vertical plate with temperature dependent viscosity are presented by Kafoussias and Williams [14].

The mass flux created by temperature gradient is known as thermal diffusion (Soret) effect. The energy flux caused by concentration differences is known as diffusion thermo (Dufour) effect. The Soret effect has been utilized for medium molecular weight (eg. N_2 and air) and isotope separation. These effects have several applications in engineering problems. Here a survey conducted by Dursunkaya and Worek [15] on Soret and Dufour effects in transient and steady natural convection from a vertical surface. The combined effects of thermal diffusion and diffusion thermo over stretching surface with free convective heat and mass transfer are investigated by Afify [16]. T. Hayat [17] describes the effects of Soret and Dufour on magneto hydrodynamic flow of Casson fluid. Alam and Rahman [18] study the Dufour and Soret effects on mixed convection flow with variable suction.

Several researchers have studied the effects of thermal radiation, and these include that Pramanik [19] investigates Casson fluid flow and heat transfer over an exponentially stretching surface in the presence of thermal radiation and porous medium. Numerical analysis of magneto hydrodynamic stagnation point flow of a Casson fluid includes thermal radiation discussed by Shateyi and Marewo [20]. Anur Ishak [21] present the magneto hydrodynamic boundary layer flow over an exponentially stretching sheet with the effect of radiation. Swathi Mukhopadhyay [22] investigate Casson fluid flow and heat transfer on an unsteady stretching surface subjected to suction/blowing with the effect of thermal radiation. Recently effects of mass transfer on magneto hydrodynamic flow of Casson fluid with chemical reaction are studied by S.A. Shehzad and T. Hayat [23].

This paper provides the solution to the problem of fluid flow, heat and mass transfer of a Casson fluid over an exponentially stretching surface by considering the effect of Soret and Dufour, thermal radiation, chemical reaction by adopting the Keller Box method.

2. Mathematical Formulation

Consider two-dimensional steady, laminar, incompressible MHD flow of a Casson fluid over a vertical exponential surface stretching with velocity $U_w(x) = ae^{\frac{x}{L}}$, temperature distribution $T_w(x)$ and concentration distribution $C_w(x)$ moving through a fluid of constant temperature T_∞ and concentration C_∞ as shown in **Figure 1**. The x-axis is oriented about the vertical surface in the upward direction and y-axis is normal to it. A magnetic field is applied in the direction perpendicular to the stretching surface. The magnetic Reynolds number is taken to be small enough such that the induced magnetic field is negligible. The rheological equation of state for an isotropic and incompressible flow of a Casson fluid can be written as:

$$\tau_{ij} = \left(\mu_b + \frac{P_y}{\sqrt{2\pi}}\right) 2e_{ij} \quad \text{when} \quad \pi > \pi_c \tag{1}$$

$$\tau_{ij} = \left(\mu_b + \frac{P_y}{\sqrt{2\pi_c}}\right) 2e_{ij} \quad \text{when} \quad \pi < \pi_c \tag{2}$$

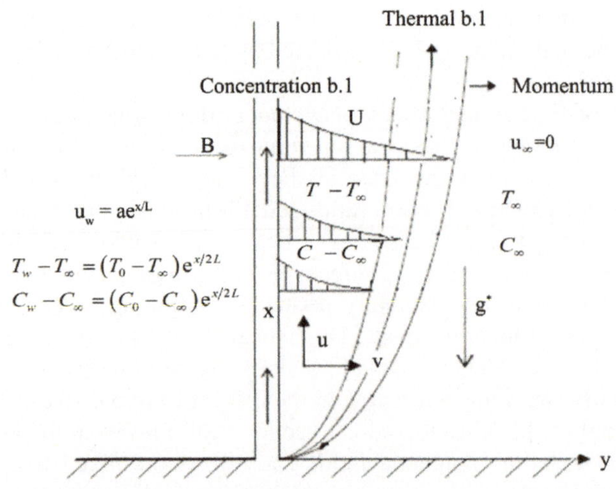

Figure 1. Physical model and coordinate system.

P_y is known as yield stress of the fluid, mathematically expressed as $P_y = \frac{\mu_b\sqrt{2\pi}}{\beta}$, μ_b is known as plastic dynamic viscosity of the non Newtonian fluid, π is the product of the component of deformation rate with itself (*i.e.* $\pi = e_{ij}e_{ij}$) and τ_{ij} is the (i, j)th component of the stress tensor.

Considering u, v as velocity components in the directions of x and y respectively in the flow field. The governing equations of continuity, momentum, energy and concentration are given by

$$\frac{\partial u}{\partial x}+\frac{\partial v}{\partial y}=0 \tag{3}$$

$$u\frac{\partial u}{\partial x}+v\frac{\partial u}{\partial y}=\upsilon\left(1+\frac{1}{\beta}\right)\frac{\partial^2 u}{\partial y^2}+g^*\left(\beta_T\left(T-T_\infty\right)+\beta_C\left(C-C_\infty\right)\right)-\frac{\sigma B^2 u}{\rho} \tag{4}$$

$$u\frac{\partial T}{\partial x}+v\frac{\partial T}{\partial y}=\alpha\frac{\partial^2 T}{\partial y^2}-\frac{1}{\rho c_p}\frac{\partial q_r}{\partial y}+\frac{v}{C_p}\left(1+\frac{1}{\beta}\right)\left(\frac{\partial u}{\partial y}\right)^2+\frac{Dk_T}{C_pC_s}\frac{\partial^2 C}{\partial y^2} \tag{5}$$

$$u\frac{\partial C}{\partial x}+v\frac{\partial C}{\partial y}=D\frac{\partial^2 C}{\partial y^2}+\frac{DK_T}{T_m}\frac{\partial^2 T}{\partial y^2}-k\left(C-C_\infty\right). \tag{6}$$

where ρ is the density of the fluid, σ is the electrical conductivity, υ is the kinematic viscosity, B is the magnetic induction, α is the thermal diffusivity, D is the solutal diffusivity of the medium, β_T and β_C are the coefficient of thermal and solutal expansions, μ is the dynamic co-efficient of viscosity, C_p and C_s are the specific heat capacity and concentration susceptibility, T_m is the mean fluid temperature, K_T is the thermal diffusion ratio, γ is the chemical reaction parameter and $\beta = \mu_b\sqrt{2\pi_c}/P_y$ is the Casson parameter.

The boundary conditions considered are defined as:

$$u=U_w\left(x\right)=ae^{\frac{x}{L}},\ v=0,\ T=T_w=T_\infty+\left(T_0-T_\infty\right)e^{\frac{x}{2L}}$$

$$C=C_w=C_\infty+\left(C_0-C_\infty\right)e^{\frac{x}{2L}}\ \text{ at }\ y=0 \tag{7}$$

$$u=0,\ T\to T_\infty,\ C\to C_\infty\ \text{ as }\ y\to\infty \tag{8}$$

where subscripts w and ∞ indicate that conditions at the wall and at the external(outer) edge of the boundary layer respectively and $U_w(x)$ is the stretching velocity, $T_w(x)$ and $C_w(x)$ are the exponential temperature and concentration distribution respectively.

In above conditions "*a*" is velocity parameter of the stretching surface, T_0 is temperature distribution parameter, C_0 is the parameter of the concentration distribution in the stretching surface.

We introduce stream function ψ as $u=\frac{\partial \Psi}{\partial y}$, $v=-\frac{\partial \psi}{\partial x}$ and define variables.

$$u=ae^{\frac{x}{L}}f'(\eta),\ \eta=\sqrt{\frac{a}{2\upsilon L}}e^{\frac{x}{2L}}y,\ v=-\sqrt{\frac{a\upsilon}{2L}}e^{\frac{x}{2L}}\left[f(\eta)+\eta f'(\eta)\right]$$

$$T(x,y)=T_\infty+(T_0-T_\infty)e^{\frac{x}{2L}}\theta(\eta),\ C(x,y)=C_\infty+(C_0-C_\infty)e^{\frac{x}{2L}}\phi(\eta)$$

By using the Rosseland approximation for radiative heat flux, q_r is defined as:

$$q_r=-\frac{4\sigma^*}{3K^*k}\frac{\partial T^4}{\partial y} \tag{9}$$

where σ^* is the Stephen Boltzman constant and K^* is the mean absorption coefficient. Since the temperature differences within the flow field are assumed to be small, and then we linearize and expand T^4 into the Taylor series about T_∞, which after neglecting higher order forms take the form.

$$T^4\cong 4T_\infty^3T-3T_\infty^4$$

To obtain similarity solutions, we assumed that the magnetic field is of the form $B(x)=B_0e^{\frac{x}{L}}$ where B_0 is the constant magnetic field and $K(x)=K_0e^{\frac{x}{L}}$ is the exponential reaction, K_0 is constant.

With the help of above similarity transformations the governing equations are transformed in to

$$\left(1+\frac{1}{\beta}\right)f'''+ff''-2(f')^2+2\lambda e^{\frac{-3X}{2}}(\theta+N\phi)-Mf'=0 \tag{10}$$

$$\left(1+\frac{4}{3}Nr\right)\theta''+Pr(f\theta'-f'\theta)+PrEce^{\frac{3X}{2}}\left(1+\frac{1}{\beta}\right)(f'')^2+PrD_f\phi''=0 \tag{11}$$

$$\phi''+Sc(f\phi'-f'\phi)+ScSr\theta''-Sc\gamma\phi=0 \tag{12}$$

The corresponding boundary conditions are

$$\eta=0: f(0)=1,\ f'(0)=1,\ \theta(0)=1,\ \phi(0)=1$$

$$\eta\to\infty: f'(\infty)=0,\ \theta(\infty)=0,\ \phi(\infty)=0$$

where $X=\frac{x}{L}$ is the X-location, observing Equations (10), (11) in MHD mixed convection due to Casson fluid, the velocity and temperature profiles are not similar because the x coordinate cannot be deleted from these equa tions. It is hard to proceed in this case. To reduce the difficulty of analysis, it was considered $X=\frac{x}{L}$ to find local similarity solutions for the governing equations. Now one can study the effects of different parameters on various profiles at any given X-location.

$M=\frac{2\sigma B_0^2L}{\rho a}$ is the magnetic parameter, $\lambda=\frac{Gr}{Re^2}$ is the mixed convection parameter, $Re=\frac{aL}{\nu}$ is the Reynolds number, $Gr=\frac{g^*\beta_T(T_0-T_\infty)L^3}{\nu^2}$ is the Grashoff number, $N=\frac{\beta_c(C_0-C_\infty)}{\beta_T(T_0-T_\infty)}$ is the Buoyancy ratio, $Pr=\frac{\nu}{\alpha}$ is the Prandtl number, $Ec=\frac{a^2}{C_p(T_0-T_\infty)}$ is the Eckert number, $Nr=\frac{4\sigma^*T_\infty^3}{3kk^*}$ is the Thermal radiation parameter, $D_f=\frac{DK_T(C_0-C_\infty)}{\nu C_sC_p(T_0-T_\infty)}$ is the Dufour number, $Sr=\frac{DK_T(T_0-T_\infty)}{\nu T_m(C_0-C_\infty)}$ is the Soret number, $Sc=\frac{\nu}{D}$ is the Schmidt number, $\gamma=\frac{K_0L}{a}$ is the chemical reaction parameter.

The wall shear stress, heat and mass transfer coefficients with constant density are given by $\tau_w = \left(\mu_b + \frac{P_y}{\sqrt{2\pi}}\right)\left(\frac{\partial u}{\partial y}\right)_{y=0}$, $q_w = -k\left(\frac{\partial T}{\partial y}\right)_{y=0}$ and $q_m = -D\left(\frac{\partial C}{\partial y}\right)_{y=0}$ where k is the thermal conductivity. The dimensionless form of skin friction $C_f = \frac{2\tau_w}{\rho U_*^2}$. Where U_* is the characteristic velocity, the local Nusselt number $Nu_x = \frac{xq_w}{k(T_w - T_\infty)}$, and local Sherwood number $Sh_x = \frac{xq_w}{D(C_w - C_\infty)}$ are given by

$$Cf_x = \frac{\sqrt{2}X}{\sqrt{Re_x}}\left(1+\frac{1}{\beta}\right)f''(0),\ Nu_x = -\left(1+\frac{4}{3}Nr\right)\sqrt{\frac{Re_x X}{2}}\theta'(0) \text{ and } Sh_x = -\sqrt{\frac{Re_x}{2}}\phi'(0).$$

3. Method of Solution

The governing equations with boundary equations are solved numerically by using finite difference scheme known as Keller Box method which is described by Cebeci and Bradshaw [24]. This method involves the following steps:

Step 1: Reducing higher order ODEs (systems of ODES) in to systems of first order ODEs;
Step 2: Writing the systems of first order ODEs into difference equations using central difference scheme;
Step 3: Linearizing the difference equations using Newton's method and writing it in vector form;
Step 4: Solving the system of equations using block elimination method.

4. Numerical Discussions

To solve the above differential equations numerically, we adopt Matlab software. In this process the step size, $\Delta\eta = 0.01$ is used to obtain numerical solution with five decimal place accuracy as the criterion of convergence. **Table 1** shows that the comparison between wall temperature gradient $\theta'(0)$ calculated by the present method and of Magyari and Keller for $\lambda = X = D_f = Nr = Sr = \gamma = N = M = Sc = Ec = 0$ and $\beta \to \infty$. The numerical vales of $\left(1+\frac{1}{\beta}\right)f''(0)$, $-\theta'(0)$, and $-\phi'(0)$ for the various values of λ, X, D_f, Nr, Sr, γ, β are shown in **Table 2**.

5. Results and Discussion

In the present study, the following values are adopted for the numerical computations $N = 0.5$, $D_f = 0.03$, $Sr = 0.5$, $\lambda = 1.0$, $Pr = 1.0$, $M = 1.0$, $Sc = 0.22$, $X = 0.5$, $\gamma = 0.1$, $\beta = 0.5$. These values are used throughout the calculations unless otherwise mentioned.

Figure 2 shows that the non-dimensional velocity profile for different values of the mixed convection parameter λ. We have observed the results for both aiding flow and apposing flow. It illustrates that the dimensionless velocity increases, as the value of λ increases. As λ increases, the buoyancy effects increase and hence the fluid flow accelerates. In the case of mixed convection, temperature is low for aiding flow and more for

Table 1. Comparison of results for $-\theta'(0)$ with previous published work.

Pr	Magyari and Keller [4]	Present
0.5	−0.59434	−0.5946
1.0	−0.95478	−0.9548
3.0	−1.86908	−1.8691
5.0	−2.50014	−2.5002
8.0	−3.24213	−3.2423
10.0	−3.66038	−3.6606

Table 2. Effects of skin friction coefficient, heat and mass transfer coefficients for various values of λ, X, D_f, Nr, Sr, γ and β.

λ	X	Nr	Sr	D_f	γ	β	$(1+1/\beta)''(0)$	$-\theta'(0)$	$-\Phi'(0)$
1.0	0.5	1.0	0.5	0.03	0.1	0.5	−1.9167	0.6505	0.4246
2.0	0.5	1.0	0.5	0.03	0.1	0.5	−1.1259	0.7017	0.4560
3.0	0.5	1.0	0.5	0.03	0.1	0.5	−0.3918	0.7383	0.4797
1.0	0.2	1.0	0.5	0.03	0.1	0.5	−1.4607	0.6870	0.4431
1.0	1.0	1.0	0.5	0.03	0.1	0.5	−2.3684	0.5841	0.4044
1.0	2.0	1.0	0.5	0.03	0.1	0.5	−2.7091	0.3124	0.4054
1.0	0.5	1.5	0.5	0.03	0.1	0.5	−1.8828	0.5637	0.4371
1.0	0.5	3.0	0.5	0.03	0.1	0.5	−1.8174	0.4204	0.4593
1.0	0.5	4.6	0.5	0.03	0.1	0.5	−1.7766	0.3429	0.4722
1.0	0.5	1.0	1.0	0.03	0.1	0.5	−1.9071	0.6527	0.3779
1.0	0.5	1.0	3.0	0.03	0.1	0.5	−1.8687	0.6612	0.1894
1.0	0.5	1.0	5.0	0.03	0.1	0.5	−1.8309	0.6692	−0.0019
1.0	0.5	1.0	0.5	1.0	0.1	0.5	−1.8774	0.5596	0.4387
1.0	0.5	1.0	0.5	2.0	0.1	0.5	−1.8372	0.4601	0.4530
1.0	0.5	1.0	0.5	3.0	0.1	0.5	−1.7967	0.3535	0.4675
1.0	0.5	1.0	0.5	0.03	0.15	0.5	−1.9203	0.6496	0.4394
1.0	0.5	1.0	0.5	0.03	0.25	0.5	−1.9266	0.6480	0.4670
1.0	0.5	1.0	0.5	0.03	0.35	0.5	−1.9326	0.6466	0.4927
1.0	0.5	1.0	0.5	0.03	0.1	0.8	−1.6085	0.6370	0.4171
1.0	0.5	1.0	0.5	0.03	0.1	1.0	−1.4976	0.6315	0.4143
1.0	0.5	1.0	0.5	0.03	0.1	1.5	−1.3415	0.6233	0.4101

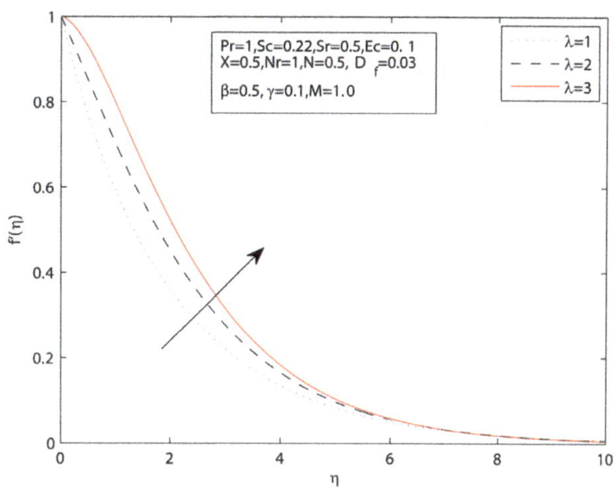

Figure 2. Velocity profiles for various values of mixed convection parameter λ.

opposing flow. As λ increases, the convection cooling effect increases and therefore temperature decreases. **Figure 3**, **Figure 4** reveal that the dimensionless temperature and concentration reduces with increase of λ.

Figure 5 depicts the effects of Casson parameter β on dimensionless velocity profiles for impermeable stretching surface. The velocity is found to decrease with the increasing Casson parameter β. **Figure 6**, **Figure 7** present the temperature and concentration profiles for Casson parameter β. It shows that the both temperature and concentration profiles decreasing with increasing of β. **Figure 8** illustrates the velocity profile for different values of the magnetic parameter M. The presence of magnetic field reduces the momentum boundary layer

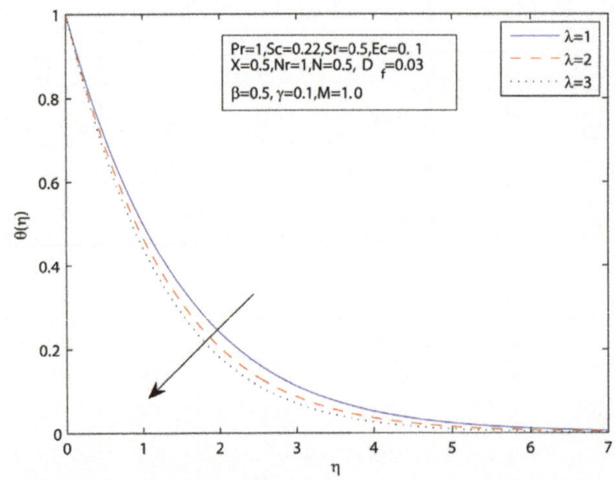

Figure 3. Temperature profiles for various values of mixed convection parameter λ.

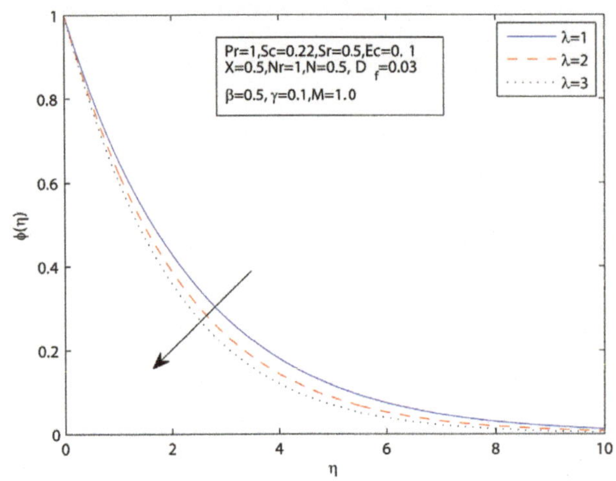

Figure 4. Concentration profiles for various values of mixed convection parameter λ.

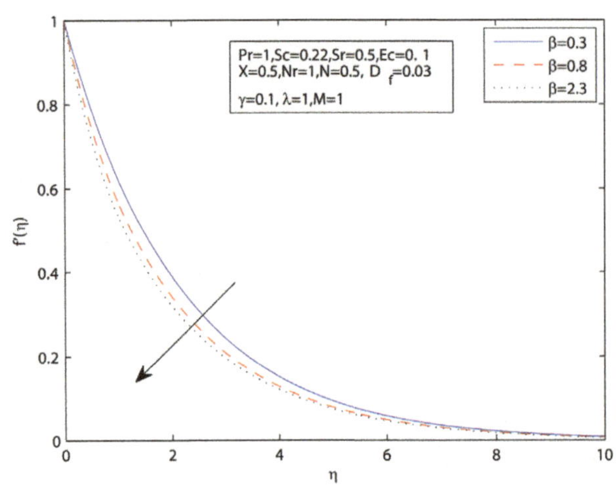

Figure 5. Velocity profiles for various values of Casson parameter β.

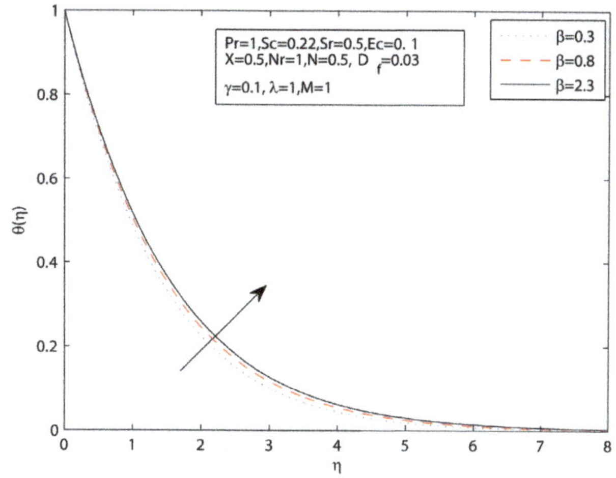

Figure 6. Temperature profiles for various values of Casson parameter β.

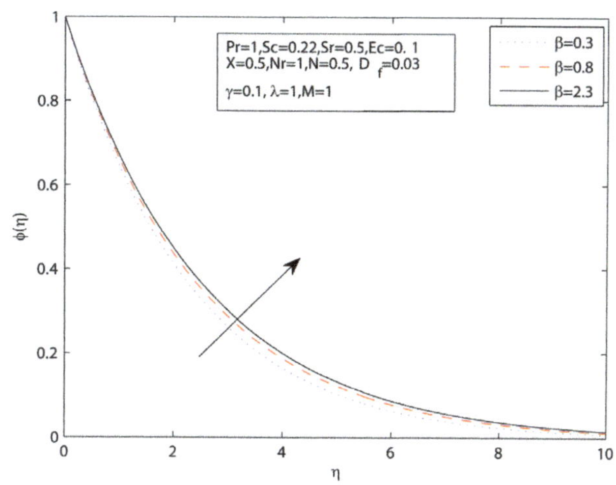

Figure 7. Concentration profiles for various values of Casson parameter β.

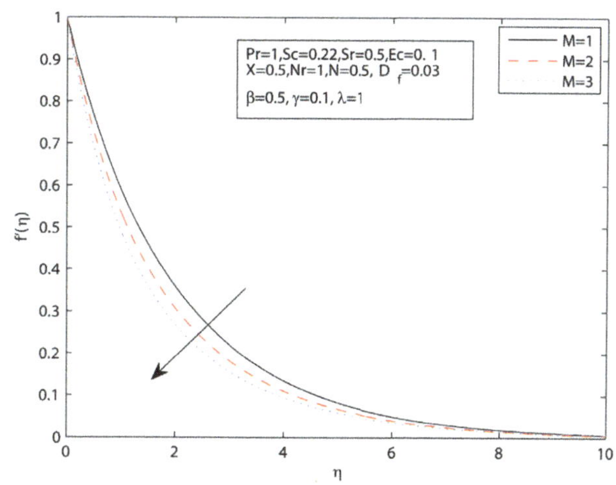

Figure 8. Velocity profile for various values of magnetic parameter M.

thickness. Thus the velocity decreases as the magnetic parameter increases.

Increasing the values of the *Ec* generates heat in the fluid due to frictional heating. **Figure 9** demonstrates that influence of the *Ec* on the profile of temperature. It is observed that the temperature profile increases as the value of the *Ec* increases. The effects of Prandtl number on temperature profiles are shown in **Figure 10**. Prandtl number can be used to increase the rate of cooling conducting fluids. Prandtl number signifies the ratio of momentum diffusivity to thermal diffusivity. Temperature is found to decrease with increasing *Pr*. $Pr = 1$, indicates that the velocity boundary layer and thermal boundary layer are approximately equal.

Figure 11 displays the effect of thermal radiation of the temperature profiles. From this we observe that the temperature increases with increasing value of the parameter *Nr*. **Figure 12** shows that the effect of Soret on concentration profile. An increase in the Soret number increases the concentration and the boundary layer thickness. It is observed that the concentration profile increases as the *Sr* increases.

Figures 13-15 explain the effects of the *X*-location on the velocity, temperature and concentration profiles. From **Figure 13**, it is observed that the velocity profile decreases with an increase in *X* in the boundary layer. **Figure 14** depicts that as the value of *X* increases, the temperature increases. The concentration profile increases with as increasing the value of *X*.

The effect of D_f on temperature profile is depicted in **Figure 16**. The temperature profile increases as D_f increases. **Figure 17** displays the nature of concentration profiles for various values of chemical reaction

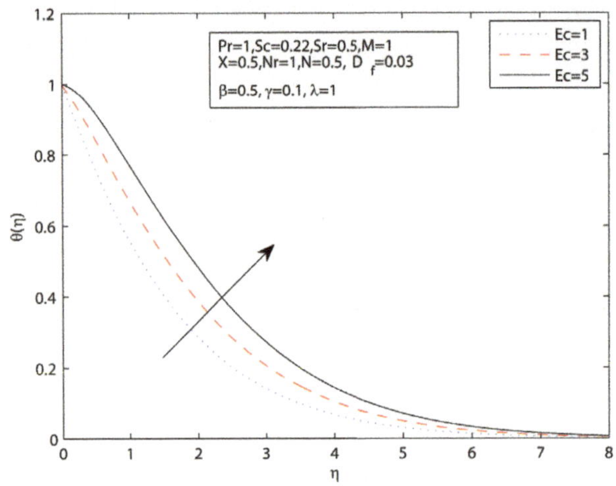

Figure 9. Temperature profile for various values of Eckert number *Ec*.

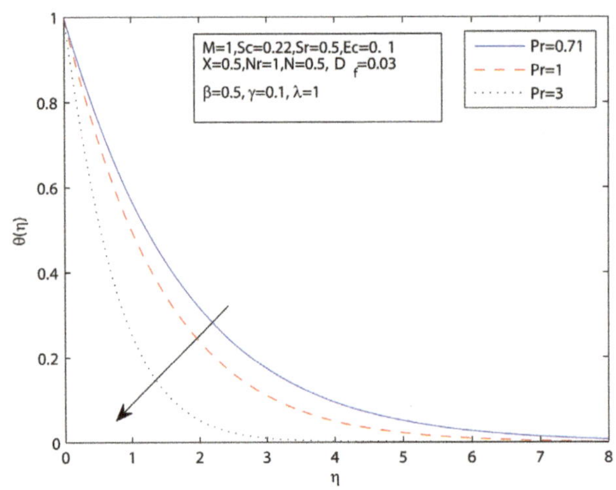

Figure 10. Temperature profile for various values of number *Pr*.

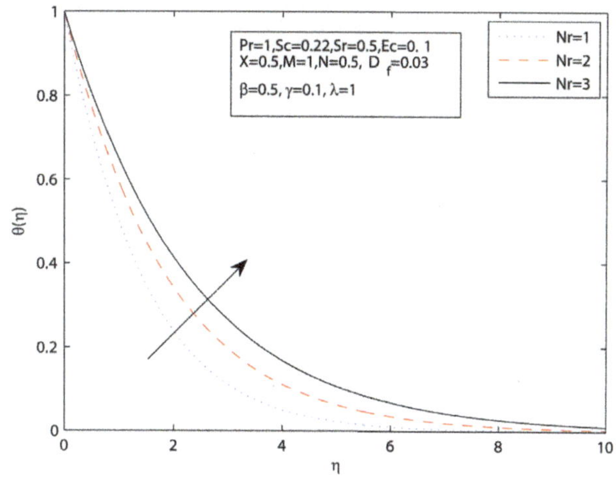

Figure 11. Temperature profile for various values of radiation parameter *Nr*.

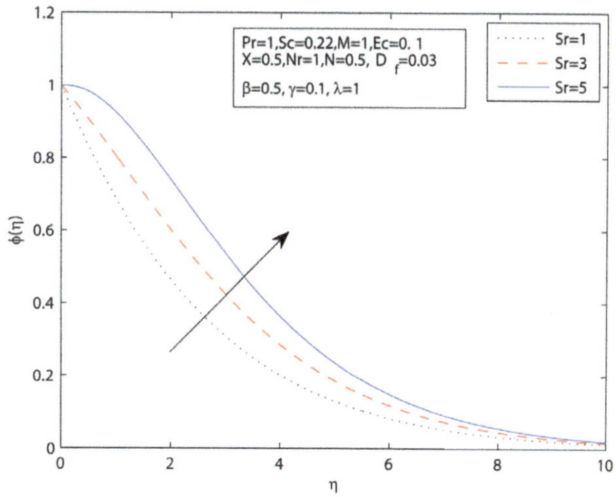

Figure 12. Concentration profile for various values of Soret number *Sr*.

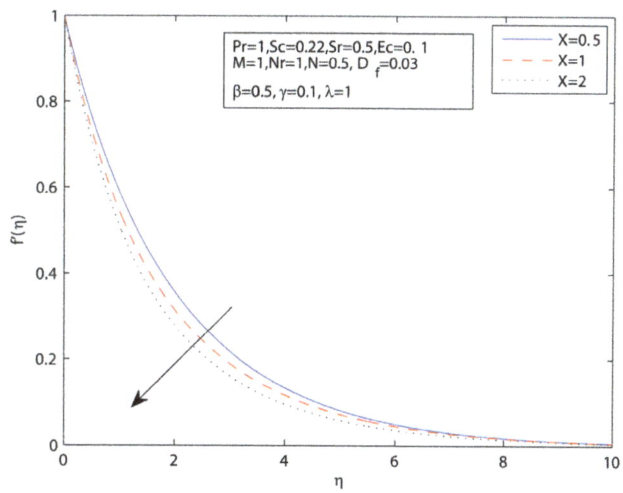

Figure 13. Velocity profile for various values of *X*-location.

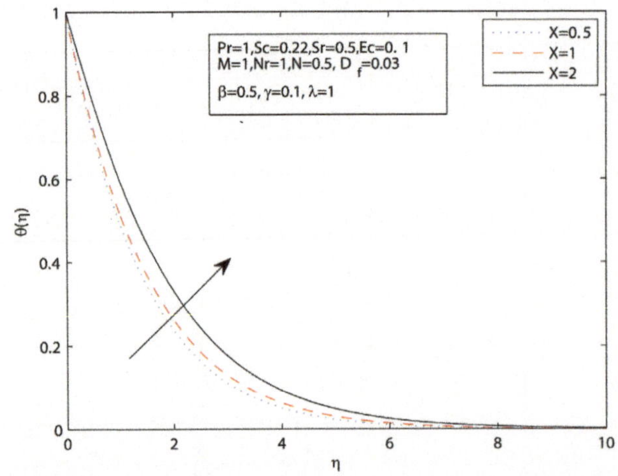

Figure 14. Temperature profile for various values of *X*-location.

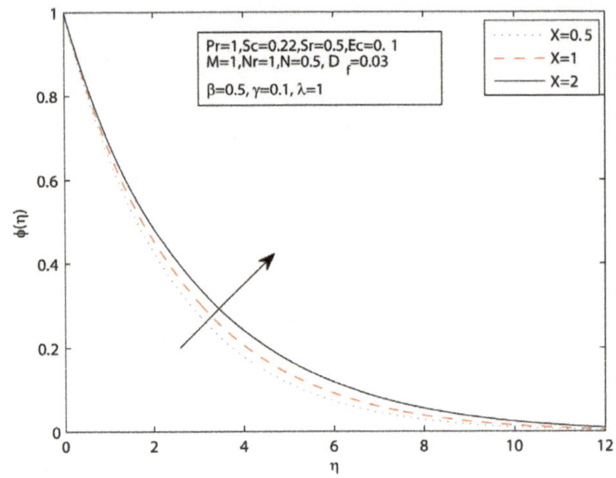

Figure 15. Concentration profile for various values of *X*-location.

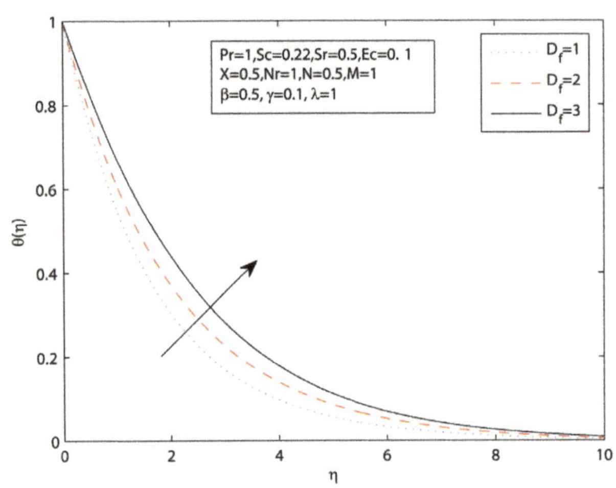

Figure 16. Temperature profile for various values of Dufour number D_f.

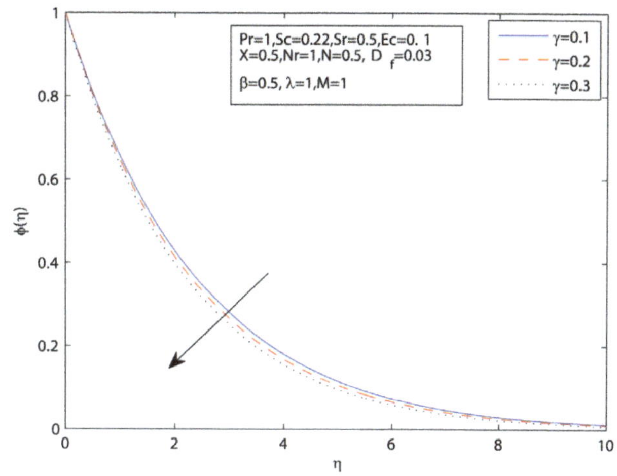

Figure 17. Concentration profile for various values of chemical reaction parameter γ.

parameter γ. Chemical reaction increases the rate of interfacial mass transfer and reduces the local concentration. Therefore the concentration profile decreases with an increase of the parameter γ.

6. Conclusions

The numerical solutions for MHD mixed convection flow of a Casson fluid over an exponentially stretching surface with Soret and Dufour, thermal radiation, chemical reaction are analyzed. The results are presented graphically with various parameters and the values are agreed with previously published work. From the graphical representations, we have the following observations:

- The effect of Casson fluid parameter when treated as fluid with variable plastic dynamic viscosity, the velocity profile decreases, temperature and concentration distributions are increased.
- The temperature increases with increasing values of the radiation parameter and Dufour number D_f.
- An increase in the mixed convection parameter λ increases the velocity profiles, and decreases the temperature profiles and concentration profiles.
- An increase in the X-location reduces the non-dimensional velocity increase in the temperature and concentration distribution.
- With increasing values of chemical reaction parameter γ, the velocity profile decreases.

An increase in the Soret number increases the concentration profile and the boundary layer thickness.

References

[1] Garnier, M. (1962) Magneto Hydrodynamics in Materials Processing. *Philosophical Transactions of the Royal Society A*, **344**, 249-263. http://dx.doi.org/10.1098/rsta.1993.0090

[2] Sakiadas, B.C. (1961) Boundary-Layer Behavior on Continuous Solid Surfaces. *AIChE Journal*, **7**, 26-28. http://dx.doi.org/10.1002/aic.690070108

[3] Crane, L.J. (1970) Flow Past a Stretching Plate. *Zeitschrift für Angewandte Mathematik und Physik*, **21**, 590-595. http://dx.doi.org/10.1007/BF01587695

[4] Magyari, E. and Keller, B. (1999) Heat and Mass Transfer in the Boundary Layers on an Exponentially Stretching Continuous Surface. *Journal of Physics D*, **32**, 577-585. http://dx.doi.org/10.1088/0022-3727/32/5/012

[5] Bidin, B. and Nazar, R. (2009) Numerical Solution of the Boundary Layer Flow over an Exponentially Stretching Sheet with Thermal Radiation. *European Journal of Scientific Research*, **33**, 710-717.

[6] Sajid, M. and Hayat, T. (2008) Influence of Thermal Radiation on the Boundary Layer Flow Due to an Exponentially Stretching Sheet. *International Communications in Heat and Mass Transfer*, **35**, 347-356. http://dx.doi.org/10.1016/j.icheatmasstransfer.2007.08.006

[7] Merrill, F.W. (1965) Rheology of Human Blood and Some Speculations on Its Role in Vascular Homeostatics Biomechanical Mechanisms in Vascular Homeostatics and Intravascular Thrombosis. In: Sawyer, P.N., Ed., *Appleton Cen-*

tury Crofts, New York, 127-137.

[8] MacDonald, D.A. (1979) On Study Flow through Modelled Vascular Stenoses. *Journal of Biomechanics*, **12**, 13-20. http://dx.doi.org/10.1016/0021-9290(79)90004-6

[9] Nadeem, S., Haq, R.U. and Lee, C. (2012) MHD Flow of a Casson Fluid over an Exponentially Shrinking Sheet. *Scientia Iranica*, **19**, 1550-1553. http://dx.doi.org/10.1016/j.scient.2012.10.021

[10] Mukhopadhyay, S., De Ranjan, P., Bhattacharyya, K. and Layek, G.C. (2013) Casson Fluid Flow over an Unsteady Stretching Surface. *Ain Shams Engineering Journal*, **4**, 933-938.

[11] McGregor, J.L. (1970) The Application of the Minimal Energy Hypothesis to a Casson Fluid. *Bulletin of Mathematical Bio Physics*, **32**, 249-262. http://dx.doi.org/10.1007/BF02476889

[12] Partha, M.K., Murthy, P.V.S.N. and Rajasekhar, G.P. (2005) Effect of Viscous Dissipation on the Mixed Convection Heat Transfer from an Exponentially Stretching Surface. *Heat and Mass Transfer*, **41**, 360-366. http://dx.doi.org/10.1007/s00231-004-0552-2

[13] Pal, D. (2010) Mixed Convection Heat Transfer in the Boundary Layers on an Exponentially Stretching Surface with Magnetic Field. *Applied Mathematics and Computation*, **217**, 2356-2369. http://dx.doi.org/10.1016/j.amc.2010.07.035

[14] Kafoussias, N.G. and Williams, E.M. (1995) Thermal-Diffusion and Diffusion-Thermo Effects on Mixed Free-Forced Convective and Mass Transfer Boundary Layer Flow with Temperature Dependent Viscosity. *International Journal of Engineering Science*, **33**, 1369-1384.

[15] Dursunkaya, Z. and Worek, Z.W. (1992) Diffusion Thermo and Thermal Diffusion Effects in Transient and Steady Natural Convection from a Vertical Surface. *International Journal of Heat and Mass Transfer*, **35**, 2060-2065. http://dx.doi.org/10.1016/0017-9310(92)90208-A

[16] Afify, A.A. (2009) Similarity Solution in MHD Effects of Thermal Diffusion and Diffusion-Thermo on Free Convective Heat and Mass Transfer over a Stretching Surface Considering Suction or Injection. *Communication in Non-Linear Science and Numerical Simulation*, **14**, 2202-2214.

[17] Hayat, T., Shehzad, S.A. and Alsaed, A. (2012) Soret and Dufour Effects on MHD Flow of Casson Fluid. *Applied Mathematics and Mechanics*, **33**, 1301-1312. http://dx.doi.org/10.1007/s10483-012-1623-6

[18] Alam, M.S. and Rahman, M.M. (2006) Dufour and Soret Effects on Mixed Convection Flow Past a Vertical Porous Flat Plate with Variable Suction. *Nonlinear Analysis, Modeling and Control*, **11**, 3-12.

[19] Pramanik, S. (2014) Casson Fluid Flow and Heat Transfer Past an Exponentially Porous Stretching Surface in the Presence of Thermal Radiation. *Ain Shams Engineering Journal*, **5**, 205-212. http://dx.doi.org/10.1016/j.asej.2013.05.003

[20] Shateyi, S. and Marewo, G.T. (2014) Numerical Analysis of MHD Stagnation Point Flow of Casson Fluid, Heat and Mass Transfer over a Stretching Sheet. *Proceedings of the 7th International Conference on Finite Differences, Finite Elements, Finite Volumes, Boundary Elements*, Gdansk, 15-17 May 2014, 128-132.

[21] Ishak, A. (2011) MHD Boundary Layer Flow Due to an Exponentially Stretching Sheet with Radiation Effect. *Sains Malaysiana*, **40**, 391-395.

[22] Mukhopadhyay, S. (2013) Effects of Thermal Radiation on Casson Fluid Flow and Heat Transfer over an Unsteady Stretching Surface Subjected to Suction/Blowing, *Chinese Physics B*, **22**, Article ID: 114702. http://dx.doi.org/10.1088/1674-1056/22/11/114702

[23] Shehzad, S.A. and Hayat, T. (2013) Effects of Mass Transfer on MHD Flow of Casson Fluid with Chemical Reaction. *Brazilian Journal of chemical Engineering*, **30**, 187-195. http://dx.doi.org/10.1590/S0104-66322013000100020

[24] Cebeci, T. and Bradshaw, P. (1984) Physical and Computational Aspects of Convective Heat Transfer. Springer, Berlin Heidelberg. http://dx.doi.org/10.1007/978-3-662-02411-9

13

A Quaternion Solution of the Motion in a Central Force Field Relative to a Rotating Reference Frame

Ioan-Adrian Ciureanu[1], Daniel Condurache[2]

[1]Department of Medical Informatics and Biostatistics, University of Medicine and Pharmacy "Gr.T. Popa", Iasi, Romania

[2]Department of Theoretical Mechanics, Technical University of Iasi, Iasi, Romania

Email: adrian.ciureanu@umfiasi.ro, daniel.condurache@gmail.com

Abstract

The paper presents a quaternion approach of giving a closed form solution of the motion in a central force field relative to a rotating reference frame. This new method involves two quaternion operators: the first one transforms the motion from a non-inertial reference frame to a inertial one with a very significant consequence of vanishing all the non-inertial terms (Coriolis and centripetal forces); the second quaternion operator provides the solution of the motion in the non-inertial reference frame by applying it to the solution in the inertial reference frame. This process will govern the inverse transformation of the motion and is proved on two particular cases, the Foucault Pendulum and Keplerian motions problems relative to rotating reference frames.

Keywords

Quaternion, Rotating Reference Frame, Foucault Pendulum Motion, Keplerian Motion

1. Introduction

The present paper presents a quaternion solution of the motion in a central force field relative to a rotating reference frame. It starts from the main Cauchy problem stated below:

$$\begin{aligned} &\ddot{\boldsymbol{r}} + 2\boldsymbol{\omega}\times\dot{\boldsymbol{r}} + \boldsymbol{\omega}\times(\boldsymbol{\omega}\times\boldsymbol{r}) + \dot{\boldsymbol{\omega}}\times\boldsymbol{r} + f(r)\frac{\boldsymbol{r}}{r} = 0, \quad \boldsymbol{\omega}\in V_3^{\mathbb{R}} \\ &\boldsymbol{r}(0) = \boldsymbol{r}_0, \quad \dot{\boldsymbol{r}}(0) = \boldsymbol{v}_0 \end{aligned} \tag{1.1}$$

where $\boldsymbol{\omega}$ is a differentiable vectorial map and r is the magnitude of vector $\boldsymbol{r}$.

The quaternion method which will be presented in this paper involves two quaternion operators from which the first one transforms the non-linear with variable coefficients initial value problem (1.1) in another one without the coefficients and the second quaternion operator, applied to the solution of the last problem, will provide the time-explicit closed form solutions for two specific cases, Foucault Pendulum and Keplerian motion problem when $\boldsymbol{\omega}$ has a fixed direction.

The structure of this paper consists of the following four parts. Section 2 starts with a brief presentation of the quaternion algebra and continues with the presentation of Darboux problem in quaternion form in order to prepare the defining of the quaternion operators.

The next section represents the core of the paper because there the quaternion operators are defined, but not before the transformation in the quaternion form of the Equation (1.1) to be done.

Section 4 proves the accuracy of the method of using quaternion operators for computing the time-explicit closed form solutions for two particular cases, the Foucault Pendulum and Keplerian motions problems in rotating reference frame.

2. Mathematical Preliminaries

2.1. Algebra of Quaternions

The quaternions were invented by William Rowan Hamilton in 1843 [1]. A quaternion can be written as a linear combination:

$$\boldsymbol{q} = w + x\boldsymbol{i} + y\boldsymbol{j} + z\boldsymbol{k} \tag{2.1}$$

where w, x, y, z are the *constituents* of the quaternion and $\boldsymbol{i}$, $\boldsymbol{j}$, $\boldsymbol{k}$ are the *imaginary units*. The multiplication of two quaternions satisfies the fundamental rules introduced by Hamilton:

$$\boldsymbol{i}^2 = \boldsymbol{j}^2 = \boldsymbol{k}^2 = -1; \quad \boldsymbol{ij} = -\boldsymbol{ji} = \boldsymbol{k}; \quad \boldsymbol{jk} = -\boldsymbol{kj} = \boldsymbol{i}; \quad \boldsymbol{ki} = -\boldsymbol{ik} = \boldsymbol{j} \tag{2.2}$$

For the quaternion $\boldsymbol{q}$, w is the first constituent and it's named "*the real part*" and *x*, *y*, *z* form the vector part of the same quaternion. We can use the quaternions when we need to model rotations, especially in the case of the motion of the rigid body around a fixed point. A quaternion can also be noted as:

$$\boldsymbol{q} = (a_0, \boldsymbol{a}) \tag{2.3}$$

where a_0 is a real number and $\boldsymbol{a}$ is a vector. In this case, a_0 is named *the real part* of $\boldsymbol{q}$ and $\boldsymbol{a}$ is *the vector part* of $\boldsymbol{q}$. A quaternion with zero real part called *vector quaternion.*

The set of quaternions is denoted by $\mathbb{H}$ and is a noncommutative; associative four dimensional division algebra with respect to the scalar multiplication, quaternion sum and quaternion product, defined as:

$$\begin{cases} \lambda(a_0, \boldsymbol{a}) = (\lambda a_0, \lambda \boldsymbol{a}); \\ \boldsymbol{q}_1 = (a_0, \boldsymbol{a}); \quad \boldsymbol{q}_2 = (b_0, \boldsymbol{b}); \\ \boldsymbol{q}_1 + \boldsymbol{q}_2 = (a_0 + b_0, \boldsymbol{a} + \boldsymbol{b}); \\ \boldsymbol{q}_1\boldsymbol{q}_2 = (a_0 b_0 - \boldsymbol{a}\cdot\boldsymbol{b}, a_0\boldsymbol{b} + b_0\boldsymbol{a} + \boldsymbol{a}\times\boldsymbol{b}). \end{cases} \tag{2.4}$$

with $\cdot$ being the vector dot product and $\times$ representing the vector cross product.

We already know that an algebra is a vector space where the product may be defined as an additional internal operation. Also, the dimension of an algebra is the algebraic dimension of the vector space. We will define a division algebra as an algebra where the division operation is possible. So, for any $\boldsymbol{a}$ and $\boldsymbol{b}$, with $\boldsymbol{b} \neq \boldsymbol{0}$, there are two unique elements $\boldsymbol{x}$ and $\boldsymbol{y}$ in the algebra, as:

$$\boldsymbol{a} = \boldsymbol{xb}; \quad \boldsymbol{a} = \boldsymbol{by}. \tag{2.5}$$

We will denote with $\boldsymbol{q}^*$ the conjugate of the quaternion $\boldsymbol{q}$ from (2.3), the conjugate being defined as $\boldsymbol{q}^* = (a_0, -a)$. The norm of the quaternion $\boldsymbol{q} = (a_0, \boldsymbol{a})$ is given by:

$$\|\boldsymbol{q}\| \overset{\text{def}}{=} \sqrt{\boldsymbol{qq}^*} = \sqrt{a_0^2 + |\boldsymbol{a}|^2} \tag{2.6}$$

when $|\boldsymbol{a}|$ is the magnitude of vector $\boldsymbol{a}$. We will denote with $\boldsymbol{a}=(0,\boldsymbol{a})$ and $\boldsymbol{b}=(0,\boldsymbol{b})$ two vectors and their corresponding vector quaternion form. We also know that the vector dot product and cross product may be expressed in a quaternion way as below:

$$\begin{cases} \boldsymbol{a}\cdot\boldsymbol{b} = -\frac{1}{2}(\boldsymbol{ab}+\boldsymbol{ba}); \\ \boldsymbol{a}\times\boldsymbol{b} = \frac{1}{2}(\boldsymbol{ab}-\boldsymbol{ba}). \end{cases} \tag{2.7}$$

We can describe the motion of a particle on a sphere with a constant radius with the help of time-depending quaternions such as:

$$\boldsymbol{r}(t)=\boldsymbol{q}(t)\boldsymbol{r}_0\boldsymbol{q}^*(t) \tag{2.8}$$

where $\boldsymbol{r}=\boldsymbol{r}(t)$ is the vector quaternion that models the motion, $\boldsymbol{r}_0$ is a constant vector quaternion and $\boldsymbol{q}(t)$ is a time-depending quaternion with $\|\boldsymbol{q}(t)\|=1$. The next equation will describe the finite rotation with an angle $\alpha\in[0,2\pi)$ of the vector $\boldsymbol{r}_0$ around the axis whose orientation is modeled by the vector quaternion $\boldsymbol{u}$ with $\|\boldsymbol{u}\|=1$:

$$\boldsymbol{r}=\boldsymbol{q}(\boldsymbol{u},\alpha)\boldsymbol{r}_0\boldsymbol{q}^*(\boldsymbol{u},\alpha) \tag{2.9}$$

where $\boldsymbol{q}(\boldsymbol{u},\alpha)$ has the form as:

$$\boldsymbol{q}(\boldsymbol{u},\alpha)=\left(\cos\frac{\alpha}{2},\boldsymbol{u}\sin\frac{\alpha}{2}\right). \tag{2.10}$$

2.2. Darboux Equation in Quaternion Shape

It is well known that in rigid body kinematics, we need to describe the instantaneous rotation when we know the angular velocity [2]. The common solution is to use the Riccati differential equation which describes the instantaneous rotation of a rigid body when the instantaneous angular velocity is given [3].

If $\boldsymbol{R}$ is the rotation matrix, the rotation with angular velocity $\boldsymbol{\omega}$ of a constant vector $\boldsymbol{r}_0$ is described by [4]

$$\boldsymbol{r}=\boldsymbol{R}\boldsymbol{r}_0 \tag{2.11}$$

If a vector $\boldsymbol{\omega}$ is represented in Cartesian coordinates with respect to the orthonormal right oriented basis $\{\boldsymbol{e}_1,\boldsymbol{e}_2,\boldsymbol{e}_3\}$,

$$\boldsymbol{\omega}=\omega_1\boldsymbol{e}_1+\omega_2\boldsymbol{e}_2+\omega_3\boldsymbol{e}_3 \tag{2.12}$$

and if the matrix $\tilde{\boldsymbol{\omega}}$ is related to the vector $\boldsymbol{\omega}$ as below

$$\tilde{\boldsymbol{\omega}}=\begin{bmatrix} 0 & -\omega_3 & \omega_2 \\ \omega_3 & 0 & -\omega_1 \\ -\omega_2 & \omega_1 & 0 \end{bmatrix} \tag{2.13}$$

the instantaneous angular velocity vector $\boldsymbol{\omega}$ associated to the proper orthogonal valued function is defined by

$$\tilde{\boldsymbol{\omega}}=\dot{\boldsymbol{R}}\boldsymbol{R}^{\mathrm{T}}. \tag{2.14}$$

The rotation matrix that models the rotation with a given instantaneous angular velocity $\boldsymbol{\omega}$ is given by the solution to the Darboux equation represented below in the matrix shape:

$$\dot{\boldsymbol{R}}=\tilde{\boldsymbol{\omega}}\boldsymbol{R};\quad \boldsymbol{R}(0)=\boldsymbol{I}_3 \tag{2.15}$$

where $t=0$ is the initial moment of time and $\boldsymbol{R}$ is a 3×3 matrix whose elements are differentiable scalar functions.

The rotation matrix $\boldsymbol{R}$ associated with vector $\boldsymbol{\omega}$ is the solution of the initial value problem (2.15) and it is a proper orthogonal matrix function with the following properties:

$$\boldsymbol{R}\boldsymbol{R}^{\mathrm{T}}=\boldsymbol{I}_3;\quad \det\boldsymbol{R}=1. \tag{2.16}$$

Consider $\boldsymbol{\omega}$ the vector quaternion corresponding to the instantaneous angular velocity vector and $\boldsymbol{q}$ the unit quaternion that models the rotation. The quaternion operator defined as below rotates any constant vector $\boldsymbol{r}_0$ with instantaneous angular velocity $\boldsymbol{\omega}$.

$$\boldsymbol{F}_\omega \overset{\text{def}}{=} \boldsymbol{q}(\ )\boldsymbol{q}^*. \tag{2.17}$$

From Equation (2.14), it results that:

$$\tilde{\boldsymbol{\omega}} \times \boldsymbol{r}_0 = \dot{\boldsymbol{F}}_\omega\left(\boldsymbol{F}_\omega^{-1}\boldsymbol{r}_0\right) \tag{2.18}$$

and using vector quaternions property (2.7) we will rewrite (2.18) as

$$\frac{1}{2}(\boldsymbol{\omega r}_0 - \boldsymbol{r}_0\boldsymbol{\omega}) = \dot{\boldsymbol{q}}\left(\boldsymbol{q}^*\boldsymbol{r}_0\boldsymbol{q}\right)\boldsymbol{q}^* + \boldsymbol{q}\left(\boldsymbol{q}^*\boldsymbol{r}_0\boldsymbol{q}\right)\dot{\boldsymbol{q}}^* = \dot{\boldsymbol{q}}\boldsymbol{q}^*\boldsymbol{r}_0 + \boldsymbol{r}_0\boldsymbol{q}\dot{\boldsymbol{q}}^*. \tag{2.19}$$

Due to the fact that $\boldsymbol{r}_0$ is an arbitrary constant vector quaternion, from (2.19) results that the unit quaternion $\boldsymbol{q}$, which describes the rotation with angular velocity $\boldsymbol{\omega}$, is the solution to the following Darboux-like equation:

$$\dot{\boldsymbol{q}} = \frac{\boldsymbol{\omega}}{2}\boldsymbol{q}; \quad \boldsymbol{q}(0) = \hat{\mathbf{1}} \tag{2.20}$$

where $\hat{\mathbf{1}} = (1,0)$ is a unit quaternion. In this case, from Equation (2.15), $\boldsymbol{\omega}$ is the vector quaternion associated with vector $\boldsymbol{\omega}$.

Using (2.15) and the expression of $\boldsymbol{F}_\omega$ from (2.17), it follows that the continuous rotation with instantaneous angular velocity modeled by the vector quaternion $-\boldsymbol{\omega}$ is the solution to the quaternion initial value problem:

$$\dot{\boldsymbol{q}} + \frac{\boldsymbol{\omega}}{2}\boldsymbol{q} = 0; \quad \boldsymbol{q}(0) = \hat{\mathbf{1}} \tag{2.21}$$

3. The Solutions of the Motion in a Central Force Field Relative to a Rotating Reference Frame

In order to find the solutions of the equations specific to the motions in a central force field relative to a rotating reference frame, two reciprocal transformations will be done: first, the motion in the non-inertial reference frame will be transformed in a inertial one through the quaternion operator $\boldsymbol{F}_\omega$. Then will be proved that the solution of the equation specific to the non-inertial reference frame results very easy by applying the quaternion operator $\boldsymbol{R}_{-\omega}$ to the solution specific to the inertial reference frame where $\boldsymbol{R}_{-\omega} = \left(\boldsymbol{F}_\omega\right)^{-1}$.

Quaternionic Operator

In this section, a quaternion operator $\boldsymbol{F}_\omega$ will be defined in order to determine the solution of the below non-linear initial value problem which describes the motion in a central force field relative to a rotating reference frame. The first step is to recall the Cauchy problem specific to the motion in a central force field:

$$\begin{cases} \ddot{\boldsymbol{r}} + 2\boldsymbol{\omega}\times\dot{\boldsymbol{r}} + \boldsymbol{\omega}\times(\boldsymbol{\omega}\times\boldsymbol{r}) + \dot{\boldsymbol{\omega}}\times\boldsymbol{r} + f(r)\dfrac{\boldsymbol{r}}{r} = \mathbf{0}; \quad \boldsymbol{\omega}\in V_3^{\mathbb{R}}; \\ \boldsymbol{r}(0) = \boldsymbol{r}_0, \\ \dot{\boldsymbol{r}}(0) = \boldsymbol{v}_0. \end{cases} \tag{3.1}$$

Knowing that $\boldsymbol{\omega}$ is a differentiable vectorial value map, r is the magnitude of vector $\boldsymbol{r}$, and $f:\mathbb{R}_+ \to \mathbb{R}$ is a continous real valued map. Using (2.7), the last equation becomes:

$$\ddot{\boldsymbol{r}} + 2\frac{1}{2}(\boldsymbol{\omega}\dot{\boldsymbol{r}} - \dot{\boldsymbol{r}}\boldsymbol{\omega}) + \frac{1}{2}\boldsymbol{\omega}\times(\boldsymbol{\omega r} - \boldsymbol{r\omega}) + \frac{1}{2}\left(\dot{\boldsymbol{\omega}}\boldsymbol{r} - \boldsymbol{r}\dot{\boldsymbol{\omega}}\right) + f(r)\frac{\boldsymbol{r}}{r} = \mathbf{0}; \quad \boldsymbol{\omega}\in V_3^{\mathbb{R}} \tag{3.2}$$

and further,

$$\ddot{\boldsymbol{r}} + (\boldsymbol{\omega}\dot{\boldsymbol{r}} - \dot{\boldsymbol{r}}\boldsymbol{\omega}) + \frac{1}{4}\left(\boldsymbol{\omega}^2\boldsymbol{r} + \boldsymbol{r}\boldsymbol{\omega}^2 - 2\boldsymbol{\omega r \omega}\right) + \frac{1}{2}\left(\dot{\boldsymbol{\omega}}\boldsymbol{r} - \boldsymbol{r}\dot{\boldsymbol{\omega}}\right) + f(r)\frac{\boldsymbol{r}}{r} = \mathbf{0}; \quad \boldsymbol{\omega}\in V_3^{\mathbb{R}} \tag{3.3}$$

Now, the following quaternion operator $\boldsymbol{F}_\omega$ is defined as:

$$\boldsymbol{F}_\omega \overset{\text{def}}{=} \boldsymbol{q}^*(\ )\boldsymbol{q} \tag{3.4}$$

where $\boldsymbol{q}$ is the solution of the following equation:

$$\dot{\boldsymbol{q}} + \frac{\boldsymbol{\omega}}{2}\boldsymbol{q} = 0; \quad \boldsymbol{q}(0) = \hat{1} \tag{3.5}$$

If $\boldsymbol{\omega} \in V_3^*$, then the Equations (3.4) and (3.5) determines the following properties:

1. For any quaternions $\boldsymbol{a}$ and $\boldsymbol{b}$ and scalars λ_1 and λ_2, the operator $\boldsymbol{F}_\omega$ is linear, *i.e.*:

$$\boldsymbol{F}_\omega(\lambda_1\boldsymbol{a} + \lambda_2\boldsymbol{b}) = \lambda_1\boldsymbol{F}_\omega\boldsymbol{a} + \lambda_2\boldsymbol{F}_\omega\boldsymbol{b}; \tag{3.6}$$

2. For any quaternions $\boldsymbol{a}$ and $\boldsymbol{b}$, the operator $\boldsymbol{F}_\omega$ preserves the quaternionic product *i.e.*

$$\boldsymbol{F}_\omega(\boldsymbol{ab}) = (\boldsymbol{F}_\omega\boldsymbol{a})(\boldsymbol{F}_\omega\boldsymbol{b}) \tag{3.7}$$

3. For any quaternion $\boldsymbol{a}$, the operator $\boldsymbol{F}_\omega$ preserves the quaternionic norm *i.e.*

$$\|\boldsymbol{F}_\omega\boldsymbol{a}\| = \|\boldsymbol{a}\| \tag{3.8}$$

4. If $\boldsymbol{r}$ is a quaternion-valued function of a real variable, the derivative with respect to time of $\boldsymbol{F}_\omega\boldsymbol{r}$ is:

$$\dot{\boldsymbol{F}}_\omega(\boldsymbol{r}) = \boldsymbol{F}_\omega\left(\dot{\boldsymbol{r}} + \frac{1}{2}(\boldsymbol{\omega r} - \boldsymbol{r\omega})\right); \quad \forall \boldsymbol{r} \in V_3^* \tag{3.9}$$

5. If $\boldsymbol{r}$ is a quaternion-valued function of a real variable, the second derivative with respect to time of $\boldsymbol{F}_\omega\boldsymbol{r}$ is:

$$\ddot{\boldsymbol{F}}_\omega(\boldsymbol{r}) = \boldsymbol{F}_\omega\left(\ddot{\boldsymbol{r}} + (\boldsymbol{\omega}\dot{\boldsymbol{r}} - \dot{\boldsymbol{r}}\boldsymbol{\omega}) + \frac{1}{4}(\boldsymbol{\omega}^2\boldsymbol{r} + \boldsymbol{r}\boldsymbol{\omega}^2 - 2\boldsymbol{\omega r \omega}) + \frac{1}{2}\left(\dot{\boldsymbol{\omega}}\boldsymbol{r} - \boldsymbol{r}\dot{\boldsymbol{\omega}}\right)\right); \quad \forall \boldsymbol{r} \in V_3^* \tag{3.10}$$

6. $\boldsymbol{F}_\omega$ is invertible and it's inverse is denoted with $\boldsymbol{R}_{-\omega}$ *i.e.*:

$$\boldsymbol{R}_{-\omega} = (\boldsymbol{F}_\omega)^{-1} = \boldsymbol{q}(\ )\boldsymbol{q}^* \tag{3.11}$$

where $\boldsymbol{R}_{-\omega}$ describes the rotation with the angular velocity $-\boldsymbol{\omega}$ which corresponds to the vector quaternion $-\boldsymbol{\omega}$ and $\boldsymbol{q}$ is the solution of Equation (3.5).

Theorem 3.1.

The solution of the Cauchy problem:

$$\begin{cases} \ddot{\boldsymbol{r}} + 2\boldsymbol{\omega}\times\dot{\boldsymbol{r}} + \boldsymbol{\omega}\times(\boldsymbol{\omega}\times\boldsymbol{r}) + \dot{\boldsymbol{\omega}}\times\boldsymbol{r} + f(r)\frac{\boldsymbol{r}}{r} = \boldsymbol{0}; \quad \boldsymbol{\omega} \in V_3^{\mathbb{R}}, \\ \boldsymbol{r}(0) = r_0, \\ \dot{\boldsymbol{r}}(0) = v_0 \end{cases} \tag{3.12}$$

will be obtained by applying the quaternion operator $\boldsymbol{R}_{-\omega}$, *to the solution of the Cauchy problem*:

$$\begin{cases} \ddot{\boldsymbol{r}} + f(r)\frac{\boldsymbol{r}}{r} = 0 \\ \boldsymbol{r}(0) = \boldsymbol{r}_0 \\ \dot{\boldsymbol{r}}(0) = \boldsymbol{v}_0 + \boldsymbol{\omega}_0\times\boldsymbol{r}_0 \end{cases} \tag{3.13}$$

Proof. If we apply $\boldsymbol{F}_\omega$ to the Equation (3.3), it results:

$$\boldsymbol{F}_\omega\left[\ddot{\boldsymbol{r}} + (\boldsymbol{\omega}\dot{\boldsymbol{r}} - \dot{\boldsymbol{r}}\boldsymbol{\omega}) + \frac{1}{4}(\boldsymbol{\omega}^2\boldsymbol{r} + \boldsymbol{r}\boldsymbol{\omega}^2 - 2\boldsymbol{\omega r\omega}) + \frac{1}{2}\left(\dot{\boldsymbol{\omega}}\boldsymbol{r} - \boldsymbol{r}\dot{\boldsymbol{\omega}}\right)\right] + \boldsymbol{F}_\omega\left[f(r)\frac{\boldsymbol{r}}{r}\right] = \boldsymbol{0}; \quad \boldsymbol{\omega} \in V_3^{\mathbb{R}}; \quad \omega_0 \in \mathbb{R}_+^*$$

$$\text{with} \begin{cases} \boldsymbol{F}_\omega(\boldsymbol{r})\big|_{t=0} = \boldsymbol{r}_0 \\ \dot{\boldsymbol{F}}_\omega(\boldsymbol{r})\big|_{t=0} = \boldsymbol{v}_0 + \boldsymbol{\omega}_0\times\boldsymbol{r}_0 \end{cases} \quad \text{where } \boldsymbol{r}_0 = \boldsymbol{r}(0);\ \boldsymbol{v}_0 = \dot{\boldsymbol{r}}(0);\ \boldsymbol{\omega}_0 = \boldsymbol{\omega}(0) \tag{3.14}$$

Using the Equation (3.10), it results that:

$$\ddot{\boldsymbol{F}}_{\omega}(\boldsymbol{r})+\frac{f(r)}{r}\boldsymbol{F}_{\omega}(\boldsymbol{r})=0 \tag{3.15}$$

Replacing $\boldsymbol{F}_{\omega}(\boldsymbol{r})$ with $\boldsymbol{r}$ and using the Equation (3.11), it results:

$$\begin{aligned}&\ddot{\boldsymbol{r}}+f(r)\frac{\boldsymbol{r}}{r}=0\\&\boldsymbol{r}(0)=\boldsymbol{F}_{\omega}(\boldsymbol{r})\big|_{t=0}=\boldsymbol{r}_0\\&\dot{\boldsymbol{r}}(0)=\dot{\boldsymbol{F}}_{\omega}(\boldsymbol{r})\big|_{t=0}=\boldsymbol{v}_0+\boldsymbol{\omega}_0\times\boldsymbol{r}_0\end{aligned} \tag{3.16}$$

Consequently, by using the quaternion operator $\boldsymbol{F}_{\omega}$, the complex problem given by the non-linear initial value problem with variable coefficients described by Equation (3.1) is reduced to the finding the solution of Equation (3.16) which describes the motion in a central force field, with $\boldsymbol{\omega}$ being the instantaneous angular velocity of the rotating reference frame. Thereby, the movement in the non-inertial reference frame is transformed to an inertial one and all non-inertial coefficients within Equation (3.1) are canceled. The solution of Equation (3.1) will be obtained by applying the quaternion operator $\boldsymbol{R}_{-\omega}$, to the solution of the problem (3.16).

In the next sections will be studied two particular cases of motions in central force field: the Foucault Pendulum and the Kepler's motions relative to a rotating reference frame problems.

4. Study of Particular Cases: Foucault Pendulum and Keplerian Motion Problems in Rotating Reference Frames

This section presents the methods adequate to the very known two topics: the Foucault Pendulum and Keplerian motion problems relative to a rotating reference frame problems. In order to achieve the goal of this paper, the motion in central force field Equation (1.1) will be particularized for these two specific cases giving for each of them the characteristic eqaution of $f(r)$ and the quaternion operator $\boldsymbol{R}_{-\omega}$, will be used as presented in last section.

4.1. Foucault Pendulum Problem

The Foucault Pendulum motion is described by the below initial value problem which is a particular form of the Equation (1.1) that coresponds to a spatial harmonic oscillator relative to a rotating reference frame, with

$$f(r)=\omega_*^2\boldsymbol{r},$$

$$\begin{cases}\ddot{\boldsymbol{r}}+2\boldsymbol{\omega}\times\dot{\boldsymbol{r}}+\boldsymbol{\omega}\times(\boldsymbol{\omega}\times\boldsymbol{r})+\dot{\boldsymbol{\omega}}\times\boldsymbol{r}+\omega_*^2\boldsymbol{r}=\boldsymbol{0}; \quad \boldsymbol{\omega}\in V_3^{\mathbb{R}}; \quad \omega_*>0,\\ \boldsymbol{r}(0)=r_0,\\ \dot{\boldsymbol{r}}(0)=v_0\end{cases} \tag{4.1}$$

where $\boldsymbol{r}$ is the position vector, $\boldsymbol{\omega}$ represents the angular velocity of the rotating reference frame and is a diferential vectorial map and at last but not the least important, ω_* is the pulsation of the pendulum which depends on both the gravitational acceleration at the place of the experiment and the length of the pendulum.

Applying the quaternion operator $\boldsymbol{F}_{\omega}$, the Equation (4.1), we will produce the below initial value problem:

$$\begin{aligned}&\ddot{\boldsymbol{r}}+\omega_*^2\boldsymbol{r}=0,\\&\boldsymbol{r}(0)=\boldsymbol{r}_0,\\&\dot{\boldsymbol{r}}(0)=\boldsymbol{v}_0+\boldsymbol{\omega}_0\times\boldsymbol{r}_0.\end{aligned} \tag{4.2}$$

The Equation (4.2) models the spatial harmonic oscillator and it's solution is:

$$\boldsymbol{r}(t)=\boldsymbol{r}_0\cos\omega_*t+\frac{\boldsymbol{v}_0+\boldsymbol{\omega}_0\times\boldsymbol{r}_0}{\omega_*}\sin\omega_*t \tag{4.3}$$

Due to the Theorem 3.1., the solution of the initial value problem

$$\begin{cases} \ddot{\boldsymbol{r}} + 2\boldsymbol{\omega}\times\dot{\boldsymbol{r}} + \boldsymbol{\omega}\times(\boldsymbol{\omega}\times\boldsymbol{r}) + \dot{\boldsymbol{\omega}}\times\boldsymbol{r} + \omega_*^2\boldsymbol{r} = \boldsymbol{0}, \\ \boldsymbol{r}(0) = \boldsymbol{r}_0, \quad \dot{\boldsymbol{r}}(0) = \boldsymbol{v}_0. \end{cases} \tag{4.4}$$

results from applying the the quaternion operator $\boldsymbol{R}_{-\omega}$ to the solution (4.3) of the Cauchy problem (4.2) as below:

$$\boldsymbol{r}(t) = (\boldsymbol{R}_{-\omega}\boldsymbol{r}_0)\cos\omega_* t + \frac{1}{\omega_*}\boldsymbol{R}_{-\omega}(\boldsymbol{v}_0 + \boldsymbol{\omega}_0\times\boldsymbol{r}_0)\sin\omega_* t \tag{4.5}$$

The solution of Equation (4.5) coresponds to a harmonic planar oscillation (with ω_* being the pulsation of the pendulum) composed with a precession of $-\boldsymbol{\omega}$ angular velocity of the oscillation plane [5].

In order to compute the closed form solutions of Equation (4.1), we must recall that we've assumed that the direction of the vector $\boldsymbol{\omega}$ associated with the quaternion $\boldsymbol{\omega}$ is considered to be fixed $\boldsymbol{\omega}(t) = \omega(t)\boldsymbol{u}$ where $\boldsymbol{u}$ is a constant unit vector with $\omega : [0, +\infty) \to \mathbb{R}$ and, from Eqaution (3.11), that the quaternion operator $\boldsymbol{R}_{-\omega} = \boldsymbol{q}(\ )\boldsymbol{q}^*$. Consequently,

$$\begin{aligned} \boldsymbol{r}(t) = {} & \cos\omega_* t \times \left\{ \frac{\boldsymbol{r}_0\cdot\boldsymbol{\omega}}{\omega^2}\boldsymbol{\omega} - \frac{\sin\varphi(t)}{\omega}\tilde{\boldsymbol{\omega}}\boldsymbol{r}_0 - \frac{\cos\varphi(t)}{\omega^2}\tilde{\boldsymbol{\omega}}^2\boldsymbol{r}_0 \right\} + \frac{\sin\omega_* t}{\omega_*} \times \left\{ \frac{\boldsymbol{v}_0\cdot\boldsymbol{\omega}}{\omega^2}\boldsymbol{\omega} - \frac{\sin\varphi(t)}{\omega}\tilde{\boldsymbol{\omega}}(\boldsymbol{v}_0 + \boldsymbol{\omega}_0\times\boldsymbol{r}_0) \right\} \\ & - \frac{\sin\omega_* t}{\omega_*}\frac{\cos\varphi(t)}{\omega^2}\tilde{\boldsymbol{\omega}}^2(\boldsymbol{v}_0 + \boldsymbol{\omega}_0\times\boldsymbol{r}_0). \end{aligned} \tag{4.6}$$

with $\varphi(t) = \int_0^t \omega(\xi)\,\mathrm{d}\xi$.

If we'll note:

$$\boldsymbol{\Phi}_* = \boldsymbol{r}_0\cos\omega_* t + \frac{\boldsymbol{v}_0 + \boldsymbol{\omega}_0\times\boldsymbol{r}_0}{\omega_*}\sin\omega_* t \tag{4.7}$$

than the Equation (4.6) can be rewritten as following:

$$\boldsymbol{r}(t) = \frac{\boldsymbol{\omega}\cdot\boldsymbol{\Phi}_*}{\omega^2}\boldsymbol{\omega} - \frac{\sin\varphi(t)}{\omega}\boldsymbol{\omega}\times\boldsymbol{\Phi}_* - \frac{\cos\varphi(t)}{\omega^2}\boldsymbol{\omega}\times(\boldsymbol{\omega}\times\boldsymbol{\Phi}_*) \tag{4.8}$$

In conclusion, when the direction of the vector $\boldsymbol{\omega}$ associated with the quaternion $\boldsymbol{\omega}$ is considered to be fixed, the motion is a harmonic oscillation described by the Equation (4.3) in a plane that has a fixed point and a precession with the angular velocity $-\boldsymbol{\omega}$.

4.2. Kepler's Problem in Rotating Reference Frame

The Keplerian motion in a rotating reference frame that rotates with the angular velocity $\boldsymbol{\omega}$ is described by the following linear initial value problem which is a particular form of the Equation (3.1) with $f(r) = \frac{\mu}{r^2}$

$$\begin{cases} \ddot{\boldsymbol{r}} + 2\boldsymbol{\omega}\times\dot{\boldsymbol{r}} + \boldsymbol{\omega}\times(\boldsymbol{\omega}\times\boldsymbol{r}) + \dot{\boldsymbol{\omega}}\times\boldsymbol{r} + \frac{\mu}{r^3}\boldsymbol{r} = \boldsymbol{0}; \quad \boldsymbol{\omega}\in V_3^{\mathbb{R}}; \quad \mu > 0, \\ \boldsymbol{r}(0) = \boldsymbol{r}_0, \\ \dot{\boldsymbol{r}}((0) = \boldsymbol{v}_0. \end{cases} \tag{4.9}$$

where $\boldsymbol{r}$ is the position vector of the body related to the attraction center, $\boldsymbol{\omega}$ represents the angular velocity of the rotating reference frame and is a differential vectorial map and μ is a constant with $\mu = kM$ where k is the universal attraction constant and M is the mass of the attraction center.

It was proved in the second section that the solution to the Cauchy problem is obtained by applying the quaternion operator $\boldsymbol{R}_{-\omega}$ to the solution of the following Cauchy problem:

$$\begin{aligned}&\ddot{\boldsymbol{r}}+\frac{\mu}{r^3}\boldsymbol{r}=0,\\&\boldsymbol{r}(0)=\boldsymbol{r}_0,\\&\dot{\boldsymbol{r}}(0)=\boldsymbol{v}_0+\boldsymbol{\omega}_0\times\boldsymbol{r}_0.\end{aligned}\tag{4.10}$$

The Equation (4.10) describes a typical Keplerian motionunder certain conditions.
In the particular case of negative specific energy, the solution of (4.11) is: [6] [7]

$$\boldsymbol{r}(t)=\left[\cos E(t)-e\right]\boldsymbol{a}_0+\sin E(t)\boldsymbol{b}_0\tag{4.11}$$

where $t\geq 0$

In the Equation (4.11), the coefficients $\boldsymbol{a}_0$ and $\boldsymbol{b}_0$ are the vectorial semimajor and, respectively, the semi-minor axes of the elliptical inertial trajectory, $\boldsymbol{e}_0$ is the vectorial eccentricity of the trajectory with e being its magnitude and constant n is named mean motion as below:

$$\begin{aligned}\boldsymbol{a}_0&=\begin{cases}\dfrac{\mu}{2e|h|}\boldsymbol{e}_0, & \text{if } \boldsymbol{e}_0\neq 0.\\ \boldsymbol{r}_0, & \text{if } \boldsymbol{e}_0=0.\end{cases}\\ \boldsymbol{b}_0&=\begin{cases}\dfrac{1}{e\sqrt{2|h|}}\boldsymbol{\Omega}_0\times\boldsymbol{e}_0, & \text{if } \boldsymbol{e}_0\neq 0.\\ \dfrac{1}{n}\left(\boldsymbol{v}_0+\boldsymbol{\omega}_0\times\boldsymbol{r}_0\right), & \text{if } \boldsymbol{e}_0=0.\end{cases}\end{aligned}\tag{4.12}$$

where the specific energy is noted with h and is equal with:

$$h=\frac{1}{2}\left(\boldsymbol{v}_0+\boldsymbol{\omega}_0\times\boldsymbol{r}_0\right)^2-\frac{\mu}{r_0}<0\tag{4.13}$$

and the specific angular momentum of the inertial trajectory is noted with $\boldsymbol{\Omega}_0$ and equal to:

$$\boldsymbol{\Omega}_0=\boldsymbol{r}_0\times\left(\boldsymbol{v}_0+\boldsymbol{\omega}_0\times\boldsymbol{r}_0\right).\tag{4.14}$$

The eccentricity of the trajectory is given by:

$$\boldsymbol{e}_0=\frac{\left(\boldsymbol{v}_0+\boldsymbol{\omega}\times\boldsymbol{r}_0\right)\times\left[\boldsymbol{r}_0\times\left(\boldsymbol{v}_0+\boldsymbol{\omega}_0\times\boldsymbol{r}_0\right)\right]}{\mu}-\frac{\boldsymbol{r}_0}{r_0}\tag{4.15}$$

with

$$e=|\boldsymbol{e}|=\sqrt{1-\frac{2\Omega_0^2|h|}{\mu^2}}\tag{4.16}$$

and the mean motion is:

$$n=\frac{\left(2|h|\right)^{3/2}}{\mu}\tag{4.17}$$

$$\text{where } t\in[0,+\infty).\tag{4.18}$$

The function $E(t)$ is the eccentric anomaly defined by:

$$E(t)-e\sin E(t)=nt+E_0-e\sin E_0 \quad \text{for } t\in[0,+\infty)\tag{4.19}$$

with $E_0\in[0,2\pi)$ given by:

$$\begin{aligned}\cos E_0&=\frac{1}{e}\left(1-n\frac{r_0}{\sqrt{2|h|}}\right),\\ \sin E_0&=n\frac{\left(\boldsymbol{v}_0\cdot\boldsymbol{r}_0\right)}{2e|h|}\left[1-\frac{\boldsymbol{\omega}_0\cdot\boldsymbol{\Omega}_0}{\mu}r_0\right].\end{aligned}\tag{4.20}$$

Now, in order to find the solution to the Cuchy problem (4.21), the quaternion operator $\boldsymbol{R}_{-\omega}$ has to be applied to the solution of the Equation (4.11) resulting:

$$\boldsymbol{r}(t)=\boldsymbol{R}_{-\omega}\left\{\left[\cos E(t)-e\right]\boldsymbol{a}_0+\sin E(t)\boldsymbol{b}_0\right\} \tag{4.22}$$

Using the properties of the quaternion operator $\boldsymbol{R}_{-\omega}$, Equation (4.22) becomes:

$$\boldsymbol{r}(t)=\left[\cos E(t)-e\right]\boldsymbol{R}_{-\omega}\boldsymbol{a}_0+\sin E(t)\boldsymbol{R}_{-\omega}\boldsymbol{b}_0 \tag{4.23}$$

Again, the direction of the vector $\boldsymbol{\omega}$ associated with the quaternion $\boldsymbol{\omega}$ is considered to be fixed $\boldsymbol{\omega}(t)=\omega(t)\boldsymbol{u}$ where $\boldsymbol{u}$ is a constant unit vector with $\omega:[0,+\infty)\to\mathbb{R}$ and, from Eqaution (3.11), the quaternion operator $\boldsymbol{R}_{-\omega}=\boldsymbol{q}(\ )\boldsymbol{q}^*$ transforms the Equation (4.23) as below:

$$\begin{aligned}\boldsymbol{r}(t)=&\left[\cos E(t)-e\right]\left\{\boldsymbol{a}_0-\sin\varphi(t)\frac{\boldsymbol{\omega}\times\boldsymbol{a}_0}{\omega}+\left[1-\cos\varphi(t)\right]\frac{\boldsymbol{\omega}\times(\boldsymbol{\omega}\times\boldsymbol{a}_0)}{\omega^2}\right\}\\&+\sin E(t)\left\{\boldsymbol{b}_0-\sin\varphi(t)\frac{\boldsymbol{\omega}\times\boldsymbol{b}_0}{\omega}+\left[1-\cos\varphi(t)\right]\frac{\boldsymbol{\omega}\times(\boldsymbol{\omega}\times\boldsymbol{b}_0)}{\omega^2}\right\}.\end{aligned} \tag{4.24}$$

with $\varphi(t)=\int_0^t\omega(\xi)\mathrm{d}\xi$.

Consequenly, similar to the Foucault pendulum case, the Keplerian motion relative to a rotating reference frame consists of two motions: a Keplerian elliptical motion described by the Equation (4.11) and a rotation with the angular velocity $-\boldsymbol{\omega}$.

5. Conclusion

The quaternion method described in this work presents a new perspective to the clasical problem of motion in central force field relative to the rotating reference frames and provides us a very powerfull tool to solve the similar problems. Throughout the paper, two quaternion operators are defined in order to reveal the closed form solution to the two particular problems of the Foucault Pendulum and Keplerian motions in rotating reference frame.

References

[1] Hamilton, W.R. (2000) On Quaternions, or on a New System of Imaginaries in Algebra. *The London, Edinburgh and Dublin Philosophical Magazine and Journal of Science*, Vols. xxv-xxxvi, No. 3rd Series, 92 p.

[2] Darboux, G. (1887) Leçons sur la théorie générale des surfaces et les applications géométriques du calcul infinitésimal. Gauthier-Villars, Paris.

[3] Condurache, D. and Martinusi, V. (2010) Quaternionic Exact Solution to the Relative Orbital Motion Problem. *Journal of Guidance, Control, and Dynamics*, **33**, 1035-1047. http://dx.doi.org/10.2514/1.47782

[4] Angeles, J. (1988) Rational Kinematics. (Springer Tracts in Natural Philosophy, Vol. 34). Springer-Verlag, New York.

[5] Condurache, D. and Martinusi, V. (2008) Foucault Pendulum-Like Problems: A Tensorial Approach. *International Journal of Non-Linear Mechanics*, **43**, 743-760. http://dx.doi.org/10.1016/j.ijnonlinmec.2008.03.009

[6] Condurache, D. and Martinusi, V. (2007) Kepler's Problem in Rotating Reference Frames; Part 1: Prime Integrals, Vectorial Regularization. *Journal of Guidance, Control and Dynamics*, **30**, 192-200. http://dx.doi.org/10.2514/1.20466

[7] Condurache, D. and Martinusi, V. (2007) A Complete Closed Form Vectorial Solution to the Kepler Problem. *Meccanica*, **42**, 465-476. http://dx.doi.org/10.1007/s11012-007-9065-7

14

Viscosity Transient Phenomenon during Drop Impact Testing and Its Simple Dynamics Model

Chen Liu, Yoshikazu Tanaka, Yukio Fujimoto

Department of Transportation and Environmental Engineering, Graduate School of Engineering Science, Hiroshima University, Higashi-Hiroshima, Japan

Email: fujimoto@naoe.hiroshima-u.ac.jp

Abstract

Most soft materials behave as if they were hardened when subjected to an impact force. The strain rate dependence of viscosity resistance is the reason for this behavior. The authors carried out drop impact tests on several types of soft materials under the condition of a flat frontal impact. The impact force waveform of soft materials was found to consist of a thorn-shaped waveform and a succeeding mountain-shaped waveform. Based on our experimental observations, we believe that a large viscosity resistance is rapidly changed to a small resistance in the course of the impact. In the present study, the cause of this distinct waveform is discussed based on a dynamics model. The study applies a standard linear solid (SLS) model in which the viscosity transient phenomenon is considered is applied. Three types of impact force waveforms of actual soft materials are simulated using the SLS model. Some features of the impact force waveform of soft materials can be explained using the SLS model.

Keywords

Impact Force, Soft Material, Standard Linear Solid Model, Viscosity Transient, Strain Rate Dependence

1. Introduction

Most soft materials have the properties of both viscosity and elasticity. Elastic properties are only slightly affected by the strain rate dependence. On the other hand, viscosity properties (including plastic deformation) are strongly affected by the strain rate dependence. A soft material with a strong viscosity resistance behaves as if it were hardened under a fast load, such as an impact load. For example, in a static state, agar jelly is softer than rubber. However, its impact force waveform has a shorter impact period than that of rubber.

In order to observe the features of impact force waveforms of soft materials, we conducted drop impact experiments for several types of soft materials [1]. The examined materials are agar jelly, sponge, sponge rubber, nitrile rubber, oil clay, paper clay, low-rebound urethane foam, ham, eligible meat, konjac, and cork. The experiments were carried out under a flat frontal impact condition.

The experiments reveal that the impact force waveform of a soft material consists of a thorn-shaped waveform, followed by a mountain-shaped waveform. The thorn-shaped waveform is a spike-like waveform, which is observed in the rising segment of the waveform. This type of thorn-shaped waveform appears in all the results for all of the soft materials except cork sheet. The existence of the thorn-shaped (spike-like) waveform is also reported in a previous study based on a Kolsky bar experiment [2].

From the experimental observations, we believe that the thorn-shaped waveform is caused primarily by the viscosity resistance, and the mountain-shaped waveform is caused primarily by the elastic resistance. In other words, a transient occurs from the viscosity-dominant waveform to the elasticity-dominant waveform during the impact period.

In the present study, we attempt to simulate the impact force waveforms of soft materials by applying a dynamics model considering the viscosity transient mechanism. The standard linear solid (SLS) model is used in the simulation [3] [4]. The soft material is modeled as a combination of a spring and a dashpot. The viscosity transient is defined by the rapid change in viscosity resistance of the dashpot. The simulated waveforms are compared with the actual waveforms measured in the experiment.

2. Impact Force Waveform of a Soft Material

Figure 1(a) shows a soft material subjected to a flat frontal impact. The impact object is a free drop hammer. **Figure 1(b)** shows typical thorn-shaped waveforms of the impact force. The features of the thorn-shaped waveform include a steep slope θ on the rising segment and a thorn-shaped waveform similar to the yield point of low-carbon steel. After that, at a certain point, the slope α becomes gentler and generates a second mountain-shaped waveform. The rising segment of the thorn is usually similar to a straight line.

Figure 1(c) summarizes the occurrence mechanism of thorn-shaped waveforms obtained by the experimental observation. The impact force waveform of soft materials is a mixture of two types of waveforms. The first (thorn-shape) waveform exhibits a large viscosity resistance (red curve), and the second (mountain-shape) waveform exhibits a small viscosity resistance (dashed black curve). This resistance change occurs suddenly at the thorn peak. In other words, during the impact period, the phenomenon changes from a viscosity-dominant waveform to anelasticity-dominant waveform. The compressive stress at the thorn peak is considered as the trigger of this transient. The soft material surface is struck twice by the relative movement of collision objects. The reason is that the slope α becomes smaller than the initial slope θ because the viscosity discontinuity is induced by the first strike.

3. Condition of Flat Frontal Impact and Compact Drop Test Equipment

Through a number of experiments, we found that the thorn-shaped waveform could be clearly measured under the condition of a flat frontal impact [1] [5]. In **Figure 2**, the impact object is solid and has a flat contact surface. The soft material that receives the impact force is a plate-like material with a uniform thickness without surface irregularities. The impact object strikes the soft material in the normal direction. The thorn-shaped waveform can be steadily measured under the above condition. When the bottom surface of the impact object is curved (**Figure 2(b)**), when the impact object is tilted (**Figure 2(c)**), or when the upper surface of the soft material is curved (**Figure 2(d)**), the thorn-shaped waveform becomes unclear. Here, "unclear" does not mean that thorn-shaped waveform does not occur. Rather, the thorn-shaped waveform is not obvious in the contact area that expands gradually from the local area, and the phenomenon is diluted.

Figure 3 shows the free drop test equipment used in this study. The device is braced by vertical columns on the left and right of the bottom plate. The top of the side column is connected with a metal plate. The drop hammer is made of an aluminum circular disk plate and a stainless cylinder (mass of 1.9 kg). The contact surface of the hammer bottom is a disk plate with a diameter of $D = 60$ mm. The upper sensor is an orbicular pad sensor and is attached near the drop hammer bottom. The lower sensor is a pad sensor of 100 mm square and is

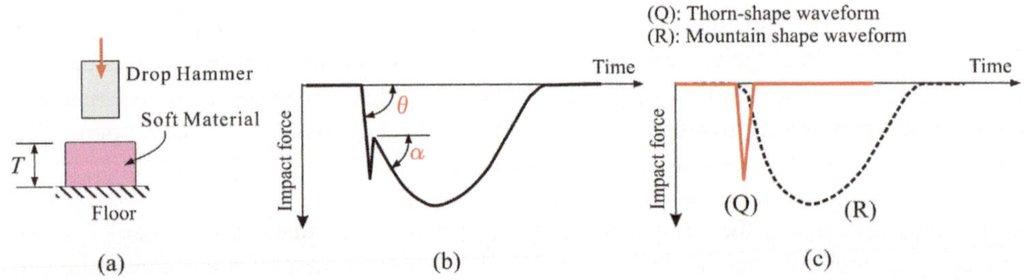

Figure 1. Impact force waveform of a soft material. (a) Flat frontal impact by a free drop hammer for a plate-like soft material; (b) Typical impact force waveform of a soft material including a thorn-shaped waveform; (c) Two types of waveforms contained in the impact force waveform of a soft material.

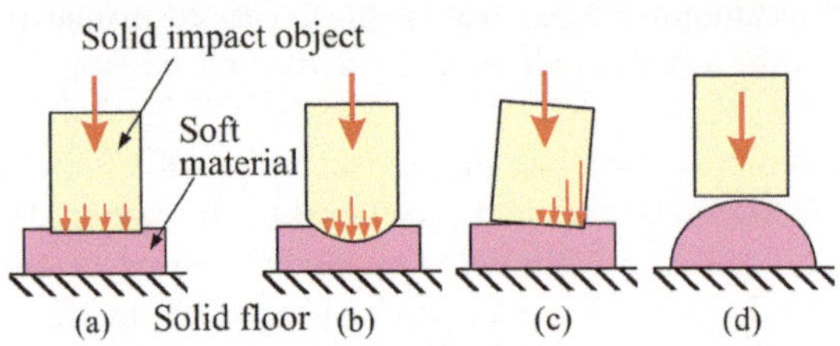

Figure 2. (a) Flat frontal impact condition for which the thorn appears clearly; (b), (c), and (d) Flat frontal impact condition for which the thorn is indistinct.

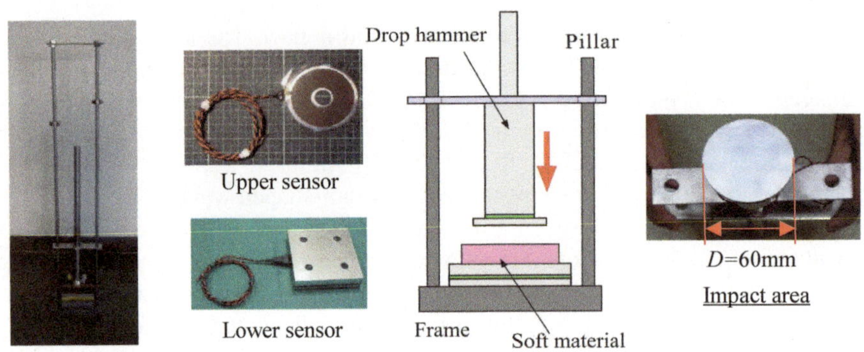

Figure 3. Drop test equipment and the upper and lower sensors used in the device.

attached to the frame bottom. Both sensors are made from piezoelectric film. A high-speed voltage recorder with a sampling rate of 20 - 200 kHz is used for recording the sensor output. The output of the upper sensor is modified by 8% to account for the mass of the disk plate.

Based on rigorous tests using a servo-controlled testing machine and a high-rigidity load cell, it was confirmed that both sensors could measure the impact force very accurately when the contact time was in the range between 0.1 msec and 2 sec. The drop height h is the height from the upper surface of the soft material to the bottom surface of the drop hammer. The impact velocity is V_0 = 1.4 m/sec when h = 0.1 m, V_0 = 2.73 m/sec when h = 0.38 m, and V_0 = 3.13 m/sec when h = 0.5 m.

4. Examples of Measured Impact Force Waveforms

Figure 4(a) shows the impact force waveforms of a gel sheet (product name: a gel) with a T = 20 mm thickness. **Figure 4(b)** is an enlarged view of the thorn part. The drop height is h = 380 mm ($\dot{\varepsilon}_0$ = 136 sec^{-1}). The thorn-shape waveform can be seen clearly in both the upper and lower sensor outputs. In the enlarged view, we can observe that the slope θ of the rising segment is larger than the slope α of the segment past the thorn. The impulse of the thorn part is quite small compared with the succeeding mountain-shape waveform.

Figure 5(a) is the impact force of a nitrile rubber (NBR) plate with a T = 30 mm thickness. The drop height is h = 380 mm ($\dot{\varepsilon}_0$ = 91 sec^{-1}). The upper sensor output shows a relatively large thorn, and the lower sensor out-

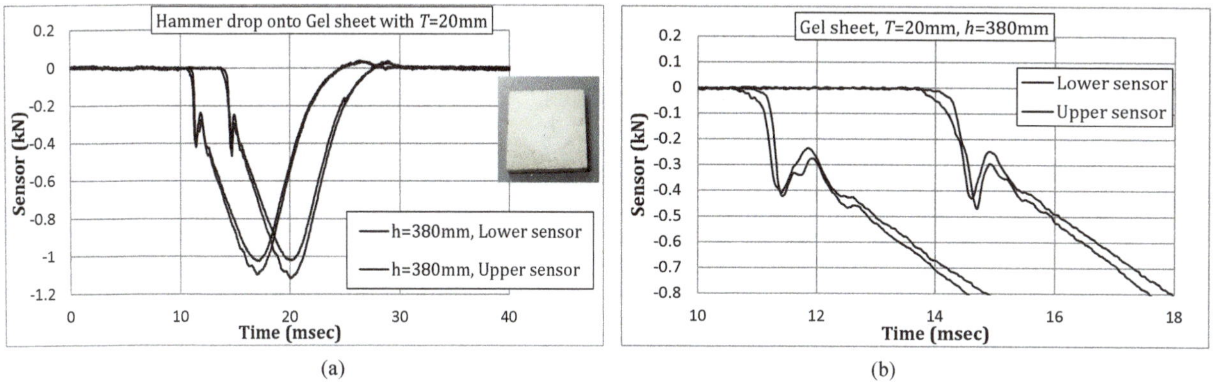

Figure 4. Impact force waveform induced by the hammer drop onto gel sheet with 20 mm thickness.

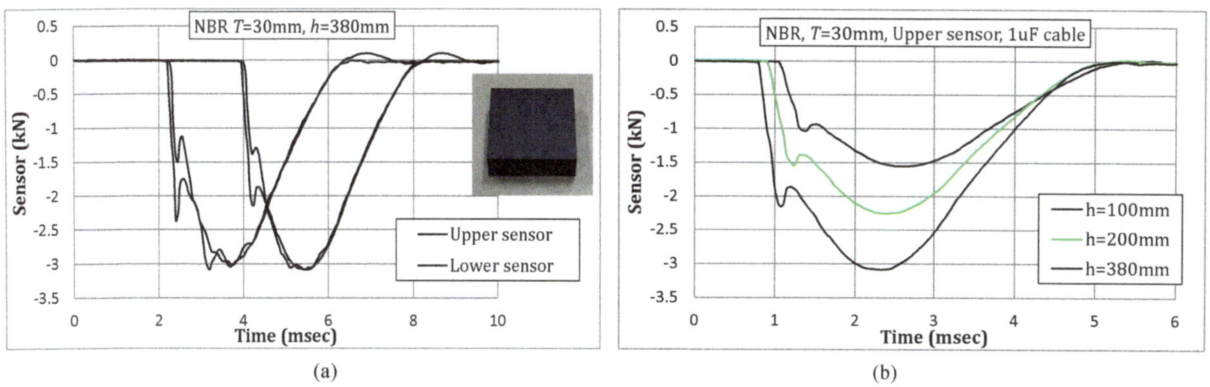

Figure 5. Impact force induced by the hammer drop onto NBR with $T = 30$ mm.

put shows a smaller thorn. The slopes θ of the upper and lower sensor outputs are almost the same. **Figure 5(b)** compares the impact force waveforms of the upper sensor for drop heights of $h = 100$, 200, and 380 mm. In the figure, when the impact velocity V_0 is faster, the slope θ of the rising segment and the height of the thorn peak are larger. This means that the slope θ and the height of the thorn peak are dependent on the impact velocity or strain rate.

Figure 6 shows the impact force waveform of an oil clay plate with a $T = 50$ mm thickness. The drop height is $h = 100$ and 530 mm. A large thorn is seen by the upper sensor, and a small thorn is seen by the lower sensor. As the drop height increases (V_0 becomes faster), the slope θ becomes larger, and the thorn peak becomes taller. The thorns are more obvious than those of other materials. From the enlarged view, the thorn shape is close to an isosceles triangle.

5. Viscosity Transient Phenomenon of a Soft Material under the Drop Test

In the drop impact experiment, strain rate varies with time. The strain rate is large at the beginning of the collision. The strain rate is reduced with the compression of the material and becomes zero at the time of maximum compression.

Therefore, the strain rate dependency of the material changes at every moment during the impact. If it is assumed that the viscosity resistance decreases gradually along with the decrease of the strain rate, the thorn-shaped waveform will not be generated. From the experimental observations, however, we believe that the viscosity resistance changes rapidly or discontinuously (viscosity transient) in the course of the collision process.

Figure 7 shows a schematic diagram of the relationship between the viscosity resistance and the strain rate. We believe that two phases of viscosity resistance appear in the impact force results for soft materials. One phase is an excessively large viscosity resistance (red curve), which appears at the first stage, and the other phase is a small viscosity resistance (black dashed curve), which appears in the second stage. This viscosity transient (green dashed arrow) occurs at the peak of the thorn.

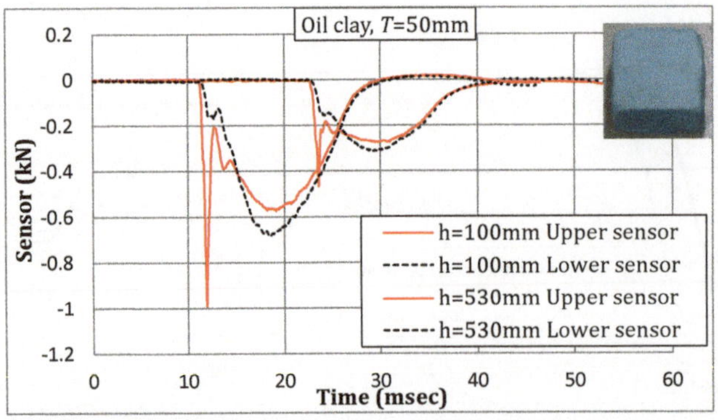

Figure 6. Impact force induced by the hammer drop onto oil clay with 50 mm thickness.

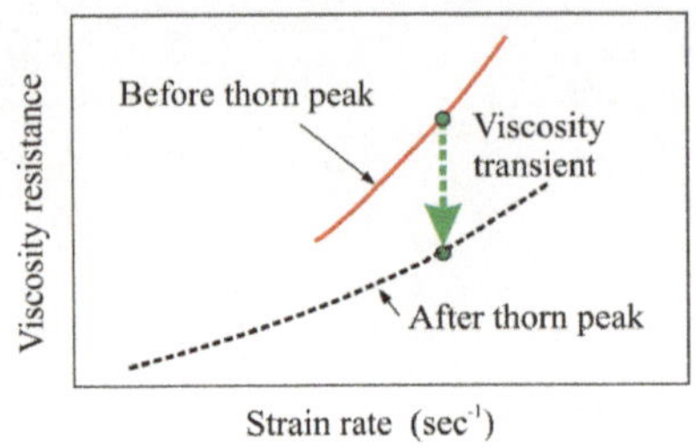

Figure 7. Viscosity resistance vs. strain rate relationship.

In the case of some colloidal suspensions, the viscosity changes rapidly at a certain shear rate. This phenomenon is known as the viscosity discontinuity [6]. This discontinuity also occurs under a reversible load [7]. Based on the experimental observations, we believe that this viscosity discontinuity phenomenon also occurs in most soft materials (soft solids). The impact force waveform of the soft material takes on a distinctive shape as a result of the viscosity discontinuity. In the next chapter, the impact force waveform is simulated considering the viscosity transient.

6. Simulation of the Impact Force Waveform by the Voigt Model and the Standard Linear Solid (SLS) Model

Figure 8 shows the condition under which the drop hammer (mass: m = 1.9 kg) collides with the plate-like soft material at a speed V_0. The thickness of the soft material is T. The soft material is modeled by the Voigt model or the SLS model. The mass (drop hammer) is not connected to the spring and dashpot. The Voigt model can be represented by a viscous damper c and an elastic spring k_1 connected in parallel, as shown in the figure. The SLS model adds an elastic spring k_1 to the Maxwell model (c, k_2) in parallel. The impact velocity V_0 = 2.73 m/s, which corresponds to the drop height h = 0.38 m of the hammer, is given to the mass as the initial condition. The differential equation of the SLS model is given as Equation (1), where x_2 is the displacement of the hammer position, and x_1 is the displacement of the dashpot. The initial strain rate is $\dot{\varepsilon}_0 = V_0/T$ (s^{-1}). The Runge-Kutta method is used for solving the equation of motion.

$$\begin{cases} m\ddot{x}_2 + k_1 x_2 + k_2\left(x_2 - x_1\right) = 0 \\ k_2\left(x_2 - x_1\right) = c\dot{x}_1 \end{cases} \tag{1}$$

First, an analysis is carried out using the Voigt model. The spring coefficient k_1 is assumed to be k_1 = 1000 kN/m, which is constant throughout the collision. As mentioned above, we believe that the viscosity transient occurs in the early stage of the impact force. This influence is taken into consideration through the damping coefficient c of the dashpot. **Figure 9** shows the assumed c as a time-dependent value. Here, c is c = 500 kg/s ($0 < t < 0.3$ ms). Then, c decreases linearly ($0.3 < t < 0.4$ ms), and finally becomes c = 200 kg/s ($t > 0.4$ ms).

Figure 10 shows the impact force waveform calculated by the Voigt model. The impact force waveform rises vertically at the beginning of the impact due to the damping coefficient of the dashpot. Although the thorn-

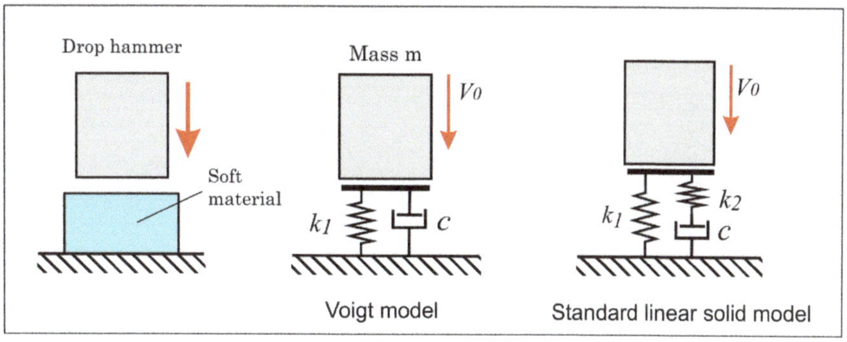

Figure 8. Voigt model and standard linear solid model for the simulation of the impact force waveform of a soft material.

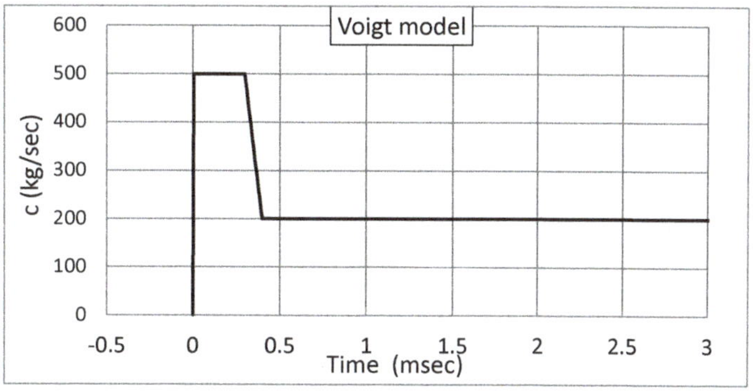

Figure 9. Damping coefficient *c* assumed in the Voigt model and the standard linear solid model.

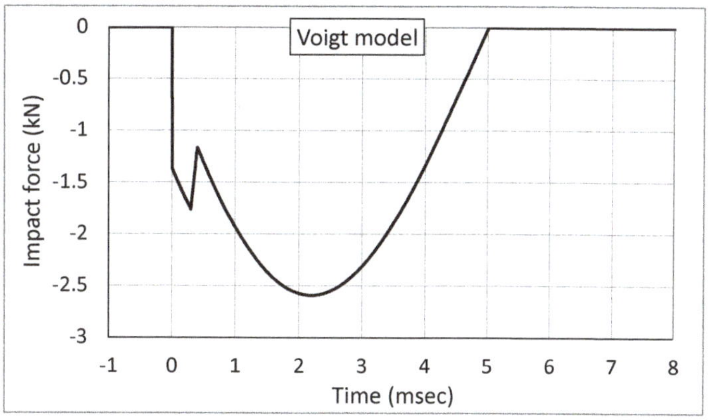

Figure 10. Impact force waveform simulated using the Voigt model.

shaped waveform caused by the viscosity transient appears, the thorn-shaped waveform is different from the measured waveform.

Next, an analysis using the SLS model is carried out. Here, k_1 and c are the same as the values used in the above mentioned Voigt model. The value of k_2 is assumed to be 5000 kN/m. **Figure 11** shows the impact force waveform calculated by the SLS model. By placing the spring k_2 in series with the damper, a small thorn-shaped waveform similar to that measured in the experiment is generated. From this figure, the SLS model is considered to be effective in the simulation of the thorn-shaped waveform.

Several types of thorn-shaped waveform are calculated using the SLS model. **Figure 12** shows the impact force waveforms calculated using the SLS model for the three cases listed in **Table 1**. The values of k_2 and c in the table are the same for these three cases, and only k_1 is changed. The figure shows that the impact period becomes longer as k_1 is reduced.

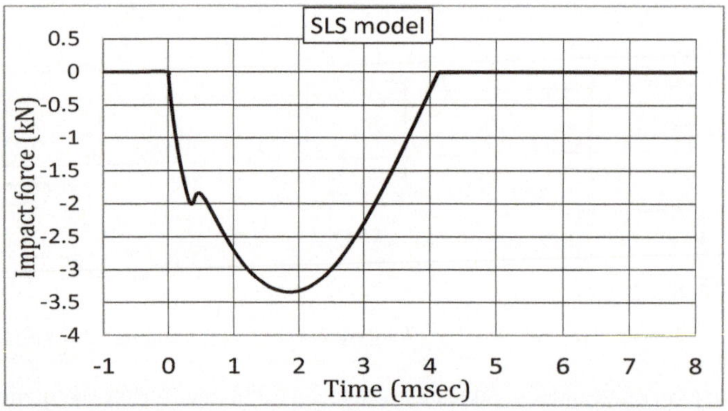

Figure 11. Impact force waveform simulated using the standard linear solid model.

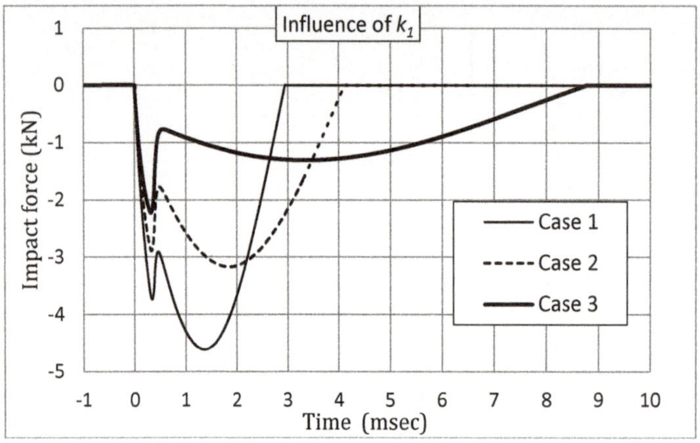

Figure 12. Impact force waveforms calculated using the standard linear solid model.

Table 1. Spring coefficients k_1, k_2 and damping coefficient c for the three cases calculated.

			c (kg/s)		
	k_1 (kN/m)	k_2 (kN/m)	$0 < t < 0.3$ ms	$0.3 < t < 0.4$ ms	$t > 0.4$ ms
Case 1	2000	5000	1000	Linearly decrease	200
Case 2	1000	5000	1000	Linearly decrease	200
Case 3	200	5000	1000	Linearly decrease	200

7. Simulation of Impact Force Waveforms of Actual Soft Materials

The SLS model is applied to the simulation of the impact force waveform of actual soft materials. Nitrile rubber is chosen as the materials for the simulation. The simulation is conducted using several values of k_1, k_2, and c in order to determine the condition in which the calculated waveform becomes close to the experimental waveform of the respective material. In other words, k_1, k_2, and c are decided such that the thorn height, the mountain height, and the total impact period of the simulated waveform match those of the experimental waveform.

Figure 13 shows the impact force waveforms of nitrile rubber simulated using the SLS model. The spring coefficients k_1 and k_2 are assumed to be constant during the impact period, and are assumed to be $k_1 = 1000$ kN/m and $k_2 = 3000$ kN/m. The damping coefficient c of the dashpot is assumed to be a time dependent value, which changes during the impact period. **Figure 14** shows the time variation of c. $c = 1500$ kg/s when $0 < t < 0.2$ ms. During the period of $0.2 < t < 0.4$ ms, c decreases linearly from 1500 kg/s to 100 kg/s. During the period of $t > 0.4$ ms, $c = 100$ kg/s. The period $0.2 < t < 0.4$ ms is the period of the viscosity transient. A thorn shape is clearly observed in the rising segment of the waveform in **Figure 14**. The peak height of the thorn, the peak

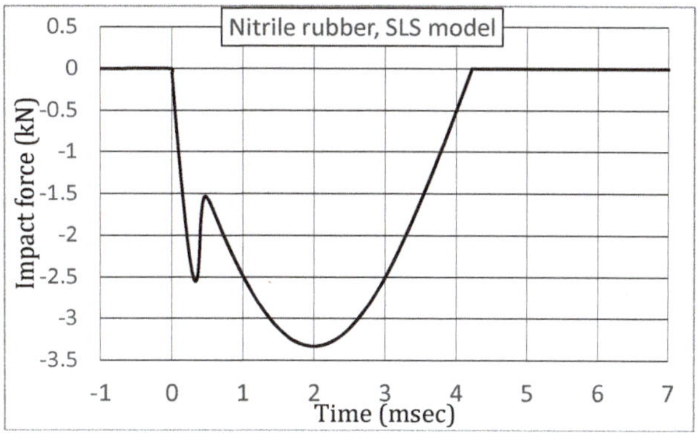

Figure 13. Impact force waveform of nitrile rubber simulated using the standard linear solid model.

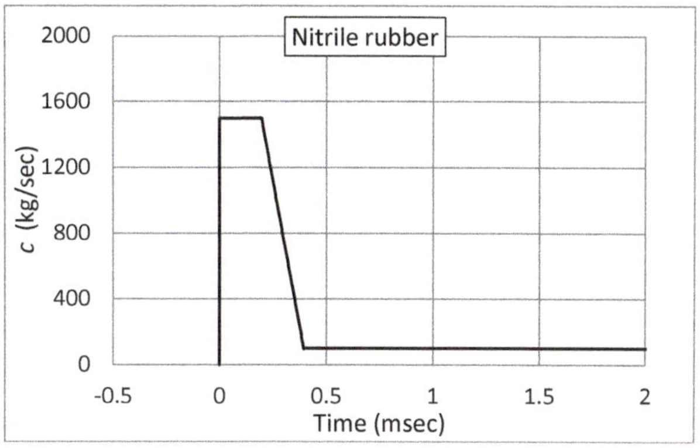

Figure 14. Damping coefficient *c* assumed for nitrile rubber.

height of the mountain-shaped waveform, and the total impact period are approximately the same as those of nitrile rubber, as shown in **Figure 13**.

8. Conclusions

When an impact test on soft material is carried out under the condition of a flat frontal impact, an impact force waveform consisting of a thorn-shaped waveform and a subsequent mountain-shaped waveform is obtained. We believe that this distinctive waveform is a result of the viscosity transient during the impact period. In this study, we attempt to simulate the impact force waveforms using a dynamics model. The obtained results are as follows.

1) When a soft material receives an impact force, a large viscous resistance is induced at the beginning of the impact. The viscous resistance then decreases rapidly with time. This viscosity transient phenomenon is thought to be the reason for the occurrence of the thorn-shaped waveform.

2) The standard linear solid (SLS) model can qualitatively explain the feature of the impact force waveforms of soft materials. In the SLS model, the occurrence mechanism of the thorn-shaped waveform is explained by treating the damping coefficient of dashpot as a time-dependent value.

3) A more sophisticated dynamics model is necessary in order to describe the details of the impact force waveform of the respective soft material. The densification mechanism of the material, the double-strike phenomenon, etc., should be considered in the model for the quantitative simulation.

References

[1] Fujimoto, Y., Liu, C., Tanaka, Y. and Shintaku, E. (2013) Measurement of Double-strike Phenomenon of Impulsive Force Using Fluctuating Load Detection Plate. *Journal of the Japanese Society for Experimental Mechanics*, **13**, 112-120.

[2] Song, B., Chen, W.W., Ge, Y. and Weerasooriya, T. (2007) Radial Inertia Effects in Kolsky Bar Testing of Extra-Soft Specimens. *Experimental Mechanics*, **47**, 659-670. http://dx.doi.org/10.1007/s11340-006-9017-5

[3] Bergstorm, J.S. and Boyce, M.C. (1998) Constitutive Modeling of the Large Strain Time-dependent Behavior of Elastomers. *Journal of the Mechanics and Physics of Solids*, **46**, 931-954. http://dx.doi.org/10.1016/S0022-5096(97)00075-6

[4] Roylance, D. (2001) Engineering Viscoelasticity. 1-37. http://ocw.mit.edu/courses/materials-science-and-engineering/3-11-mechanics-of-materials-fall-1999/modules/visco.pdf

[5] Fujimoto, Y., Liu, C., Uesugi, T., Tanaka, Y. and Shintaku, E. (2013) Pipe Surface Sensor for Impulsive Force Measurement. *Transactions of the Japan Society of Mechanical Engineers Series C*, **79**, 1949-1959. http://dx.doi.org/10.1299/kikaic.79.1949

[6] Hoffman, R.L. (1972) Discontinuous and Dilatant Viscosity Behavior in Concentrated Suspensions I. Observation of a Flow Instability. *Transactions of the Society of Rheology*, **16**, 155-173. http://dx.doi.org/10.1122/1.549250

[7] Bender, J. and Wagner, N. (1996) Reversible Shear Thickening in Monodisperse and Bidisperse Colloidal Dispersions. *Journal of Rheology*, **40**, 899-915. http://dx.doi.org/10.1122/1.550767

15

Numerical Modeling of Non-Similar Mixed Convection Heat Transfer over a Stretching Surface with Slip Conditions

A. Subba Rao[1*], V. R. Prasad[1], N. Nagendra[1], K. V. N. Murthy[1], N. Bhaskar Reddy[2], O. Anwar Beg[3]

[1]Department of Mathematics, Madanapalle Institute of Technology and Science, Madanapalle, India
[2]Department of Mathematics, Sri Venkateswara University, Tirupathi, India
[3]GortEngovation-Aerospace, Medical and Energy Engineering, Bradford, UK
Email: [*]asrsvu@gmail.com

Abstract

In this paper, the heat transfer effect on the steady boundary layer flow of a Casson fluid past a stretching surface in the presence of slip conditions was analyzed. The stretching surface is maintained at a constant temperature. The boundary layer conservation equations, which are parabolic in nature, are normalized into non-similar form and then solved numerically with the well-tested, efficient, implicit, stable Keller-box finite difference scheme. The resulting equations are solved numerically by using the Kellerbox finite-difference method, and the expressions for velocity and temperature are obtained. They satisfy all imposed initial and boundary conditions and reduce to some well-known solutions for non-Newtonian fluids. Numerical results for velocity, temperature, skin friction and Nusselt number are shown in various graphs and discussed for embedded flow parameters. It is found that both velocity and temperature decrease with an increase of the Casson fluid parameter.

Keywords

Stretching Surface, Non-Newtonian Fluid, Slip Condition, Keller-Box Numerical Method, Heat Transfer, Skin Friction Coefficient

1. Introduction

Heat transfer in non-Newtonian fluids is an important research area due to its wide applications in food

[*]Corresponding author.

processing, petroleum production, cooling of an infinite metallic plate in a cooling bath and in many industries, for example polymers melt, polymer solutions employed in the plastic processing, a long thread traveling between a feed roll and a wind-up roll etc. Flow in the boundary layer on moving solid surface was historically first investigated by Sakiadis [1] [2] who observed that the boundary layer growth was in the direction of motion of the continuous solid surface and deviates from that of the classical Blasius flow past a flat plate. Erickson *et al.* [3] extended the Sakiadis problem to include blowing or suction at the moving surface and investigated its effects on the heat and mass transfer in the boundary layer. Bijjanal *et al.* [4] obtained closed-form similarity solutions for steady two-dimensional incompressible boundary layer flow caused by a stretching sheet with Non-Uniform Heat Source/Sink. Combined forced and free convection in boundary layers adjacent to a continuous horizontal sheet maintained at a constant temperature and moving with a constant velocity it was investigated numerically by Chen and Strobel [5]. Grubka and Bobba [6] have investigated the stretching sheet problem for a surface moving with linear velocity and with a variable surface temperature. Dutta *et al.* [7] studied numerically the Temperature Field in Flow over a Stretching Surface with Uniform Heat Flux. Chen and Char [8] investigated the effects of variable surface temperature and variable heat flux on the heat transfer characteristics of a linearly stretching sheet subject to blowing or suction. Several excellent studies of stretching flows in materials processing were presented by Karwe and Jaluria [9] [10]. Patil *et al.* [11] further analysed unsteady two-dimensional mixed convection flow along a vertical semi-infinite power law stretching sheet in a parallel free stream with a power-law temperature distribution. Nath *et al.* [12] analysed the three-dimensional, time dependent stretch surface flow. Ali and Al-Yousef [13] analysed mixed convection heat transfer from a uniformly stretching vertical surface with power function form for wall temperature. Partha *et al.* [14] described the effects of viscous dissipation on mixed convection heat transfer from an exponentially stretching surface.

Non-Newtonian transport phenomena arise in many branches of chemical and material processing engineering. Such fluids exhibit shear-stress-strain relationships which diverge significantly from the Newtonian (Navier-Stokes) model. Most non-Newtonian models involve some forms of modification to the momentum conservation equations. These include power-law, and thixotropic and viscoelastic fluids (Schowalter [15]). Such rheological models however cannot simulate the microstructural characteristics of many important liquids including polymer suspensions, liquid crystal melts, physiological fluids, contaminated lubricants, etc. Several fluids in chemical engineering, multiphase mixtures, pharmaceutical formulations, china clay and coal in water, paints, synthetic lubricants, salvia, synovial fluid, jams, soups, jellies, marmalades, sewage sludge etc. are non-Newtonian. The constitutive relations for these kinds of fluids give rise to more complex and higher order equations than the Navier-Stokes equations. Considerable progress has been made on the topic by using different models of non-Newtonian fluids [16]-[25]. Previous studies indicate that not much has been presented yet regarding Casson fluid. This model (Casson [26]; Nakamura *et al.* [27]; Samir Kumar [28]) in fact is a plastic fluid that exhibits shear thinning characteristics and that quantifies yield stress and high shear viscosity. Casson fluid model is reduced to a Newtonian fluid at a very high wall shear stress, when wall stress is much greater than yield stress. This fluid has good approximations for many substances such as biological materials, foams, molten chocolate, cosmetics, nail polish, some particulate suspensions etc. The boundary layer behaviour of viscoelastic fluid has technical applications in engineering such as glass fibre, paper production, manufacture of foods, the aerodynamic extrusion of plastic sheets, the polymer extrusion in a melt spinning process and many others.

Most of the existing studies on steady boundary layer flow and heat transfer with slip conditions are limited to the non-Newtonian fluid. The considered slip conditions especially are important in the non-Newtonian fluids such as polymer melts which often exhibit wall slip. This motivates us to consider the slip conditions in the present work for non-Newtonian fluids. More exactly, our aim is to investigate steady boundary layer flow and heat transfer of a Casson fluid past a stretching sheet with slip conditions. The equations of the problem are first formulated and then transformed into their dimensionless forms where the Keller box method is applied to find the exact solutions for velocity, temperature, Skin-friction and Nusselt number.

2. Mathematical Analysis

We consider steady two-dimensional laminar mixed convection heat transfer flow along a stretching surface with partial slip. By applying two equal and opposing forces along the x-axis, the sheet is stretched with a speed proportional to the distance from the fixed origin $x = 0$ as shown in **Figure 1**. It is also assumed that the external electric field is zero and the electric field due to the polarization of chargers is negligible. The temperature is

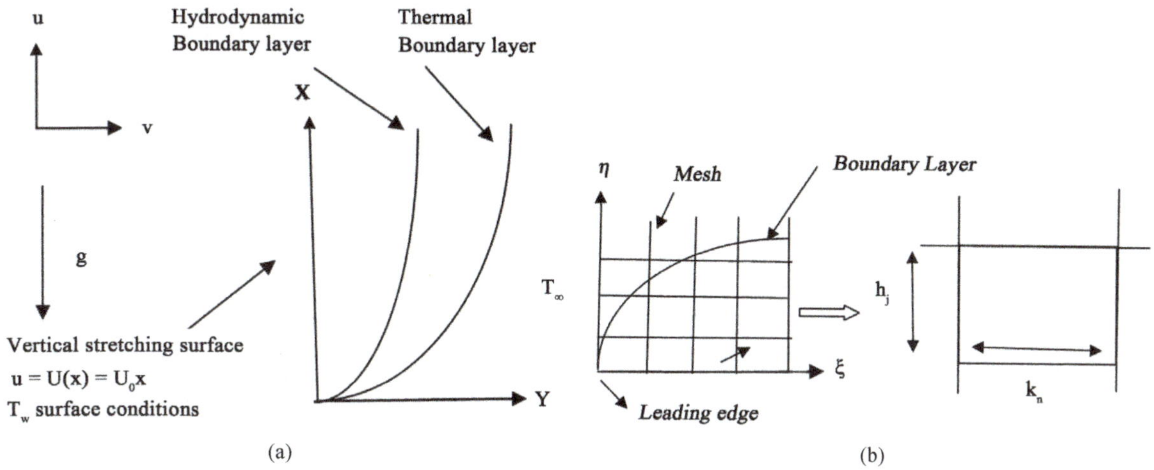

Figure 1. (a) Physical model and coordinate system; (b) Grid meshing and a Keller box computational cell.

maintained at prescribed constant value. The fluid properties are assumed to be constant except the density variation in the buoyancy force term.

The rheological equation of state for an isotropic flow of Casson fluid is (Nakamura *et al.* [27]):

$$\tau_{ij} = \begin{cases} 2\left(\mu_B + \dfrac{p_y}{\sqrt{2\pi}}\right)e_{ij}, \pi \ge \pi_c \\ 2\left(\mu_B + \dfrac{p_y}{\sqrt{2\pi_c}}\right)e_{ij}, \pi < \pi_c \end{cases} \tag{1}$$

in which $\pi = e_{ij}e_{ij}$ and e_{ij} represents the $(i,j)^{th}$ component of deformation rate, μ is the dynamic viscosity, π denotes the product of the component of deformation rate with itself, π_c shows a critical value of this product based on the non-Newtonian model, μ_B represents the plastic dynamic viscosity of non-Newtonian fluid and p_y is the yield stress of fluid.

Under the usual Boussinesq and boundary layer approximations, the equations for mass continuity, (continuity/mass conservation) momentum and energy can be written in the following form:

$$\frac{\partial u}{\partial x} + \frac{\partial v}{\partial y} = 0 \tag{2}$$

$$u\frac{\partial u}{\partial x} + v\frac{\partial u}{\partial y} = \nu\left(1 + \frac{1}{\beta}\right)\frac{\partial^2 u}{\partial y^2} + g\Omega\left(T - T_\infty\right) \tag{3}$$

$$u\frac{\partial T}{\partial x} + v\frac{\partial T}{\partial y} = \alpha\frac{\partial^2 T}{\partial y^2} \tag{4}$$

where u and v are the velocity components in the x- and y-directions, ν is the kinematic viscosity of the conducting fluid, β is the non-Newtonian Casson parameter, α is the thermal diffusivity, T is the temperature respectively.

The boundary conditions are prescribed at the stretching surface and the edge of the boundary layer regime, respectively as follows:

$$\begin{aligned} &\text{At } y = 0,\ \ u = U(x) + N_0\left(1 + \frac{1}{\beta}\right)\frac{\partial u}{\partial y},\ v = 0,\ T = T_w + K_0\frac{\partial T}{\partial y} \\ &\text{As } y \to \infty,\ \ u \to 0,\ T \to T_\infty \end{aligned} \tag{5}$$

where N_0 is the velocity slip factor and K_0 is the thermal slip factor. For $N_0 = 0 = K_0$, one can recover the no-slip case. The stream function ψ is defined by $u = \partial\psi/\partial y$ and $v = -\partial\psi/\partial x$, and therefore, the continuity equation is automatically satisfied. In order to write the governing equations and the boundary conditions in dimensionless form, the following non-dimensional quantities are introduced.

$$\psi = \nu \mathrm{Re}_x^{1/2} f(\xi,\eta),\ \xi = \frac{Gr_x}{\mathrm{Re}_x^2},\ \eta = \frac{y}{x}\mathrm{Re}_x^{1/2},$$
$$\Pr = \frac{\rho \nu c_p}{k},\ Gr = \frac{g\Omega(T_w - T_\infty)x^3}{\nu^2},\ \mathrm{Re}_x = \frac{Ux}{\nu} \tag{6}$$
$$\theta(\xi,\eta) = \frac{T - T_\infty}{T_w - T_\infty},\ \beta = \mu_B \frac{\sqrt{2\pi_c}}{p_y}$$

where ξ is the dimensionless stream wise coordinate, $f(\xi,\eta)$ is the dimensionless stream function, $\theta(\xi,\eta)$ is the temperature function, Re_x is the local Reynolds number, Gr_x is the local thermal Grash of parameter, Ω is the coefficient of thermal expansion, ρ is the density of the fluid, $Pr = \frac{\nu}{\alpha}$ is the Prandtl number, T_∞ is the free stream temperature. The local mixed convection parameter ξ is small near the leading edge where the forced convection dominates and large when the buoyancy force dominates the flow field. The stretching velocity of the surface obeys the relation:

$$U(x) = U_0(x) \tag{7}$$

where U_0 is a constant.

In view of Equations (6) and (7), Equations (2)-(4) reduce to the following coupled, nonlinear, dimensionless partial differential equations for momentum and energy for the regime

$$\left(1+\frac{1}{\beta}\right)f''' + ff'' - f'^2 + \xi\theta = \xi\left(f'\frac{\partial f'}{\partial \xi} - f''\frac{\partial f}{\partial \xi}\right) \tag{8}$$

$$\frac{\theta''}{\Pr} + f\theta' = \xi\left(f'\frac{\partial \theta}{\partial \xi} - \theta'\frac{\partial f}{\partial \xi}\right) \tag{9}$$

The transformed dimensionless boundary conditions are:

$$\text{At } \eta = 0,\ f_w = 0,\ f' = 1 + \left(1+\frac{1}{\beta}\right)S_f f''(0),\ \theta = 1 + S_T\theta'(0)$$
$$\text{As } \eta \to \infty,\ f' \to 0,\ \theta \to 0 \tag{10}$$

In the above equations, the primes denote the differentiation with respect to η, the dimensionless transverse coordinate, and ξ is the dimensionless tangential coordinate, $S_f = \frac{N_0 \mathrm{Re}_x^{1/2}}{x}$ and $S_T = \frac{K_0 \mathrm{Re}_x^{1/2}}{x}$ are the non-dimensional velocity and thermal slip parameters respectively. Here we assumed the typical values $K_0 = 0.5$, $N_0 = 0.25$ for finding the non-dimensional velocity and thermal slip parameters.

The engineering design quantities of physical interest include the skin-friction coefficient and Nusselt number, which are given by:

$$\frac{1}{2}\mathrm{Re}_x^{1/2} C_f = \left(1+\frac{1}{\beta}\right)f''(\xi,0) \tag{11}$$

$$\mathrm{Re}_x^{-1/2} Nu = -\theta'(\xi,0) \tag{12}$$

3. Numerical Solution

In this study the efficient Keller-Box implicit difference method has been employed to solve the general flow model defined by Equations (8)-(9) with boundary conditions (10). Therefore a more detailed exposition is presented here. This method, originally developed for low speed aerodynamic boundary layers by Keller [29], and has been employed in a diverse range of coupled heat transfer problems. These include Ramachandra Prasad *et al.* [30], Rao *et al.* [31] and Beg *et al.* [32].

Essentially 4 phases are central to the Keller Box Scheme.

These are

a) Reduction of the Nth order partial differential equation system to N first order equations;

b) Finite Difference Discretization;

c) Quasilinearization of Non-Linear Keller Algebraic Equations;

d) Block-tridiagonal Elimination of Linear Keller Algebraic Equations.

Phase a: ***Reduction of the Nth order partial differential equation system to N first order equations***

Equations (8)-(9) subject to the boundary conditions (10) are first written as a system of first-order equations. For this purpose, we reset Equations (8)-(9) as a set of simultaneous equations by introducing the new variables *u, v* and *t*:

$$f' = u \tag{13}$$

$$f'' = v \tag{14}$$

$$\theta' = t \tag{15}$$

$$\left(1+\frac{1}{\beta}\right)v' + fv - u^2 + \xi s = \xi\left(u\frac{\partial u}{\partial \xi} - v\frac{\partial f}{\partial \xi}\right) \tag{16}$$

$$\frac{1}{\Pr}t' + ft = \xi\left(u\frac{\partial s}{\partial \xi} - t\frac{\partial f}{\partial \xi}\right) \tag{17}$$

In terms of the dependent variables, the boundary conditions become:

$$\begin{aligned} &\text{At } \eta = 0:\ u = 1 + \left(1+\frac{1}{\beta}\right)f''(0),\ f = 0,\ s = 1 \\ &\text{As } \eta \to \infty:\ u \to 0,\ \ s \to 0 \end{aligned} \tag{18}$$

Phase b: ***Finite difference discretization***

A two dimensional computational grid is imposed on the ξ-η plane as sketched in **Figure 2**. The stepping process is defined by:

$$\eta_0 = 0, \eta_j = \eta_{j-1} + h_j, j = 1,2,\cdots,J,\ \eta_J = \eta_\infty \tag{19}$$

$$\xi^0 = 0, \xi^n = \xi^{n-1} + k_n, n = 1,2,\cdots,N \tag{20}$$

where k_n and h_j denote the step distances in the ξ and η directions respectively.

If g_j^n denotes the value of any variable at (η_j, ξ^n), then the variables and derivatives of Equations (13)-(17) at $(\eta_{j-1/2}, \xi^{n-1/2})$ are replaced by:

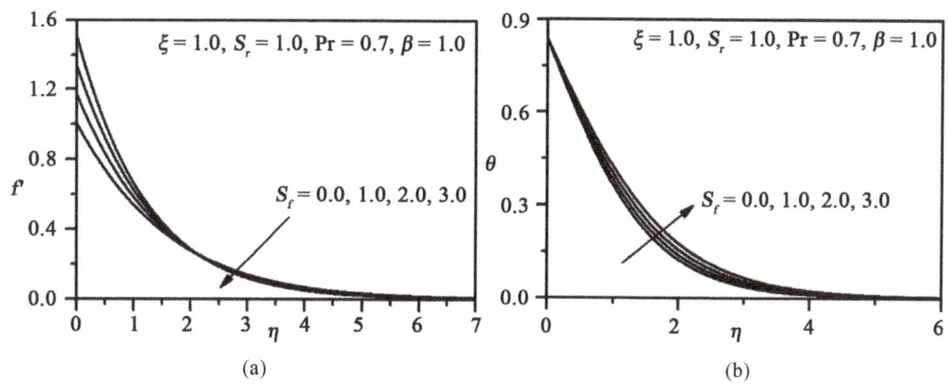

Figure 2. (a) Influence of S_f on the velocity; (b) Influence of S_f on the temperature.

$$g_{j-1/2}^{n-1/2} = \frac{1}{4}\left(g_j^n + g_{j-1}^n + g_j^{n-1} + g_{j-1}^{n-1}\right), \tag{21}$$

$$\left(\frac{\partial g}{\partial \eta}\right)_{j-1/2}^{n-1/2} = \frac{1}{2h_j}\left(g_j^n - g_{j-1}^n + g_j^{n-1} - g_{j-1}^{n-1}\right), \tag{22}$$

$$\left(\frac{\partial g}{\partial \xi}\right)_{j-1/2}^{n-1/2} = \frac{1}{2k_n}\left(g_j^n - g_{j-1}^n + g_j^{n-1} - g_{j-1}^{n-1}\right), \tag{23}$$

We now state the finite-difference approximation of Equations (13)-(17) for the *mid-point* $\left(\eta_{j-1/2}, \xi^n\right)$, below

$$h_j^{-1}\left(f_j^n - f_{j-1}^n\right) = u_{j-1/2}^n, \tag{24}$$

$$h_j^{-1}\left(u_j^n - u_{j-1}^n\right) = v_{j-1/2}^n, \tag{25}$$

$$\begin{gathered}\left(1+\frac{1}{\beta}\right)\left(v_j - v_{j-1}\right) - \frac{h_j}{4}(1+\alpha)\left(u_j + u_{j-1}\right)^2 + \frac{\alpha h_j}{2} v_{j-1/2}^{n-1}\left(f_j + f_{j-1}\right) \\ + \frac{(1+\alpha)h_j}{4}\left[\left(f_j + f_{j-1}\right)\left(v_j + v_{j-1}\right)\right] - \frac{\alpha h_j}{2} f_{j-1/2}^{n-1}\left(v_j + v_{j-1}\right) + \frac{\xi h_j}{2}\left(s_j + s_{j-1}\right) = \left[R_1\right]_{j-1/2}^{n-1}\end{gathered} \tag{26}$$

$$h_j^{-1}\left(\theta_j^n - \theta_{j-1}^n\right) = t_{j-1/2}^n, \tag{27}$$

$$\begin{gathered}\frac{1}{\Pr}\left(t_j - t_{j-1}\right) + \frac{(1+\alpha)h_j}{4}\left[\left(f_j + f_{j-1}\right)\left(t_j + t_{j-1}\right)\right] - \frac{\alpha h_j}{4}\left[\left(u_j + u_{j-1}\right)\left(s_j + s_{j-1}\right)\right] \\ + \frac{\alpha h_j}{2} s_{j-1/2}^{n-1}\left(u_j + u_{j-1}\right) - \frac{\alpha h_j}{2} u_{j-1/2}^{n-1}\left(s_j + s_{j-1}\right) - \frac{\alpha h_j}{2} f_{j-1/2}^{n-1}\left(t_j + t_{j-1}\right) \\ + \frac{\alpha h_j}{2} t_{j-1/2}^{n-1}\left(f_j + f_{j-1}\right) = \left[R_2\right]_{j-1/2}^{n-1}\end{gathered} \tag{28}$$

where we have used the abbreviations

$$\alpha = \frac{\xi^{n-1/2}}{k_n} \tag{29}$$

$$\left[R_1\right]_{j-1/2}^{n-1} = -h_j\left[\left(1+\frac{1}{\beta}\right)\left(\frac{v_j - v_{j-1}}{h_j}\right) + (1-\alpha)\left(f_{j-1/2}v_{j-1/2}\right) - (1-\alpha)\left(u_{j-1/2}\right)^2 + \xi\left(s_{j-1/2}\right)\right] \tag{30}$$

$$\left[R_2\right]_{j-1/2}^{n-1} = -h_j\left[\frac{1}{\Pr}\left(\frac{t_j - t_{j-1}}{h_j}\right) + (1-\alpha)\left(f_{j-1/2}t_{j-1/2}\right) + \alpha\left(u_{j-1/2}s_{j-1/2}\right)\right] \tag{31}$$

The boundary conditions are

$$f_0^n = u_0^n = 0,\ \theta_0^n = 1,\ u_J^n = 0,\ \theta_J^n = 0 \tag{32}$$

Phase c: *Quasilinearization of non-linear Keller algebraic equations*

If we assume $f_j^{n-1}, u_j^{n-1}, v_j^{n-1}, s_j^{n-1}, t_j^{n-1}$ to be known for $0 \le j \le J$, Equations (24)-(28) are a system of $5J+5$ equations for the solution of $5J+5$ unknowns $f_j^n, u_j^n, v_j^n, s_j^n, t_j^n$, $j = 0,1,2,\cdots,J$. This non-linear system of algebraic equations is linearized by means of Newton's method as explained in Keller [29] and Prasad *et al.* [33].

Phase d: *Block-tridiagonal elimination of linear Keller algebraic equations*

The linear system (24)-(28) can now be solved by the block-elimination method, since they possess a block-tridiagonal structure. Commonly, the block-tridiagonal structure consists of variables or constants, but here, an interesting feature can be observed, namely that it consists of block matrices. The complete linearized system is formulated as a block matrix system, where each element in the coefficient matrix is a matrix itself. Then, this system is solved using the efficient Keller-box method. The numerical results are affected by the number of mesh points in both directions. After some trials in the η-direction (radial coordinate) a larger number of mesh points are selected whereas in the ξ direction (tangential coordinate) significantly less mesh points are utilized. $\eta_{\max}$ has been set at 10 and this defines an adequately large value at which the prescribed boundary conditions are satisfied. $\xi_{\max}$ is set at 3.0 for this flow domain. *Mesh independence* is therefore achieved in the present computations. The computer program of the algorithm is executed in **MATLAB** running on a PC. The method demonstrates excellent stability, convergence and consistency, as elaborated by Keller [29] and this system is developed by Cebeci and Bradshaw [34].

4. Results and Discussions

Comprehensive solutions have been obtained and are presented in **Figures 2-7**. The numerical problem comprises 2 independent variables (ξ, η), 2 dependent fluid dynamic variables (f, θ) and 5 thermo physical and

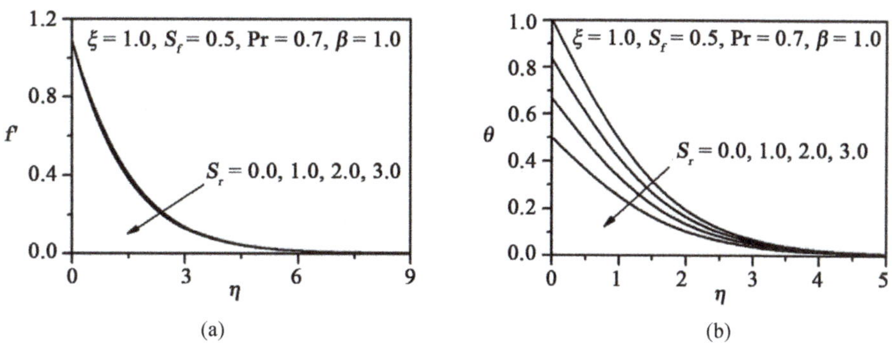

Figure 3. (a) Influence of S_T on the velocity; (b) Influence of S_T on the temperature.

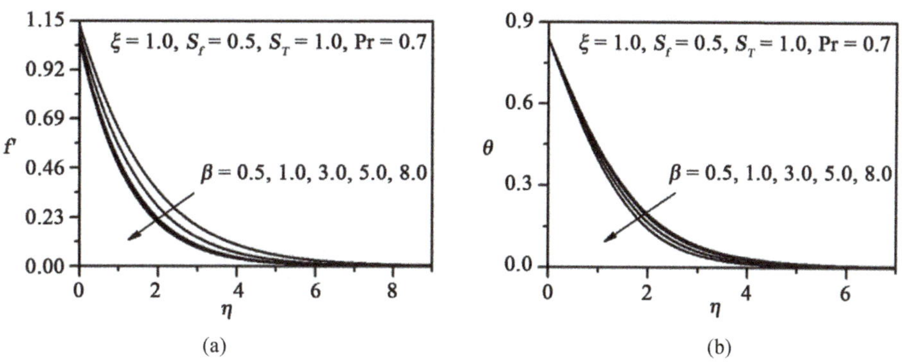

Figure 4. (a) Influence of β on the velocity; (b) Influence of β on the temperature.

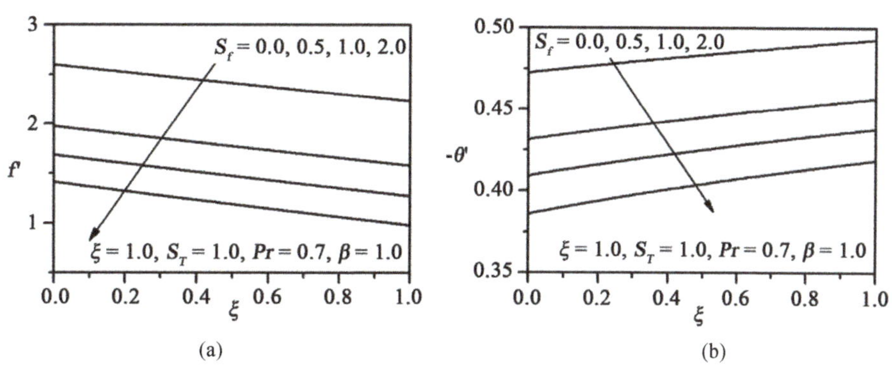

Figure 5. (a) Effect of S_f on the Skin-friction coefficient results; (b) Effect of S_f on the Nusselt number results.

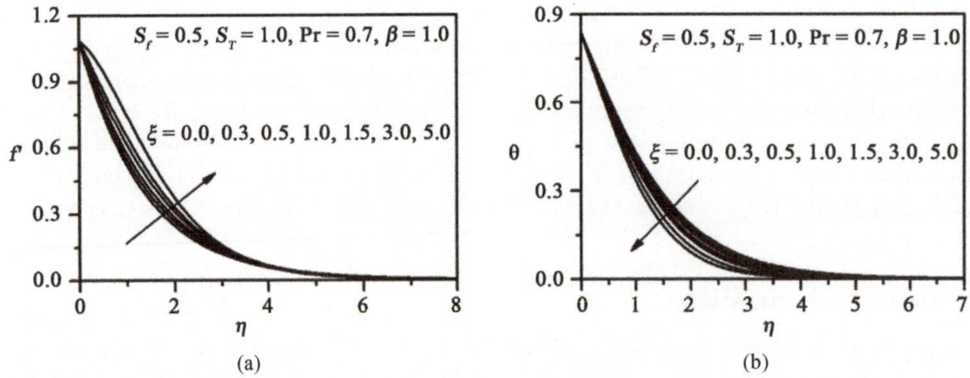

Figure 6. (a) Influence of ξ on the velocity; (b) Influence of ξ on the temperature.

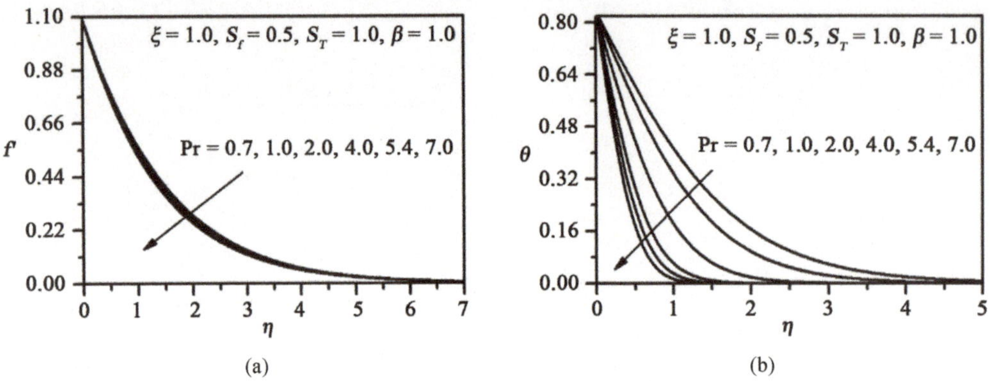

Figure 7. (a) Influence of Pr on the velocity; (b) Influence of Pr on the temperature.

body force control parameters $\Pr, S_f, S_T, \beta, \xi$. In the present computations, the following default parameters are prescribed (unless otherwise stated): Pr = 0.71, $S_f = 0.5$, $S_T = 1.0$, $\beta = 1.0$, $\xi = 1.0$.

In **Figure 2(a)** and **Figure 2(b)**, the influence of velocity slip parameter on velocity and temperature is illustrated. In **Figure 2(a)** the dimensionless velocity component at the wall reduces with increase in slip parameter and hence there will be a decrease in the boundary layer thickness. The velocity profiles damped out a bit slower for the high amount of slip parameters, because of an interception which exhibits among them. **Figure 2(b)** indicates that an increase in slip parameter tends to increase temperature in the flow field. By increasing S_f, thermal boundary layer thickness enhances.

The variation of velocity and temperature with the transverse coordinate (η), over the thermal slip parameter S_T is illustrated in **Figure 3(a)** and **Figure 3(b)**. The response of velocity is much more consistent than for the case of changing velocity slip parameter, it is strongly decreased for all locations in the radial direction. The peak velocity accompanies the case of no thermal slip ($S_T = 0$). The maximum deceleration corresponds to the case of strongest thermal slip ($S_T = 3$). Temperatures (**Figure 3(b)**) are also strongly depressed with increasing thermal slip. The maximum effect is observed at the wall. Further into the free stream, all temperature profiles converge smoothly to the vanishing value. **Figure 4(a)** and **Figure 4(b)** depict the effect of the Casson fluid parameter β on velocity and temperature. Actually, with an increase in non-Newtonian Casson parameter (β), it produces resistance in the fluid flow. An increase in β implies a decrease in yield stress P_y of the Casson fluid and increase in the value of plastic dynamic viscosity μ_B; this effect creates resistance in the flow of fluid. It is further noted that velocity decreases as the Casson fluid parameter β increases. In **Figure 4(b)** it is shown that the effect of β causes decreases in temperature. The effect of velocity slip parameter S_f on stretching surface shear stress (f''), local Nusselt number variation $(-\theta')$ are presented in **Figure 5(a)** and **Figure 5(b)**. In consistency with the earlier graphs described for velocity evolution, with an increase in S_f, wall shear stress is consistently reduced i.e. the flow is decelerated along the stretching surface. The impact of wall slip is therefore significant on the boundary layer characteristics of Casson flow from a surface. With an increasing S_f, the local Nusselt number is also considerably decreased and profiles are generally monotonic decays. Maximum local

Nusselt number always arises at the stretching surface and is minimized with proximity to the greater distance from the stretching surface. In both **Figure 5(a)** and **Figure 5(b)**, skin friction coefficient and local Nusselt number are maximized for the case of no-slip *i.e.* $S_f = 0$. In **Figure 6(a)** and **Figure 6(b)**, the variation of velocity and temperature fields with different ξ values is shown. Close to the stretching surface, velocity (f') is found to be maximized closer to the stretching surface and minimized with progressive distance away from it *i.e.* the flow is decelerated with increasing ξ. However further from the wall, a marked acceleration in the flow is generated with greater distance from the surface i.e. velocity values are higher for higher values of ξ. Temperature θ is found to noticeably decrease through the boundary layer with increasing ξ values; as such the fluid regime is cooled most efficiently at the stretching surface and heated increasingly as we progress around the stretching surface periphery upwards. The effect of Prandtl number (Pr) on the primitive flow variables of velocity and temperature is shown in **Figure 7(a)** and **Figure 7(b)**. Prandtl number signifies the ratio of viscous diffusion to thermal diffusion in the boundary layer regime. With greater Pr values, viscous diffusion rate exceeds thermal diffusion rate. An increase in Pr from 0.7 through 1.0, 2.0, 4.0, 5.4 to 7.0, strongly depresses velocities (**Figure 7(a)**) in the regime. For Pr < 1, thermal diffusivity exceeds momentum diffusivity *i.e.* heat will diffuse faster than momentum. For Pr = 1.0, both the viscous and energy diffusion rates will be the same as will the thermal and velocity boundary layer thicknesses. With increasing Pr values, temperature as shown in **Figure 7(b)**, is markedly reduced throughout the boundary layer.

To validate the present solutions, we compare the present model with the earlier Newtonian model of Merkin [35] and we observe that an excellent agreement between the previous results as shown in **Table 1**.

5. Conclusions

In this study, numerical solutions have been presented for flow and heat transfer of Casson fluid from a permeable isothermal stretching surface with partial slip. The model has been developed to simulate food stuff transport

Table 1. Values of the local heat transfer coefficient for various values of ξ with $\beta \to \infty$, $f_w = N_0 = K_0 = 0$.

ξ	$-\theta'(\xi,0)$	
	Merkin [35]	**Present results**
0.0	0.4212	0.4198
0.2	0.4204	0.4201
0.4	0.4182	0.4191
0.6	0.4145	0.4149
0.8	0.4093	0.4098
1.0	0.4025	0.4034
1.2	0.3942	0.3945
1.4	0.3843	0.3847
1.6	0.3727	0.3729
1.8	0.3594	0.3598
2.0	0.3443	0.3448
2.2	0.3270	0.3272
2.4	0.3073	0.3079
2.6	0.2847	0.2852
2.8	0.2581	0.2588
3.0	0.2252	0.2254
π	0.1963	0.1964

processes in industrial manufacturing operations. A robust, extensively-validated and implicit finite difference numerical scheme has been implemented to solve the transformed and dimensionless velocity and thermal boundary layer equations, subject to physically realistic boundary conditions. The results in summary have shown that, when increasing the velocity slip parameter, velocity, skin friction and Nusselt number decrease, but the temperature increases. A significant finding of this study is that flow separation can be controlled by increasing the value of Casson fluid parameter as well as by increasing Prandtl number.

The current study has been confined to steady-state flow *i.e.* ignored transient effects and neglected thermal radiation heat transfer effects [36]. Generally, very stable and accurate solutions are obtained with the present finite difference code and it is envisaged that other non-Newtonian flows will be studied using this methodology in the future, including Maxwell upper convected fluids [37], and couple stress fluids [38]. These aspects are also of relevance to rheological food processing simulations and will be considered in future investigations.

Acknowledgements

The authors are grateful to the reviewers for giving their constructive comments for improving this article. The work is supported by the University Grants Commission-SERO. The authors are thankful to UGC-NEWDELHI, S.V University, Tirupati and management of MITS, Madanapalle.

References

[1] Sakiadis, B.C. (1961) Boundary-Layer Behavior on Continuous Solid Surfaces: I. Boundary-Layer Equations for Two-Dimensional and Axisymmetric Flow. *AIChE Journal*, **7**, 26-28. http://dx.doi.org/10.1002/aic.690070108

[2] Sakiadis, B.C. (1961) Boundary-Layer Behavior on Continuous Solid Surfaces: II. Boundary-Layer Equations on a Continuous Flat Surface. *AIChE Journal*, **7**, 221-225. http://dx.doi.org/10.1002/aic.690070211

[3] Erickson. L.E., Fan. L.T. and Fox, V.G. (1966) Heat and Mass Transfer on Moving Continuous Flat Plate with Suction or Injection. *Industrial Engineering Chemistry Fundamentals*, **5**, 19-25.

[4] Gireesha, B.J., Roopa, G.S. and Bagewadi, C.S. (2011) Boundary Layer Flow of an Unsteady Dusty Fluid and Heat Transfer over a Stretching Sheet with Non-Uniform Heat Source/Sink. *Scientific Research*, **3**, 726-735. http://dx.doi.org/10.4236/eng.2011.37087

[5] Chen, T.S. and Strobel, F.A. (1980) Buoyancy Effects in Boundary Layer Adjacent to a Continuous, Moving Horizontal Flat Plate. *Journal of Heat Transfer*, **102**, 170-172. http://dx.doi.org/10.1115/1.3244232

[6] Grubka, L.J. and Bobba, K.M. (1985) Heat Transfer Characteristics of a Continuous, Stretching Surface with Variable Temperature. *ASME J. Heat Transfer*, **107**, 248-250. http://dx.doi.org/10.1115/1.3247387

[7] Dutta, B.K., Roy, P. and Gupta, A.S. (1985) Temperature Field in Flow over a Stretching Surface with Uniform Heat Flux. *International Communications in Heat and Mass Transfer*, **12**, 89-94. http://dx.doi.org/10.1016/0735-1933(85)90010-7

[8] Chen, C.K. and Char, M.I. (1988) Heat Transfer of a Continuous Stretching Surface with Suction or Blowing. *Journal of Mathematical Analysis and Applications*, **135**, 568-580. http://dx.doi.org/10.1016/0022-247X(88)90172-2

[9] Karwe, M.V. and Jaluria, Y. (1988) Fluid Flow and Mixed Convection Transport from a Moving Plate in Rolling and Extrusion Processes. *ASME J. Heat Transfer*, **110**, 655-661. http://dx.doi.org/10.1115/1.3250542

[10] Karwe, M.V. and Jaluria, Y. (1991) Numerical Simulation of Thermal Transport Associated With a Continuously Moving Flat Sheet in Materials Processing. *ASME J. Heat Transfer*, **113**, 612-619. http://dx.doi.org/10.1115/1.2910609

[11] Patil, P.M., Roy, S. and Pop, I. (2010) Unsteady Mixed Convection Flow over a Vertical Stretching Sheet in a Parallel Free Stream with Variable Wall Temperature. *International Journal of Heat and Mass Transfer*, **53**, 4741-4748. http://dx.doi.org/10.1016/j.ijheatmasstransfer.2010.06.018

[12] Rajeswari, V., Kumari, M. and Nath, G. (1993) Unsteady Three-Dimensional Boundary Layer Flow Due to a Stretching Surface. *Actamechanica*, **98**, 123-141.

[13] Ali, M. and Al-Yousef, F. (2002) Laminar Mixed Convection Boundary Layers Induced by a Linearly Stretching Permeable Surface. *International Journal of Heat and Mass Transfer*, **45**, 4241-4250. http://dx.doi.org/10.1016/S0017-9310(02)00142-4

[14] Partha, M.K., Murthy, P.V.S.N. and Rajasekhar, G.P. (2005) Effect of Viscous Dissipation on the Mixed Convection Heat Transfer from an Exponentially Stretching Surface. *Heat and Mass Transfer*, **41**, 360-366. http://dx.doi.org/10.1007/s00231-004-0552-2

[15] Schowalter, W.R. (1978) Mechanics of Non-Newtonian Fluids. Pergamon Press, New York.

[16] Rana, P. and Bhargava, R. (2012) Flow and Heat Transfer of a Nanofluid over a Nonlinearly Stretching Sheet: A Nu-

merical Study. *Communications in Nonlinear Science and Numerical Simulation*, **17**, 212-226. http://dx.doi.org/10.1016/j.cnsns.2011.05.009

[17] Nazar, M., Fetecau, C., Vieru, D. and Fetecau, C. (2010) New Exact Solutions Corresponding to the Second Problem of Stokes for Second Grade Fluids. *Nonlinear Analysis*: *Real World Applications*, **11**, 584-591. http://dx.doi.org/10.1016/j.nonrwa.2008.10.055

[18] Fetecau, C., Hayat, T., Zierep, J. and Sajid, M. (2011) Energetic Balance for the Rayleigh—Stokes problem of an Oldroyd-B fluid. *Nonlinear Analysis*: *Real World Applications*, **12**, 1-13. http://dx.doi.org/10.1016/j.nonrwa.2009.12.009

[19] Wang, S.W. and Tan, W.C. (2008) Stability Analysis of Double-Diffusive Convection of Maxwell Fluid in a Porous Medium Heated from Below. *Physics Letters A*, **372**, 3046-3050. http://dx.doi.org/10.1016/j.physleta.2008.01.024

[20] Tan, W.C. and Xu, M.Y. (2004) Unsteady Flows of a Generalized Second Grade Fluid with the Fractional Derivative Model between Two Parallel Plates. *Acta Mechanica Sinica*, **20**, 471-476.

[21] Zhang, Z.Y., Fu, C.J., Tan, W.C. and Wang, C.Y. (2007) On Set of Oscillatory Convection in a Porous Cylinder Saturated with a Viscoelastic Fluid. *Physics of Fluids*, **19**, 98-104.

[22] Rashidi, M.M., Chamkha, A.J. and Keimanesh, M. (2011) Application of Multi-Step Differential Transform Method on Flow of a Second Grade Fluid over a Stretching or Shrinking Sheet. *American Journal of Computational Mathematics*, **6**, 119-128. http://dx.doi.org/10.4236/ajcm.2011.12012

[23] Ali, N., Hayat, T. and Asghar, S. (2009) Peristaltic Flow of Maxwell Fluid in a Channel with Compliant Walls. *Chaos, Solitons & Fractals*, **39**, 407-416. http://dx.doi.org/10.1016/j.chaos.2007.04.010

[24] Attia, H.A. and Seddeek, M.A. (2007) On the Effectiveness of Uniform Suction or Injection on Two Dimensional Stagnation-Point Flow towards a Stretching Surface with Heat Generation. *Chemical Engineering Communications*, **194**, 553-564. http://dx.doi.org/10.1080/00986440600992537

[25] Hussain, M., Hayat, T., Asghar, S. and Fetecau, C. (2010) Oscillatory Flows of Second Grade Fluid in a Porous Space. *Nonlinear Analysis*: *Real World Applications*, **11**, 2403-2414. http://dx.doi.org/10.1016/j.nonrwa.2009.07.016

[26] Casson, N. (1959) In Reheology of Dipersed System. Peragamon Press, Oxford.

[27] Nakamura, M. and Sawada, T. (1988) Numerical Study on the Flow of a Non-Newtonian Fluid through an Axisymmetric Stenosis. *Journal of Biomechanical Engineering*, **110**, 137-143. http://dx.doi.org/10.1115/1.3108418

[28] Samir Kumar, N. (2013) Analytical Solution of MHD Stagnation-Point Flow and Heat Transfer of Casson Fluid over a Stretching Sheet with Partial Slip. *ISRN Thermodynamics*, **2013**, Article ID: 108264.

[29] Keller, H.B. (1970) A New Difference Method for Parabolic Problems. In: Bramble, J., Ed., *Numerical Methods for Partial Differential Equations*, Academic Press, New York, 327-350.

[30] Prasd, V.R., Vasu, B. and Beg, O.A. (2011) Thermo-Diffusion and Diffusion-Thermo Effects on Boundary Layer Flows. LAP Lambert Academic Publishing GmbH & Co. KG, Saarbrücken.

[31] Rao, A.S., Prasad, V.R., Reddy, N.B. and Bég, O.A. (2013) Heat Transfer in a Casson Rheological Fluid from a Semi-infinite Vertical Plate with Partial Slip. *Heat Transfer-Asian Research*, **44**, 272-291. http://dx.doi.org/10.1002/htj.21115

[32] Bég, O.A., Prasad, V.R., Vasu, B., Reddy, N.B., Li, Q. and Bhargava, R. (2011) Free Convection Heat and Mass Transfer from an Isothermal Sphere to a Micropolar Regime with Soret/Dufour Effects. *International Journal of Heat and Mass Transfer*, **54**, 9-18. http://dx.doi.org/10.1016/j.ijheatmasstransfer.2010.10.005

[33] Prasad, V.R., Rao, A.S., Reddy, N.B., Vasu, B. and Beg, O.A. (2013) Modelling Laminar Transport Phenomena in a Casson Rheological Fluid from a Horizontal Circular Cylinder with Partial Slip. *Proceedings of the Institution of Mechanical Engineers, Part E*: *Journal of Process Mechanical Engineering*, **227**, 309-326. http://dx.doi.org/10.1177/0954408912466350

[34] Cebeci, T. and Bradshaw, P. (1984) Physical and Computational Aspects of Convective Heat Transfer. Springer, New York. http://dx.doi.org/10.1007/978-3-662-02411-9

[35] Merkin, J.H. (1977) Free Convection Boundary Layers on Cylinders of Elliptic Cross Section. *Journal of Heat Transfer*, **99**, 453-457. http://dx.doi.org/10.1115/1.3450717

[36] Prasad, V.R., Vasu, B., Prashad, D.R. and Bég, O.A. (2012) Thermal Radiation Effects on Magneto-Hydrodynamic Heat and Mass Transfer from a Horizontal Cylinder in a Variable Porosity Regime. *Journal of Porous Media*, **15**, 261-281. http://dx.doi.org/10.1615/JPorMedia.v15.i3.50

[37] B´eg, O.A. and Makinde, O.D. (2011) Viscoelastic Flow and Species Transfer in a Darcian High-Permeability Channel. *Journal of Petroleum Science and Engineering*, **76**, 93-99. http://dx.doi.org/10.1016/j.petrol.2011.01.008

[38] Kairi, R.R. and Murthy, P.V.S.N. (2012) Effect of Melting on Mixed Convection Heat and Mass Transfer in a Non-Newtonian Fluid Saturated Non-Darcy Porous Medium. *Journal of Heat Transfer*, **134**, Article ID: 042601.

Nomenclature

f	non-dimensional steam function
g	acceleration due to gravity
Gr_x	Grash of number
Re_x	Local Reynolds number
C_f	skin friction coefficient
S_f	non-dimensional velocity slip parameter
S_T	non-dimensional thermal slip parameter
N_0	velocity slip factor
K_0	thermal slip factor
Nu	Local Nusselt number
Pr	Prandtl number
V	the linear (translational) fluid velocity vector
T	temperature
u, v	non-dimensional velocity components along the x- and y-directions, respectively
x	stream wise coordinate
y	transverse coordinate

Greek symbols

α	thermal diffusivity
β	the non-Newtonian Casson parameter
Ω	The coefficient of thermal expansion
η	The dimensionless transverse coordinate
μ	Dynamic viscosity
ν	Kinematic viscosity
θ	Non-dimensional temperature
ρ	Density
σ	The electrical conductivity
ξ	The dimensionless steam wise coordinate
ψ	Dimensionless stream function

Subscripts

w	conditions on the wall
∞	free stream conditions

16

Numerical Investigation of Aerodynamic Characteristics by a Rotating Thin Plate

Yoshihiro Kubota[1*], Osamu Mochizuki[2]

[1]Faculty of Science and Engineering, Toyo University, Kawagoe, Japan
[2]Department of Biomedical Engineering, Toyo University, Kawagoe, Japan
Email: *kubota548@toyo.jp

Abstract

In this study, we use a thin rotating plate to generate propulsion and lift for a paper plate. And the thin plate rotates along the spanwise axis. We numerically determine the influence on aerodynamic characteristics with a rotational velocity of the thin plate. The rotational velocity is obtained with spin parameter which is the ratio of the peripheral speed of the plate to the main flow velocity. And the numerical simulations based on the discrete vortex method show that the autorotation mode of the plate in a uniform flow appears naturally when the spin parameter is unity. Vortex formed from the backward-rotating edge is weaker than those generated from the forward-rotating edge of thin plate. The maximum lift generated at $S = 0.75$ if $S < 1$. The negative moment becomes negative for the nondimensional rotating speed $S \leq 1.75$. The most negative moment appears when $S = 1$; at that time, autorotation occurs naturally.

Keywords

Autorotation, Thinplate, Vortex, Vortexmethod, Speed of Rotation, Spin Parameter

1. Introduction

The purpose of this study is to understand the aerodynamic characteristics to generate propulsion and lift for a thin rotating paper plate. The aerodynamics of freely falling paper plate such as business cards was studied to understand the fluttering or the tumbling. We are focusing on the aerodynamics characteristics of autorotation of thin paper plate. The autorotation is free to rotate with a fixed axis of a centroid of paper plate. The displacement of paper plate during the autorotation is not only vertical but also horizontal. Moreover, the horizontal displacement is not a swing motion. This means that the paper plate has the direction of displacement with a horizontal.

*Corresponding author.

The paper plate during the autorotation generates the lift force [1]. So if we control the rotation of paper plate properly to generate the enough lift force during the autorotation, this might become a small flying plane such as a Micro Air Vehicle (MAV) Aono *et al.* (2011). This is a starting point of our study [2].

From the previous study, the relation between shape of plate and the role of vortex for thin rotating plate was studied by Kubota *et al.* [3]. They experimentally investigated the appropriate shape from the moment of inertia. The relation between the formation of lift force and the lateral vortex structure was also shown. The aerodynamics of freely falling paper plate was known with the fluttering, the tumbling, and the autorotation. Mittal *et al.* (2004) studied numerically about the flutter and tumble with a freely falling paper plate and the autorotation during paper in a uniform flow. Their result showed the effect of Reynolds number and relation between transient of fluttering to tumbling and thickness of paper. The instability on unsteady free fall of a thin solid object was computed by Bönisch *et al.* (2007) [4]. They computed the loci of unsteady free fall solid object during the fluttering and tumbling from the calculation of dynamics on slid body. Influence of density ration of paper and fluid was studied by Finn (2007) [5]. His numerical study showed that the changing of trajectory as tumbling, fluttering, and autorotation with a difference of density ratio. For the study of fluttering and tumbling, Andersen *et al.* (2005) showed the numerical study of relation between the inertia moment with geometry of paper plate and transient of fluttering to tumbling [6]. Also, they reported a comparison between numerical and experimental studies. This argued that the lift was dominated by the rotational velocity of plate. The experimental study for focusing on tumbling frequency was studied by Mahadevan *et al.* (1999) [7]. They concluded that the tumbling frequency was associated with the geometry of plate as the chord length of paper and the span length of paper. Hirata *et al.* (2011) reported the experimental study of aerodynamic characteristics of tumbling plate [8]. They were focusing on the inertia moment, rotation, lift coefficient, and drag coefficient of paper as aerodynamic characteristics. Their results proposed the empirical formulation of aerodynamic characteristics from a function of Reynolds number. Pesavento *et al.* (2004) were studied the aerodynamic lift during a tumbling [9]. The aerodynamic lift on a tumbling plate was dominated by the both falling velocity and angular velocity of stiff paper plate such as the business card. An influence of cross sectional geometry for freely falling paper was studied numerically by Jin *et al.* (2008) [10]. The influence of cross sectional geometry was the change of angular velocity. For a tumbling, an angular velocity dominated the lift force and this was studied by Pesavento *et al.* (2004) [9].

The main scope of our study is the generation of the lift force and the drag force due to the formation of vertex with the different rotational velocity of thin plate. The rotational velocity is obtained with the spin parameter which is the ratio of the peripheral speed of the plate to the main flow velocity.

In our previous study, we determined the aspect ratio of the plate by using a falling paper and observed that a plate with an aspect ratio of 7 achieved the longest flight distance, which meant it must have the highest lift/drag ratio. The numerical simulation allowed us to calculate the forces acting on the plate during rotation as a function of rotation speed. In addition, we discuss herein how to use a rotating plate to generate aerodynamic forces.

2. Simulation Model

We performed numerical simulations for nonsteady-state forces acting on the rotating plate. The phenomenon in a falling rectangular paper strip which rotates naturally along its spanwise axis is known as autorotation [1] [8]. The influence of peripheral speed of the plate relative to the main flow velocity was considered with spin parameter S, which is the ratio of the peripheral speed of the plate to the main flow velocity and is defined as follows [8]:

$$S = \pi Nc/(60U) \tag{1}$$

Here N represents the revolution per minute of the rotating plate, c is the plate chord length. Also, the Reynolds number Re was used to determine the condition of flow. The rectangular plate with an aspect ratio 7 was used on the calculation, since the effective shape of plate was determined from our previous experimental investigation. Autorotation occurs at $S = 1.0$ according to wind-tunnel tests at $Re = 9 \times 10^4$ [11] and at $S = 0.6$ according to experiments in air and water from $Re = 1.79 \times 10^3$ to 1.23×10^4 [1] [12]. To compare the effects of the spin parameter on the flow field, we preformed numerical simulations for various S values. The numerical simulations were performed using commercial software (VFS VOs-2D, College Master Hands Co.), which was based on the discrete vortex method. We investigated the vortex structure formed with rotating plate and found the effective rotating speed for a plate from the aerodynamic characteristics.

3. Results and Discussion

From the observation on our previous study, vortex formed by the autorotating plate was acting on the generation of lift force. We simulated the flow by using a vortex-blob method to observe events near the plate during rotation. The numerical simulation was implemented with the discrete vortex method. A similar simulation was performed by Iima, who simulated the flight of a butterfly by means of vortex blobs [13]. The result of our simulation for a half turn with $S = 0.6$ is shown in **Figure 1**. Darker/red dots show clockwise (cw) vortex blobs, and brighter/blue dots show counter clockwise (ccw) vortex blobs. The vortex composed of cw vortex blobs, which is generated from the A edge on the figures, travels near the upper surface and induces a negative pressure on the upper wing surface, which produces lift. However, the vortex composed of ccw vortex blobs, which is generated from the B edge on the figures, offsets the strength of ccw vortex by interfering with the cw vortex. Thus, the effect of the ccw vortex on the surface pressure of the wing is small, even if the strength of the ccw vortex at the beginning is stronger than that of the cw vortex.

By numerical simulations, we also investigated the effects of the spin parameter S on the aerodynamic characteristics. The calculated forces are shown in **Figure 2** for different values of S. Averaged values of drag, lift, moment, and overall resultant force are denoted in the figure by C_x, C_y, C_m, and R, respectively. In the range of S that we tested, the magnitude of the resultant force is a maximum at $S = 0.1$, and the extremum occurs from $S = 1.0$ to 1.25. The resultant force is not always large even if S is large. For example, the resultant force for $S = 2$ is smaller than that for $S = 1$. The force is a minimum at $S = 0.6$ if $S < 1$.

The drag shows tendencies which are similar to those of the resultant force. In contrast to the drag, the lift is a maximum at $S = 2$ and decreases to almost zero at $S = 1.25$ with decreasing S in the range from $S = 2$ to 1.25. The extremum occurs at $S = 0.75$ and becomes negative for $S \leq 0.25$. The sign of the lift changes at $S = 0.4$, which indicates that the direction of the lift force is easily controlled by changing the rotation speed around $S = 0.4$.

It is interesting that the S of flapping insects is almost 0.5. For high maneuverability, we guess that insects evolved a ratio of flapping speed to flight speed that is near 0.4. The lift-generation mechanism for $S > 1$ is the same as the Magnus effect for a rotating cylinder. However, for $S \leq 1$, lift is due to the difference of the pressures induced by vortices on both surfaces of the plate. The negative moment begins at the nondimensional rotation speed $S \leq 1.75$. The maximum value of the negative moment occurs at $S = 1$; hence, autorotation occurs naturally. The moment of the force becomes negative except for $S = 2$. Thus, the plate rotates automatically, because the negative moment assists rotation. Namely, it is called the autorotation mode. It appears naturally at $S = 1$ because of the maximum absolute value of the negative moment.

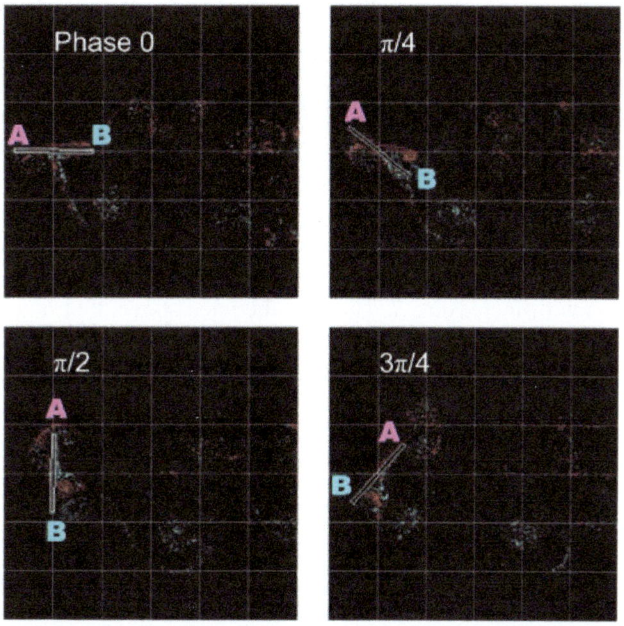

Figure 1. Process of vortex generation around plate rotating at $S = 0.6$. Data obtained by numerical simulations. Darker dots show clockwise rotating vortex blobs, and brighter dots show counter clockwise rotating vortex blobs.

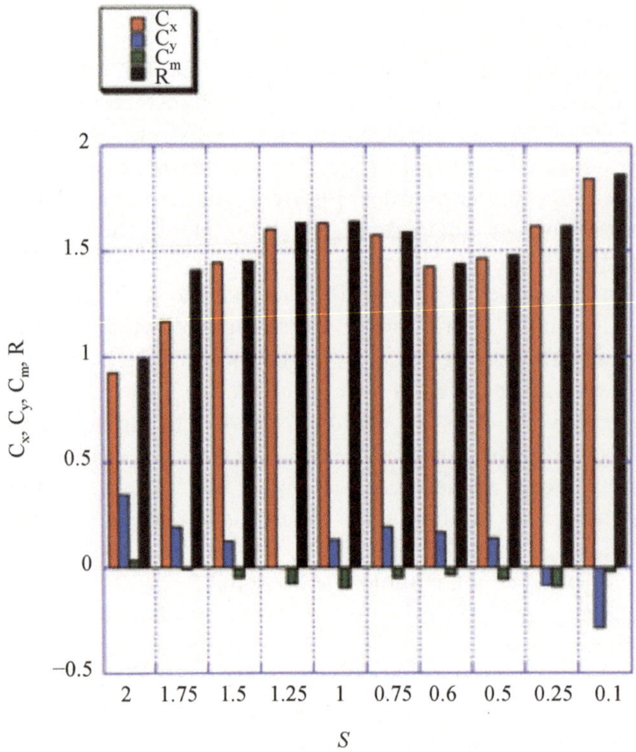

Figure 2. Relation between aerocynamic characteristics and rotationnal velocity. C_x: drag; C_y: lift; C_m: moment; R: resulting overall force; and S: spin parameter.

The ratio of lift to drag for different values of S is shown in **Figure 3**. For $S = 2$, which gives the largest ratio C_y/C_x, the flight distance in the forward direction is maximum. For $S = 1.25$, which gives $C_y/C_x = 0$, the plate falls vertically. The flight path never deflects, because it does not generate lift in this case. Moreover, for $S < 1$, the maximum ratio C_y/C_x appears at S = 0.75, and this ratio becomes increasingly negative at $S = 0.1$ because of negative lift. A test plane with this rotating plate will fly backward.

Figure 4 shows Lissajous figures on the drag and lift in the horizontal and vertical axes, respectively. One loop corresponds to a half rotation of the plate. Thus, the results in each figure correspond to many rotations. Bundled lines show average change, and the width of these bundles shows fluctuations in each loop. Lissajous shows that the change of direction on force caused by the rotation of a thin plate. For $S = 0.05$, approximately five peaks appear in a half rotation, which means that the force results from the combination of lift and drag points in five directions during a half rotation of the plate. Thus, the direction of the test plane will change five times per rotation if we choose this rotation speed, which means that we could not predict where it flies.

On the other hand, for $S = 0.5$ - 1.0, the Lissajous loops are similar to each other. The size of the loop and the fluctuation become larger for S increasing within the given range. For $S = 1.25$ - 2.0, a protuberance appearing in the second quadrant grows with increasing S. The protuberance shows that negative drag (*i.e.*, thrust) is generated at that phase. However, the drag is much larger than the thrust over a half rotation; therefore, we never notice the thrust generation at $S > 1$ when we observe the average force. The larger fluctuation results from the changes in the overall force that is produced at larger S. Thus, according to the simulations, we should adopt $S = 1$ for the effective rotating speed. We should make the plate easy to rotate so that an autorotation with $S = 1$ is naturally selected.

4. Conclusion

We investigate the forces acting on the rotating plate with the vortex method. The influence of speed on rotation of plate is considered by using spin parameter S. Vortices generated from the backward-rotating edge are weaker than those generated from the forward-rotating edge. However, the former move near the wing surface together with the plate; therefore, the induced surface pressure is larger. The lift attains a maximum at $S = 0.75$ if $S < 1$.

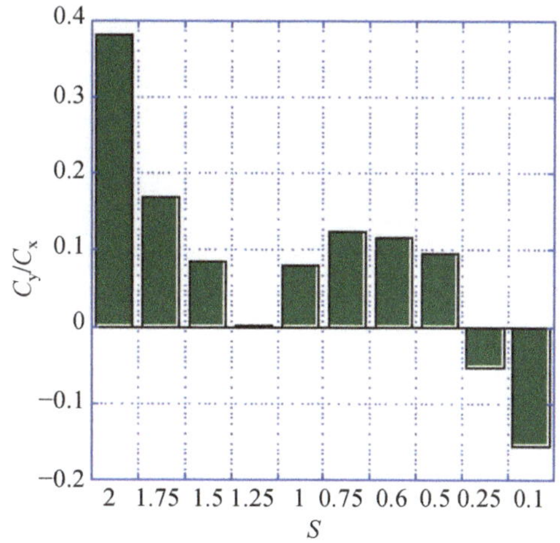

Figure 3. Ratio C_y/C_x of lift to drag versus S.

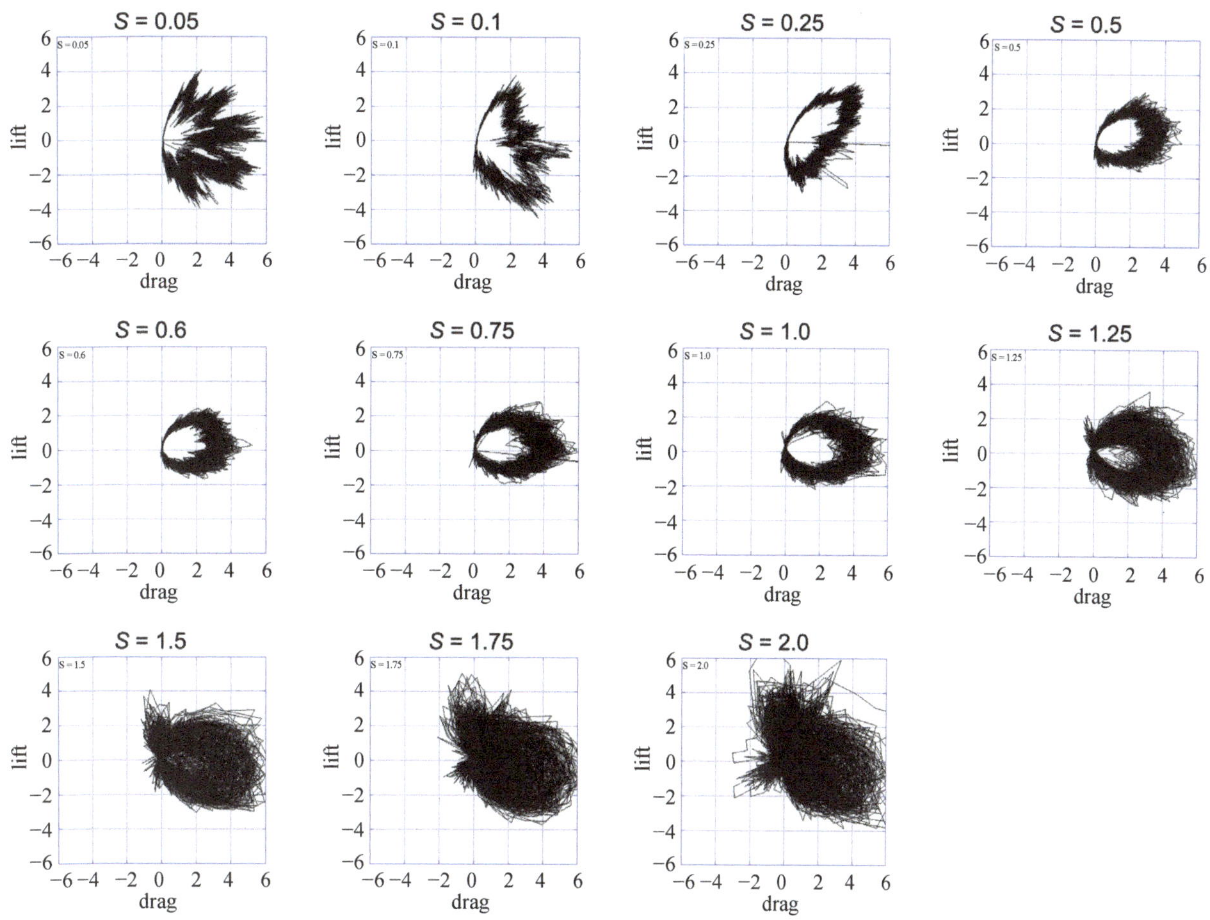

Figure 4. Lissajous figures for drag and lift with the drag on the horizontal axis and the lift on the vertical axis. Bundled lines show average change, and the width of these bundles shows fluctuations in each loop.

The moment becomes negative for the nondimensional rotating speed $S \leq 1.75$. The most negative moment occurs at $S = 1$; thus, autorotation occurs naturally. The combination of lift and drag forces shows that the effective rotational speed for plate is $S = 1$. Since, the slower rotational speed (*i.e.*, $S = 0.05$) causes the force generated unstably on the rotation cycle, and the rotating plate generates much larger drag than the thrust (or negative drag) during higher rotational speed (*i.e.*, $S \geq 1.25$).

References

[1] Mittal, R., Seshadri, V. and Udaykumar, H.S. (2004) Flutter, Tumble and Vortex Induced Autorotation. *Theoretical and Computational Fluid Dynamics*, **17**, 165-170. http://dx.doi.org/10.1007/s00162-003-0101-5

[2] Aono, H., Liang, F.Y. and Liu, H. (2008) Near- and Far-Field Aerodynamics in Insect Hovering Flight: An Integrated Computational Study. *The Journal of Experimental Biology*, **211**, 239-257. http://dx.doi.org/10.1242/jeb.008649

[3] Kubota, Y. and Mochizuki, O. (2013) Role on Moment of Inertia and Vortex Dynamics for a Thin Rotating Plate. *World Journal of Mechanics*, **3**, 270-276. http://dx.doi.org/10.4236/wjm.2013.36028

[4] Bönisch, S. and Heuveline, V. (2007) On the Numerical Simulation of the Unsteady Free Fall of a Solid in a Fluid: I. The Newtonian Case. *Computers and Fluids*, **36**, 1434-1445. http://dx.doi.org/10.1016/j.compfluid.2007.01.010

[5] Finn, D.L. (2007) Falling Paper and Flying Business Cards. *SIAM News*, **40**.

[6] Andersen, A., Pesavento, U. and Wang, Z.J. (2005) Analysis of Transitions between Fluttering, Tumbling and Steady Descent of Falling Card. *Journal of Fluid Mechanics*, **541**, 91-104. http://dx.doi.org/10.1017/S0022112005005847

[7] Mahadevan, L., Ryu, W.S. and Samuel, A.D.T. (1999) Tumbling Cards. *Physics of Fluids*, **11**, 1-3. http://dx.doi.org/10.1063/1.869919

[8] Hirata, K., Hayakawa, M. and Funaki, J. (2011) On Tumbling of a Two-Dimensional Plate under Free Flight. *Journal of Fluid Science and Technology*, **6**, 177-191. http://dx.doi.org/10.1299/jfst.6.177

[9] Pesavento, U. and Wang, Z.J. (2004) Falling Paper: Navier-Stokes Solutions, Model of Fluid Forces, and Center of Mass Elevation. *Physical Review Letters*, **93**, 144501-1-144501-4. http://dx.doi.org/10.1103/PhysRevLett.93.144501

[10] Jin, C.Q. and Xu, K. (2008) Numerical Study of the Unsteady Aerodynamics of Freely Falling Plates. *Communications in Computational Physics*, **3**, 834-851.

[11] Smith, E.H. (1971) Autorotating Wings: An Experimental Investigation. *Journal of Fluid Mechanics*, **50**, 512-534. http://dx.doi.org/10.1017/S0022112071002738

[12] Kawaguchi, D., Ymauchi, K., Funaki, J. and Hirata, K. (2008) Free-Fall Experiments on a Tumbling Plate. *JAXA-SP*-08-009, 151-156 (in Japanese).

[13] Iima, M. (2007) A Two-Dimensional Aerodynamic Model of Freely Flying Insects. *Journal of Theoretical Biology*, **247**, 657-671. http://dx.doi.org/10.1016/j.jtbi.2007.03.012

Nomenclature

S: spin parameter, $\pi Nc/(60U)$

N: revolution per minute of the rotating plate; revolution/min.

c: chord length of rotating plate; mm

U: main flow velocity; m/s

C_x: drag force acting on the rotating plate; N

C_y: lift force acting on the rotating plate; N

C_m: moment of force; Nm

R: overall resulting force; N

17

Aerodynamic Brake for Formula Cars

Roberto Capata, Leone Martellucci

Department of Mechanical and Aerospace Engineering, University of Roma "Sapienza", Rome, Italy
Email: roberto.capata@uniroma1.it

Abstract

In the last years, in formula racing cars championships, the aerodynamic had reached an ever more important stance as a performance parameter. In the last four seasons, Red Bull Racing Technical Officer had designed their Formula 1 car with the specific aim to generate the optimal downforce, in relation to the car instantaneous setup. However, this extreme research of higher downforce brings some negative effects when a car is within the wake of another car; indeed, it is well known that under these condition the aerodynamic is disturbed, and it makes difficult to overtake the leading car. To partially remedy this problem, Formula 1 regulations introduced the Drag Reduction System (DRS) in 2011, which was an adjustable flap located on the rear wing; if it is flattened, allowing to reduce the downforce, increasing significantly the velocity and, therefore, the chances to overtake the leading car. Vice versa, when the flap is closed, it ensures a higher grip, which is very useful especially in medium-slow speed turns. Keeping the focus on the rear wing, but by shifting attention from the increased top speed to increase the grip in the middle and slow speed curves, we decided to study a similar device to the DRS, but with the opposite effect. The aim is to design an aerodynamic brake integrated with the rear wing. In particular, the project idea was to sculpt, on the upper surface of the wing (pressure side), a series of "C" shaped cavity, normally covered by adequate sliding panels. These cavities, when they are discovered, at the beginning of the braking phase, produce a turbulence and additional increase downforce, lightening the load on the braking system and allowing the pilot to substantially reduce slippage and to delay the braking. Since it seems that the regulations adopted by the FIA Formula 1 Championship do not allow such a device, it has been decided to apply the concept on a Formula 4 vehicle. This paper describes the design and analyzes the effects of these details on a standard wing cavity, using a commercial CFD software.

Keywords

Aerodynamic Brake, Cavity, Dynamic Effects, Fluid Dynamic Simulation

1. Problem Formulation

In this paper, the realization of an aerodynamic brake integrated in a rear wing of a formula car has been considered. The first step consists in the choice of an appropriate aerodynamic appendix. In particular, it was decided to study an Italian Formula 4 race car [1], being a category in the first stages of development. Also, the regulation of this championship is easy to find and the car is characterized by uniformity of the mechanics and the airfoils. Therefore, taken note of the technical regulation on FIA website, it was decided to study the upper airfoil, of which was shown a dimensioned drawing (**Figure 1**). It is an aluminum alloy wing, with a chord line of 237.9 mm and a height of 54.2 mm.

Formula 4 championship will provide the use of a 4T heat engine (Otto/Beau de Rochas cycle): it can be naturally aspirated or turbocharged, with maximum power in the order of 120 kW (160 HP). Considering the weight of the car and the race tracks of the championship, it is predicted a maximum speed of 230 km/h (64 m/s). Regarding the operating conditions, an air temperature of 300K was assumed at atmospheric pressure.

Briefing Description of Airfoil Behavior

Considering an airfoil, there are several elements that have a specific nomenclature:

1) Mean camber line: locus of points halfway between the upper and lower surface as measured perpendicular to the mean chamber line itself;

2) Leading edge: the most forward point of the mean camber line;

3) Trailing edge: the rearmost point of the mean camber line;

4) Chord: the straight line joining the leading edge with the trailing edge;

5) Upper surface: the upper boundary of the profile;

6) Lower surface: the lower boundary of the profile;

7) Thickness: the distance between the lower surface and the upper surface.

The different airfoil shapes are marked by a logical numbering system which was introduced by the U.S. federal agency NACA. This system consists of four digits which have a definite meaning:

- the first digit indicates the maximum camber in hundredths of chord;
- the second digit represents the location of maximum camber along the chord from leading edge in tenths of chord;
- the third and fourth give the maximum thickness in hundredths of chord.

When an airfoil is moving relative to the air, it generates an aerodynamic force, in a rearward direction at an angle with the direction of relative motion. This aerodynamic force is commonly resolved into two components: lift and drag. Lift is the force component perpendicular to the direction of relative motion while Drag is the force component parallel to the direction of relative motion. These forces are studied at different angles of attack which is the angle at which an airfoil cleaves fluid. The experimental data show that CL varies with the angle of attack: more precisely, at low angles of attack the lift coefficient CL varies linearly with α. In a region characterized by a linear trend, the flow moves smoothly over the airfoil and is attached to the back of the wing. As soon as α increases, the flow tends to separate from the surface of the airfoil, creating a region of "dead air" behind the profile. A briefing flow analysis of the physical phenomenon in question in order to understand better what is happening in the latter case is reported. It is clear from **Figure 2** that the speed at the trailing edge tends to increase, with a strong reduction of the pressure, while in the stagnation point the speed tends to be zero and pressure rises sharply. It creates an adverse pressure gradient, thus particles of fluid move from the trailing edge to the stagnation point, and then it has a rapid separation of the boundary layer below. Stagnation point does not have a stable position in these conditions because there is not pressure recovery. The recirculation generated by

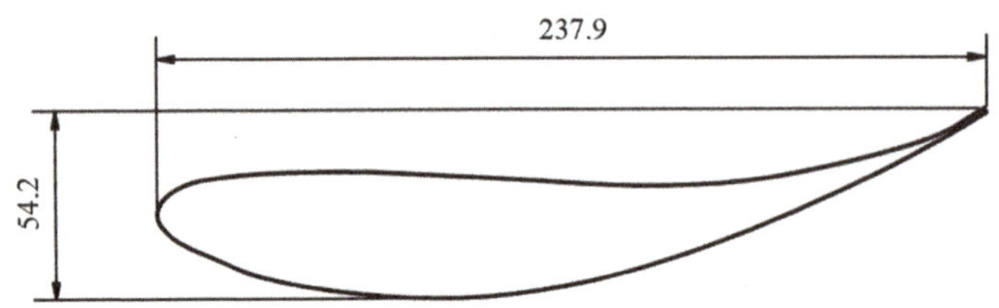

Figure 1. Dimensioned Drawing of a F4 rear wing (in mm).

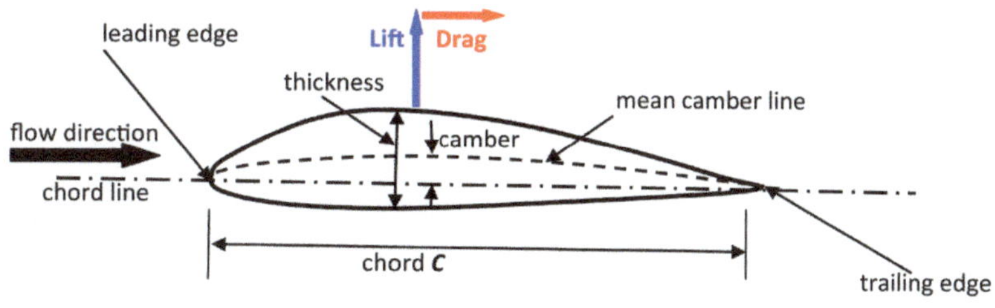

Figure 2. Airfoil characteristics with generic flow direction.

the detachment of the boundary layer creates first vortex that causes a wake vortex. It is necessary to study the turbulent behavior of the fluid that meets the wing, through the Navier-Stokes equations in order to consider the stall of the wing:

$$\rho \cdot \frac{\partial u_i}{\partial t} + \rho \cdot u_j \cdot \frac{\partial u_i}{\partial t} = \rho \cdot f_i - \frac{\partial p}{\partial x_i} + \mu \cdot \frac{\partial^2 u_i}{\partial x_j \cdot \partial x_j} \tag{1}$$

where $u(x, t)$ is the instantaneous velocity, ρ the medium density, μ the viscosity and f the applied force.

This system of equations is a system of partial differential equations that describe the behavior of a Stokesian fluid: the fluid can be considered to be continuous. There is an analytical solution only in simplified cases, while solutions in the other cases can be obtained using simplified methods of numerical analysis. The most straightforward method for the numerical simulation of turbulent flows is direct numerical simulation DNS which discretizes the Navier-Stokes equations. It resolves the entire range of turbulent length scales thus the description of the flow is so detailed that the validity of the simulation is similar to an experiment. The computational cost is proportional to Re^3, thus it is necessary to use a different solution studying turbulent flows at high Reynolds, because the computational resources required by a DNS would exceed the capacity of the most powerful computer currently available. In practical applications, the knowledge of the average quantities is enough to solve the problem of a turbulent flow; the basic idea of the technique RANS (Reynolds Averaged Navier-Stokes Equations) is to derive only the average parameters (mediated in time) from Navier-Stokes equations, reducing the enormous computational cost required by DNS. In practice, the turbulent motion consists of a mean motion and fluctuation over time. Using the decomposition of Reynolds:

$$u(x,t) = \langle u(x,t) \rangle + u'(x,t) \tag{2}$$

where $u(x,t)$ is the instantaneous velocity, $\langle u(x,t) \rangle$ is the average velocity $u'(x,t)$ is the speed fluctuating, through Navier-Stokes equations it's possible to obtain the Reynolds averaged equations. The equations for the mean motion obtained are similar to Navier-Stokes equations with the exception of the divergence of the stress tensor Reynolds: the system resulting from the Navier-Stokes equations is closed, while the system resulting from the RANS simulation is not open because Reynolds tensor introduces 6 additional unknowns. The problem mentioned is known as the problem of closure of turbulence which is solved by introducing models for the turbulent fluctuations which have to reproduce the action of fluctuating terms on mean motion.

The K-ε model is one of the most common models of turbulence, even if it is not appropriate in the case of strong adverse pressure gradients. It is a model with two equations: it includes two additional transport equations to represent properties of the turbulent flow and effects such as convection and diffusion of turbulent energy. The first variable transported is the turbulent kinetic energy, k. The second variable transported is the turbulent dissipation, ε; the second variable determines the scale of turbulence, while the first variable k determines the energy in the turbulence. There are two formulations of the K-ε models: the standard k-epsilon model and the RNG k-epsilon model.

In the standard k-epsilon model, eddy viscosity is determined by single length scale turbulence, so the turbulent diffusion is calculated only through a specified scale, whereas in reality all scales of motion will contribute to turbulent diffusion.

The approach RNG (Re-Normalisation Group), a mathematical technique that can be used to obtain a model similar to the k-epsilon turbulence, presents a modified equation ε, which attempts to explain the different scales of turbulence through changes at the term of production of turbulence. The equations used are:

a) Kinematic Eddy Viscosity

$$\nu_T = C_\mu \cdot \frac{k^2}{\varepsilon} \tag{3}$$

b) Turbulence Kinetic Energy

$$\frac{\partial k}{\partial t} + U_j \cdot \frac{\partial k}{\partial x_j} = \tau_{ij} \cdot \frac{\partial U_i}{\partial x_j} - \varepsilon + \frac{\partial}{\partial x_j} \cdot \left[\left(\nu + \frac{\nu_T}{\sigma_k} \right) \cdot \frac{\partial k}{\partial x_j} \right] \tag{4}$$

c) Dissipation Rate

$$\frac{\partial \varepsilon}{\partial t} + U_j \cdot \frac{\partial \varepsilon}{\partial x_j} = C_{\varepsilon 1} \cdot \frac{\varepsilon}{k} \cdot \tau_{ij} \cdot \frac{\partial U_i}{\partial x_j} - C_{\varepsilon 2} \cdot \frac{\varepsilon^2}{k} + \frac{\partial}{\partial x_j} \cdot \left[\left(\nu + \frac{\nu_T}{\sigma_\varepsilon} \right) \cdot \frac{\partial \varepsilon}{\partial x_j} \right] \tag{5}$$

Closure coefficient for standard k-epsilon model:

$$C_{\varepsilon 1} = 1.44,\ C_{\varepsilon 2} = 1.92,\ C_\mu = 0.09,\ \sigma_k = 1.0,\ \sigma_\varepsilon = 1.3$$

2. Project Description

The purpose of this project is to improve the race performance, reducing the breaking distance and increasing the bending speed. So, we decided to intervene on the drag generated by the wing during the breaking, and also on the grip provided by downforce, function of velocity. To explain the lift, and then the downforce, reference may be made to the wing of an airplane, observing its section. The latter is asymmetric, the top has a profile longer than the bottom: when the wing moves, it separates the relative flow in two parts, so the air layers scroll faster in the top. The outflow over the wing undergoes a boost and then is aerodynamic brake for formula cars accelerated towards the tail at a higher velocity than the air under the wing, which follows a shorter path. So the two currents are reunited in the tail after a same time interval, without creating imbalances. This is not just the facts, but as a first approximation, we can refer to this model. In reference to the Bernoulli trinomial law, since in the lower flow velocity is lower than in the upper, the pressure under the wing has to be greater than that above the wing. Therefore, the difference between the two pressures generates a resultant directed upwards, that is the lift, which holds the aircraft in the air. In detail, lift can be expressed as:

$$F_l = 1/2\, \rho V^2 A C_l \cos \alpha \tag{6}$$

where:

- ρ is the medium density;
- V is the air velocity;
- A is the reference surface;
- C_l is a **lift** dimensionless coefficient;
- α is the wing angle of attack.

In racing cars, the wing is mounted upside down and the vertical thrust towards the ground (downforce): this is correlated to the tires grip coefficient. The running resistance depends on its front section, its forward speed, the density of the medium and a drag coefficient. The drag coefficient (C_d) depends on the object shape and size of the object, the medium density and viscosity, the surface roughness, and the object velocity. The aerodynamic resistance (in general fluid dynamics), or drag, is related to a large number of factors, as shown by the formula:

$$F_d = 1/2\, \rho V^2 A C_d \cos \alpha \tag{7}$$

where:

- ρ is the medium density of the;
- V is the air velocity;
- A is the reference surface (in case of aircraft is the wing surface, the car front surface);
- C_d is a **drag** dimensionless coefficient;
- α is the wing angle of attack.

The overall resistance opposed by a fluid medium to the object forward movement is given, in first approximation, by the sum of the frictional resistance, the wake resistance and the induced resistance of lift. In particular, for a tapered body, the flow resistance is given by friction (laminar and/or turbulent), that is the rubbing of the surface against the medium. For this purpose we introduce the concept of boundary layer: it's the dynamic range, laminar or turbulent, in which internal current speed is subject to strong gradients (continuous changes), due to the viscosity of the fluid. It can be considered as the area that undergoes a disorder, and the velocity is zero on the layer surface (**Figure 3**).

The thickness of the boundary layer is very small, and it is of one order of magnitude lower than the overall dimensions of the object, that generates the viscose perturbation. Then, inside the boundary layer, the tangential shear stress is "dense". For this reason in the layer is exerted an intense dissipative braking action, converting part of the movement in thermal agitation. The dissipative action limits the relative velocity between the object and the fluid, which surrounds it. In a turbulent boundary layer, the viscous stresses are added also the stresses, due to the exchange of transverse momentum; these actions increase with the fluid density. The chaos of the turbulent fluid motions implies higher thermal dissipation, so the braking opposing force, in turbulent flow conditions, is greater than that of the laminar regime. The resistance generated, in this way, is affected by the surface roughness: moreover, the rougher surfaces ignite earlier and more easily the turbulent condition in the flow, and then, determine higher resistances. Therefore, it was decided to design some ducts, on the pressure side of the wing, initially covered by special sliding plates, for increasing the aerodynamic drag and downforce [2].

3. Wing Design

The first phase of the design is to draw the profile of the wing with a CAD software. In this way, it is possible to make a CFD simulation, to evaluate the aerodynamic performance of the wing, in terms of downforce and drag, and estimate the useful angles of attack before stall phenomenon occurs [3]. In fluid dynamics the *stall* is a reduction of the lift coefficient due to an increase of the angle of attack or due to the incident velocity decrease on an aerodynamic profile, such as an airfoil, a propeller blade or a turbomachinery rotor. The minimum value of the angle of attack for which the stall occurs is called *critical angle of attack*. This value which corresponds to the maximum lift coefficient, changes significantly, depending on the particular profile or on the considered Reynolds number [4]. Similarly, the profile of the active cavities has been reported, and appropriate simulations were performed. In this way it was possible to estimate the sizes and configurations to achieve the project target. Based on the data collected, the application of these cavities on the wing is studied, evaluating the performance on the different possible arrangements of these cavities. At this moment only 2D simulations have been performed, and a 3D series is considered as future improvement of the project. The models, the different configurations and the results obtained from all the cases mentioned above, will be shown in detail in the following paragraphs.

Geometry Modeling

To approximate the operating conditions of the wing, a control conduit with the dimensions shown in **Figure 4(a)**

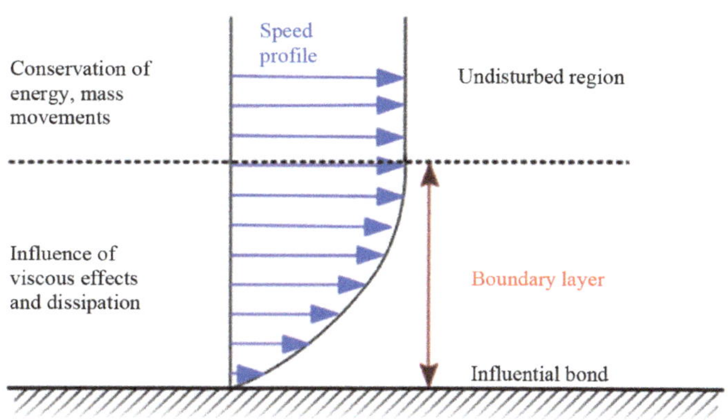

Figure 3. Boundary layer.

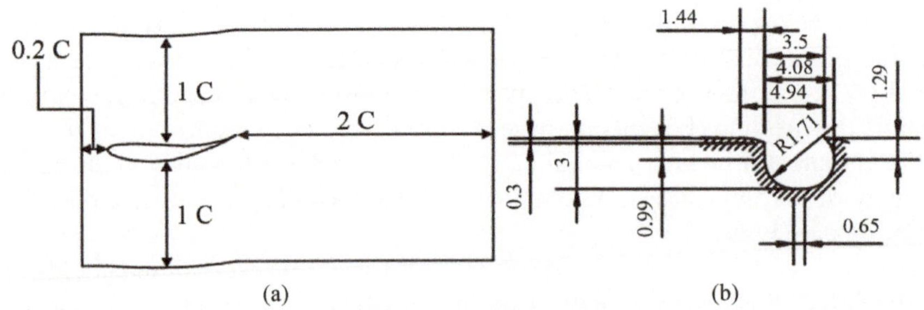

Figure 4. (a) Control conduit; (b) Active cavity dimensions (in mm).

has been chosen. Regarding to the active cavity, the geometry is illustrated in **Figure 4(b)**. The space surrounding the geometry of the aerodynamic and the cavities was discretized using a special dedicated software available as ANSYS package. Furthermore, to observe the progress of the boundary layer, it was built on a reference mesh of 5 layers, with growth factor 1.1, starting from the adjacent profiles of height 0.18 mm (**Figure 5**). To this purpose, a sizeable set of data was created by means of sufficiently accurate numerical simulations, to derive initial values. The simulations were performed on 3-D models in kinematic similarity using a commercial CFD simulation code, ANSYS/Fluent. The turbulence model was the ***k-ε* realizable**, with second order accuracy. Each model was meshed to ensure a $y^{+}_{max} \sim 5$, a necessary condition for adopting the enhanced wall treatment, since the quality of the grid has a relevant importance on the accuracy and stability of the numerical simulation.

Commercial software allows the "plastering" of cell layers to the critical boundaries of the control volume, which are obviously, in this case, the wall surfaces of the hub, casing and blades. In these zones the usual practice is that of creating a completely structured boundary layer, specifying whenever possible both the height of the first row of cells and the "growth ratio", *i.e.* the rate that determines the height of the successive cells. In this process, the height of the first row of cells is usually determined via an empirical formula that gives the value of a wall-based local Reynolds number, denoted by y^{+} ($y^{+} = u^{*} \cdot y/v$ where $u^{*} = (\tau_{wall}/\rho)^{1/2}$, with τ_{wall} being the wall shear stress). For the wing analysis control volume was split in several smaller sub-volumes, to achieve a more consistent set of faces and to better exploit the possibility of creating a locally more refined grid. The choice of the boundary conditions was made as follows: it was performed heuristically, starting from the preliminary sizing data, calibrating them by means of a first simulation, adjusting the values by iteratively resetting the outlet static pressure on the near-wake radial area downstream of the trailing edge. Through subsequent simulations the values of the inlet total pressure and temperature were refined as well in order to ensure conservation of the mass flow rate (the so-called "mass flow inlet condition" was adopted). The turbulent parameters were the turbulence intensity $I = \left(\sqrt{k}\right)/U$. Rotational periodicity was imposed on all lateral channel surfaces. The number of cells is about 65,000 elements. Finally, the starting boundary conditions are:

- fluid: it is considered air as an ideal gas at constant viscosity;
- input data: the pressure of 101325 Pa and temperature of 300 K represent the operating conditions.

boundary conditions:

- inlet → mass flow rate;
- outlet → pressure outlet;
- for both, the conditions relating to the model were set on intensity and length scales, with values of 5% and 0.03 m (≃ 1/10 of the rope), respectively;
- on the upper and lower walls of the duct it has set the periodicity condition;
- for wing, are set on the condition stationary wall and no slip;
- for the solution a simple high order term and relaxation has been chosen, by setting for all variables a relaxation factor of 0.25.

4. Simulation Results

4.1. Wing Performance

In this paragraph, the performance of the wing has been analyzed. The results obtained by using CFD simulation (see **Figures 6-8**), were used as the reference model for the subsequent tests [5]. Since, the following figure shows the results for a 0° angle of attack. In particular, for a hypothetical unitary extension wing (1 m), it is obtained:

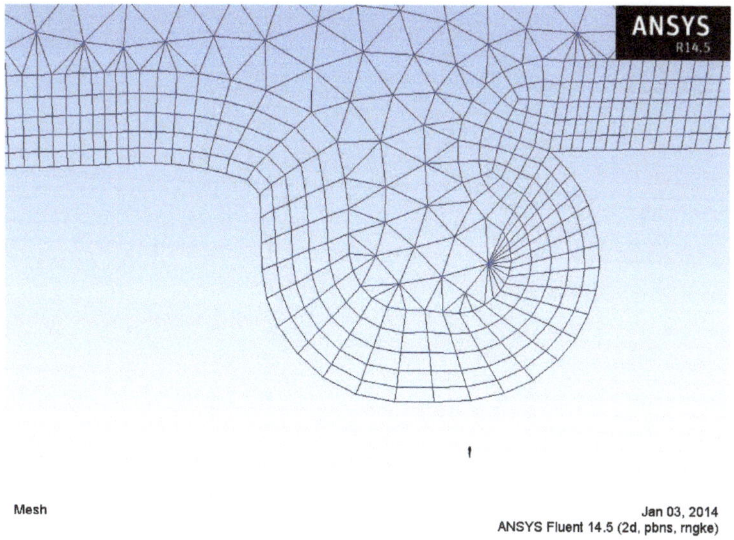

Figure 5. Cavity boundary layer mesh.

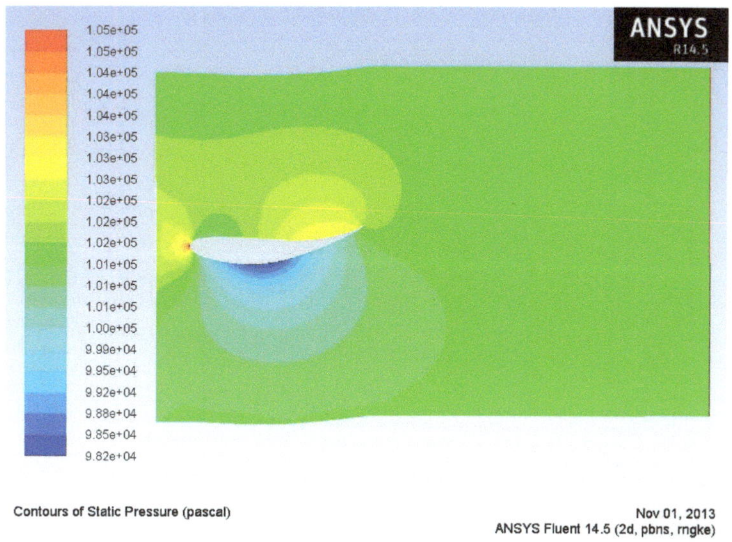

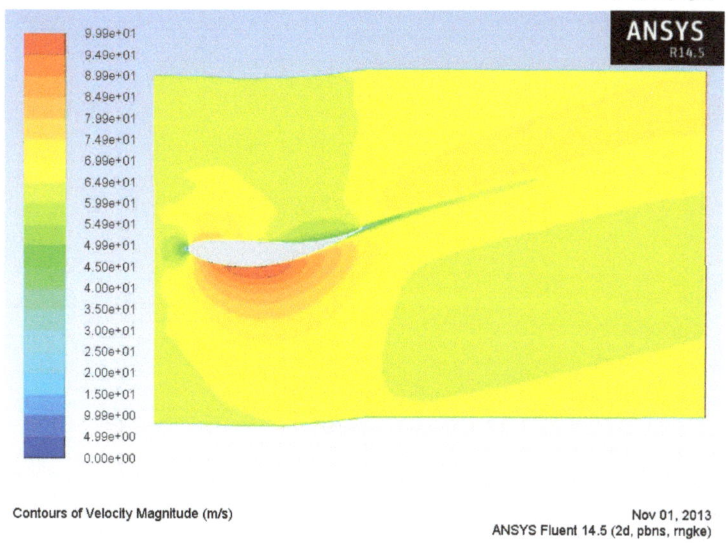

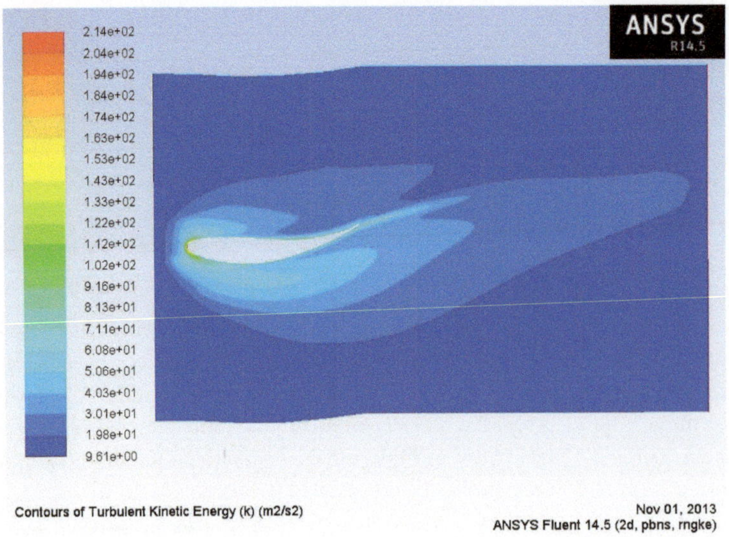

Figure 6. Pressure, velocity and turbulence plots for 0° of angle of attack.

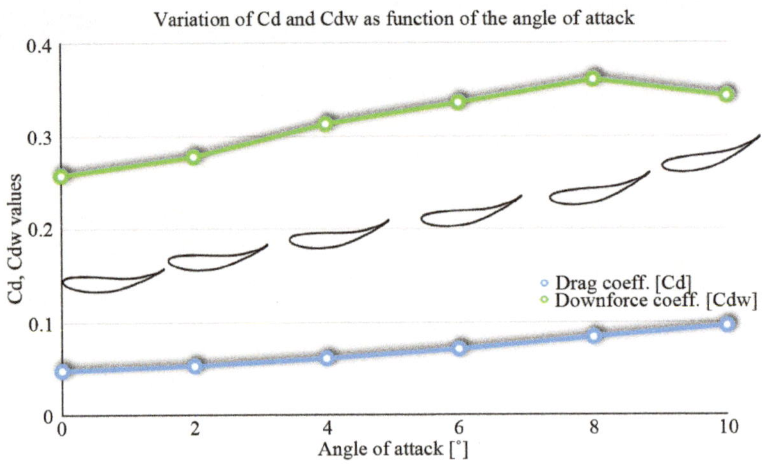

Figure 7. Plot of C_d and C_{dw} variation as function of the angle of attack.

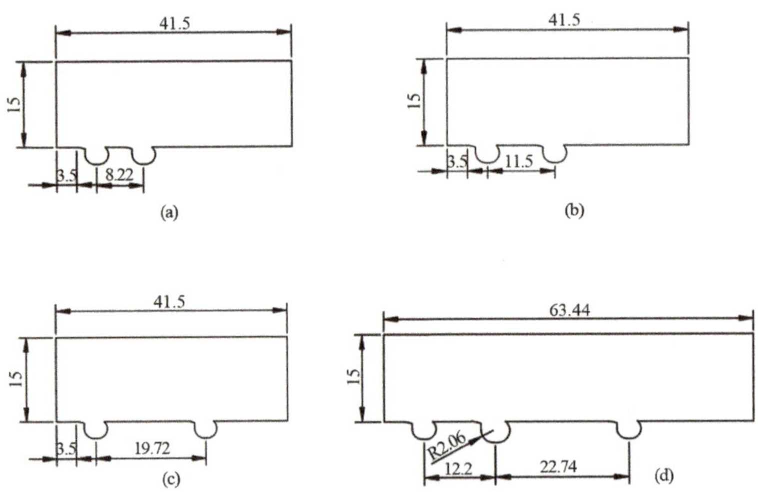

Figure 8. Different cavities configurations. (a) two cavities, (b) two cavities with space increase, (c) see case (b) (d) three cavities.

- lift coefficient $C_l = -0.25762$;
- lift $F_l = -312.66$ N;
- drag coefficient $C_d = 0.04817$;
- drag $F_d = 58.52$ N.

Finally, to individuate the stall angle, additional simulations, at different angles of attack, have been performed; precisely 2°, 4°, 6°, 8° and 10°, reporting the values obtained in **Table 1**. It can be noticed that stall occurs for angles of attack greater than 8°. The plot of C_d and C_{dw} is shown in **Figure 9**.

4.2. Active Cavity Performance Analysis

After some preliminary tests, the dimensions, that characterize the opening and depth of cavities, are 3.5 mm and 3 mm respectively. Later, built a control duct, additional simulation to find the best configuration has been carried out. In particular the configuration used were: two cavities located at 8.22 mm of distance each other, 11.5 mm, 19.72 mm and, finally, three cavities, two equal and one larger radius, interposed to 12.2 mm and 22.74 mm with respect to the remaining two (**Figure 8**). The analysis have enlightened that the quasi-optimal configuration is the one with the two equal cavities located at intermediate distance: in fact, the last configuration corresponds to the distance limit beyond which the reabsorption of the bubble pressure generated by the cavity in front occurs (circled in red in **Figure 9**). Based on these considerations, it can proceed to make the cavities on the wing to get the configurations that produce the most desired aerodynamic effects. The trend of fluid velocity is reported in **Figure 10**.

4.3. Wing with Front Cavities

According to the pressures pattern observed in the case of rear wing with zero angle of attack, in the first instance

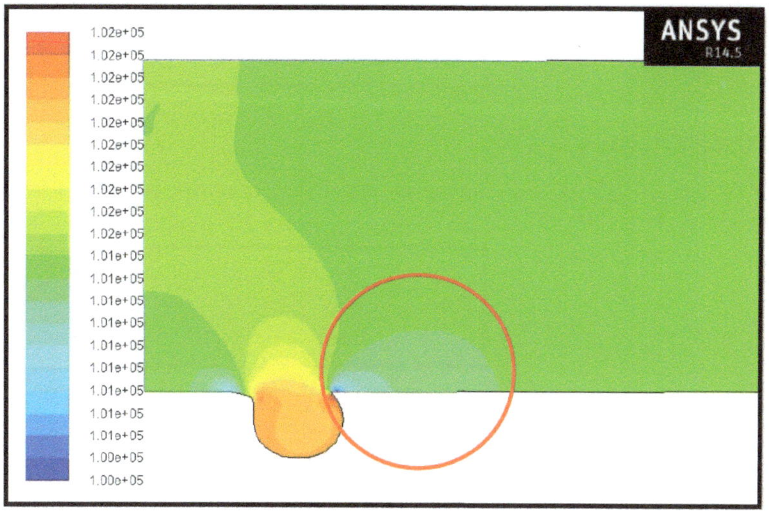

Figure 9. Pressure bubble generated by cavity. In the red circle the reabsorption of the bubble pressure generated by the cavity in front can be seen.

Table 1. Variation of C_d and C_{dw} as function of the angle of attack.

Angle of attack [°]	Drag coefficient [C_d]	Downforce coefficient [C_{dw}]
0	0.048	0.258
2	0.053	0.278
4	0.061	0.313
6	0.071	0.336
8	0.084	0.361
10	0.096	0.343

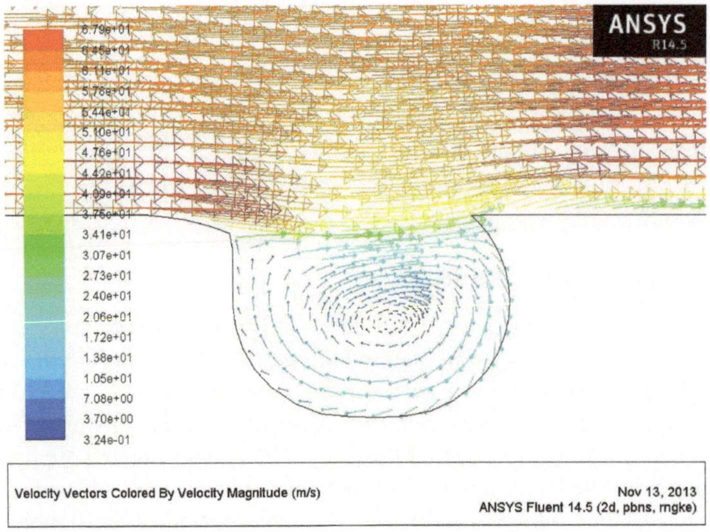

Figure 10. Velocity vectors developed inside the cavity.

it is thought to have a host of ducts on the front half of the wing, because the pressure on that section is lower than the remain aerodynamic one. After that, the CAD geometry has been modified reproducing the chosen configuration, and then we proceeded with the CFD simulation (results in **Figure 11**). The values obtained confirm the previous considerations:

$C_l = -0.26245$;
$F_l = -346.17$ N;
$C_d = 0.04777$;
$F_d = 63.00$.

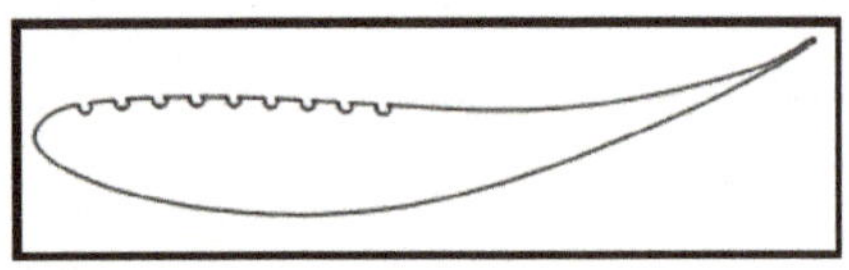

4.4. Wing with Back Cavities

For comparison, it was decided to realize the same number of cavities in the back half of the wing, leaving smooth the front half area. The CFD simulation results in **Figure 12**. In particular, it has that:

$C_l = -0.25834$;
$F_l = -341.12$ N;
$C_d = 0.04625$;
$F_d = 61.04$ N.

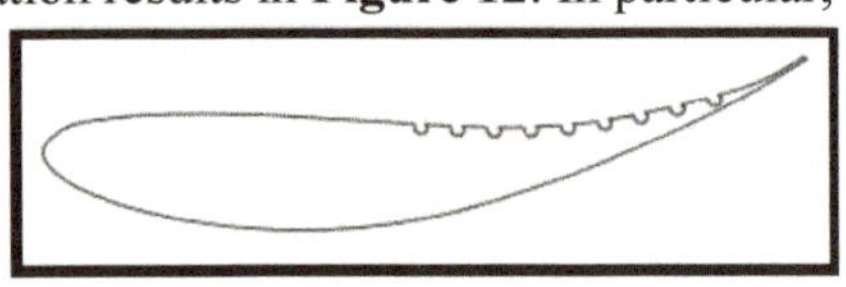

4.5. Wing with Cavities on the Whole Upper Surface

Finally (see **Figure 13**), it is decided to extend the group of active cavities over the entire pressure side. The coefficients and the forces derived are:

$C_l = -0.25611$;
$F_l = -365.17$ N;
$C_d = 0.04623$;
$F_d = 65.88$ N.

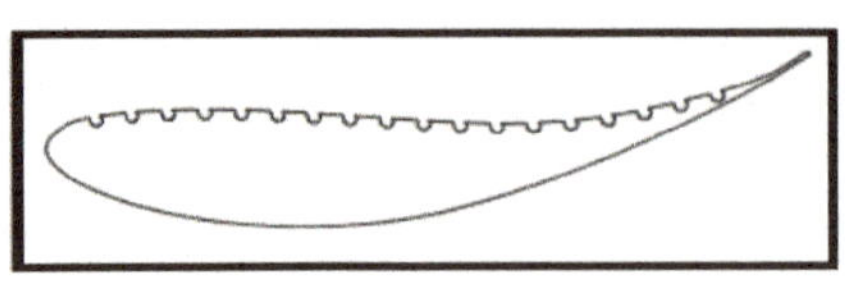

4.6. Wing with Single Cavity

For evaluating the possible influence of the single cavity on the whole group, further CFD simulations have been carried out and it is placing at different points on the back wing; appropriate pairs combinations have been also considered. Enumerating from 1 to 18 individual ducts starting from closest to the tip of the wing, the results obtained, with these additional series of simulations, are reported in **Table 2**.

It can be notice that 8 - 9 & 13 - 14 configurations denote an increase in downforce compared to the smooth wing and, at the same time, a decrease in drag.

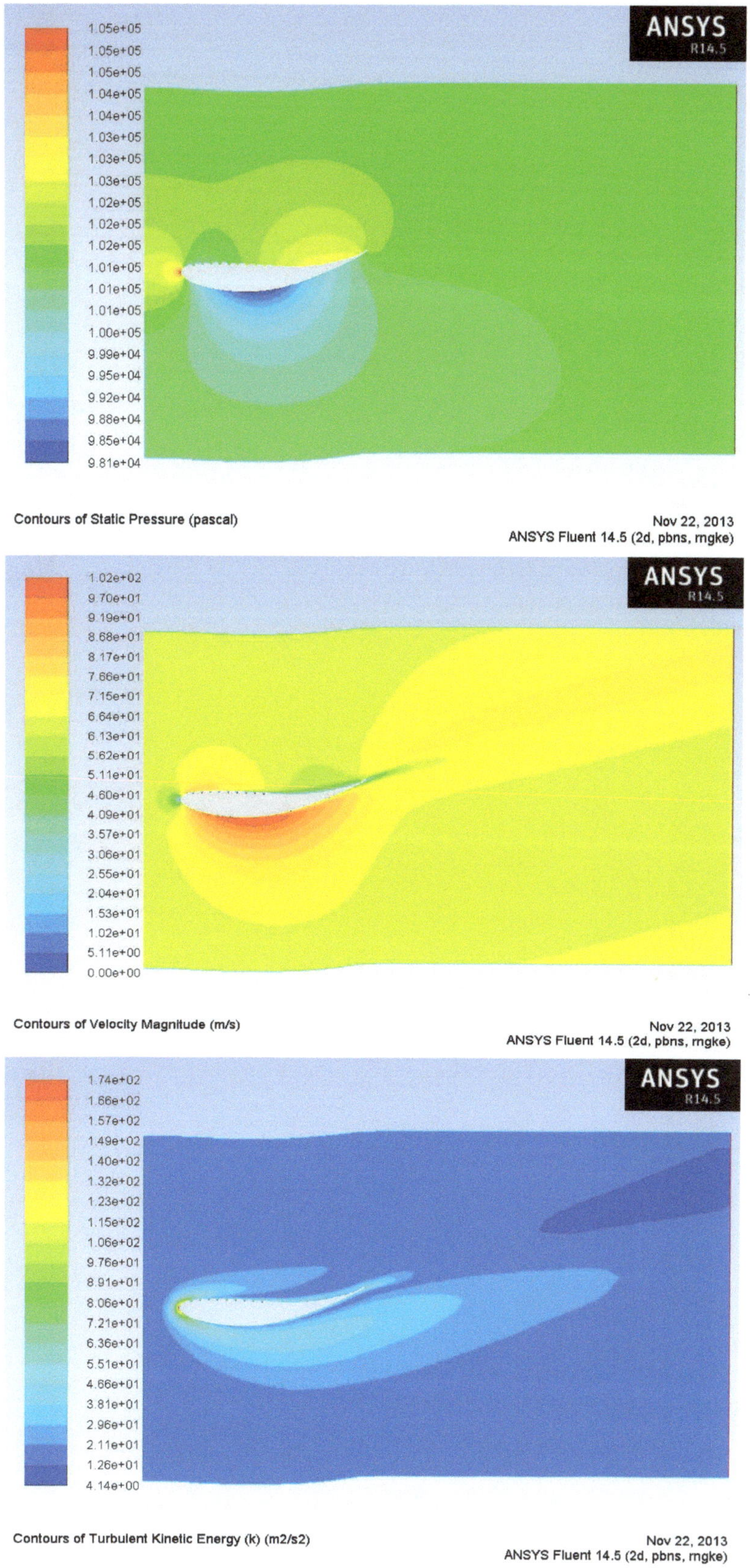

Figure 11. Pressure, velocity and turbulence plots for wing with front cavities.

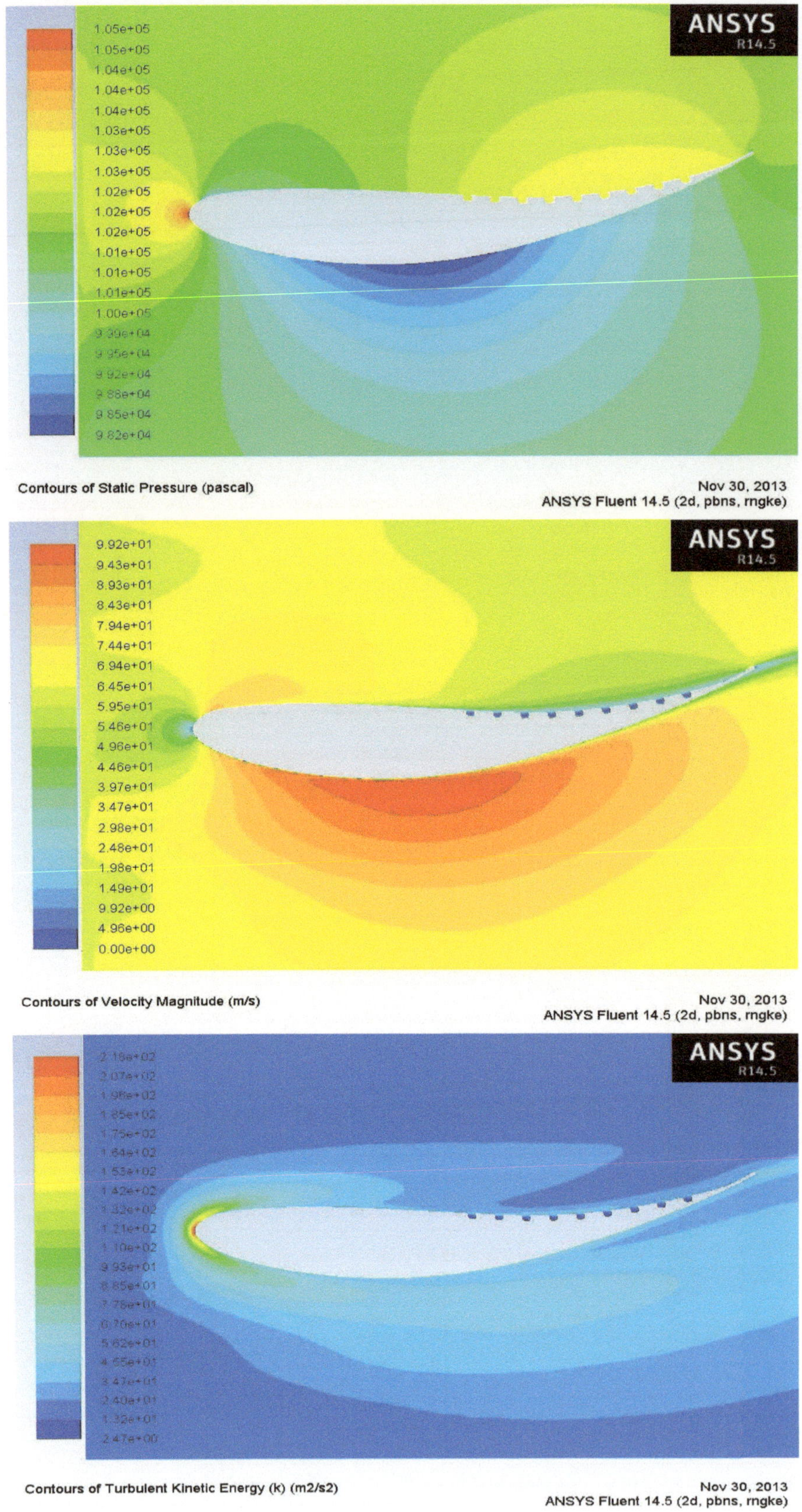

Figure 12. Pressure, velocity and turbulence plots for wing with back cavities.

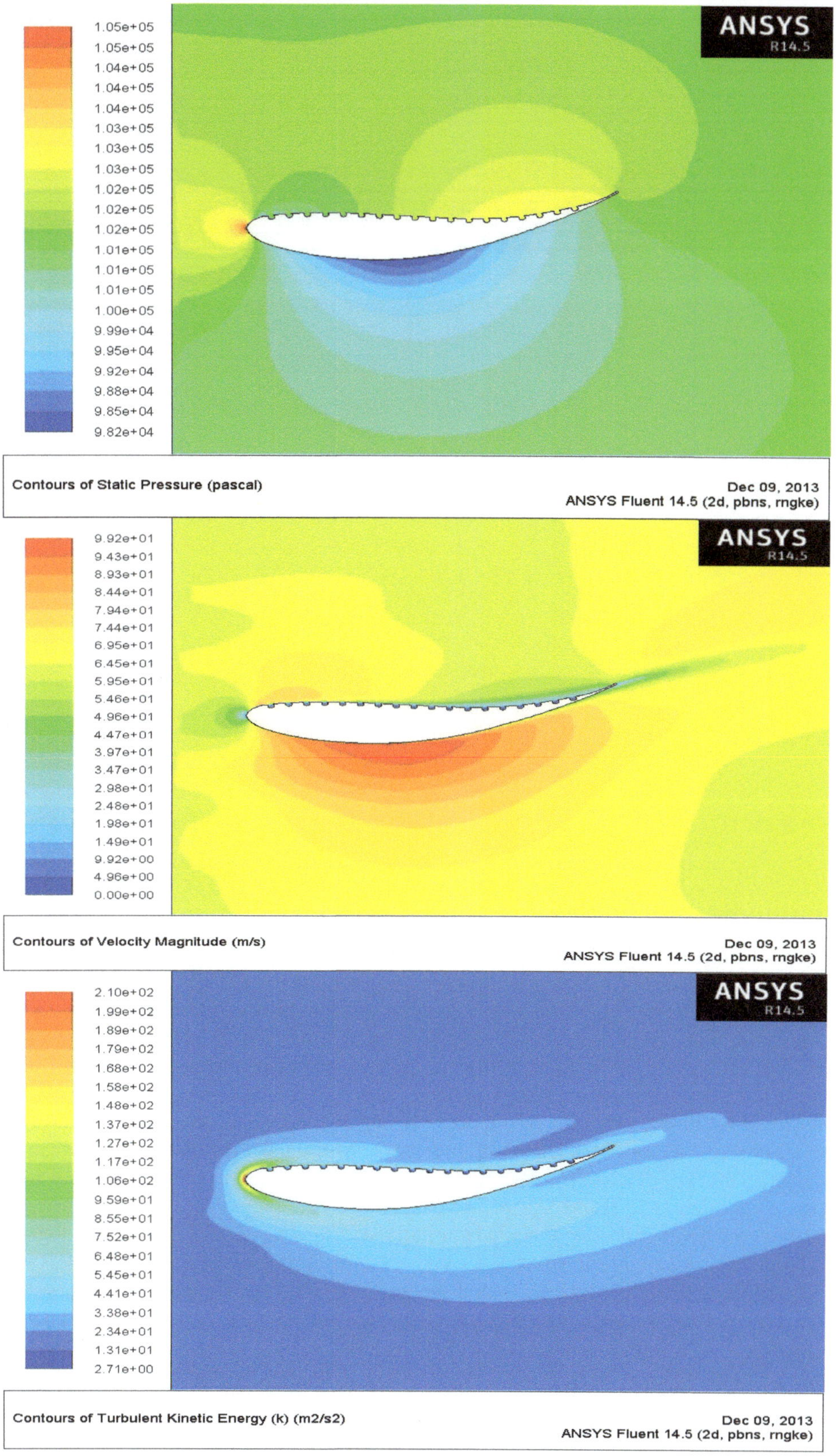

Figure 13. Pressure, velocity and turbulence plots for Wing with cavities on the whole upper surface.

Table 2. Results of CFD simulations relative to different configuration of cavities over the wing.

Cavities	Reference	C_l	F_l	C_d	F_d
3	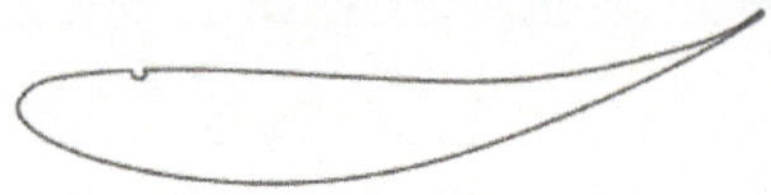	−0.25920	−317.46	0.04639	56.82
6		−0.25961	−317.96	0.04644	56.88
9		−0.25892	−317.12	0.04620	56.58
12	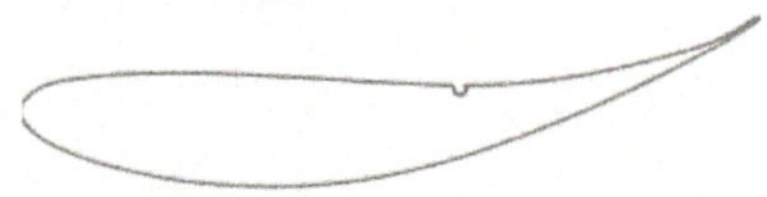	−0.25966	−318.01	0.04639	56.81
15		−0.25978	−318.16	0.04645	56.89
8 - 13		−0.25954	−320.88	0.04345	53.72
9 - 14	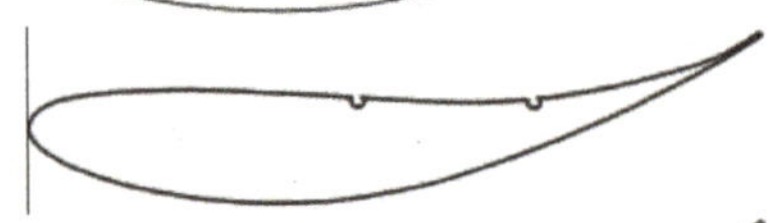	−0.25978	−321.11	0.04348	53.74
8 - 9 & 13 - 14	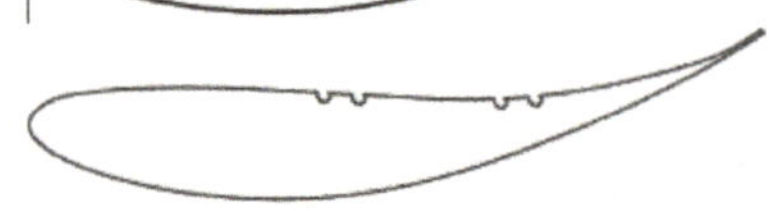	−0.25110	−316.38	0.03898	49.11
1 - 9 & 13 - 14		−0.25691	−344.88	0.04378	58.77

5. Simulation Results Analysis

The CFD simulations confirm the theoretical expectations: in particular, three of the above configurations deserve deeper analysis. In detail, it can notice that the wing with active cavities extended to the whole surface and one with the ducts only placed in the tail, show, roughly, the same value of drag (respectively 65.88 N and 61.04 N): however, in the first case, it generates greater downforce (365 N vs N 341). From this observation, it can be inferred that the rear wing with 18 ducts fulfills, in the most effective manner, a better performance, during the braking. While the same wing, but with cavities only located in the back, having a slightly lower drag, could be used during the turns: in fact, referring to the performance of the vehicle, the goal is to have a high grip to deal these turns as quickly as possible. Therefore, the choice of the most efficient set (between the configurations above) depends on the type of curve (more or less fast) and the balance between the drag and downforce forces. So, equipping the wing of pairs 8 - 9 and 13 - 14, for its characteristic to generate more downforce and a lower drag than the smooth aerodynamic profile, it could be used in starting. In fact, in this situation, to have a good sprint, it is desirable to have a low aerodynamic resistance and prevent slipping of the rear wheels: this phenomenon can be achieved by exploiting the downforce generated by the wing in this configuration. The result would be a better grip. Ultimately, you should have a wing that can satisfactory answer the two situations. The configuration must be able to provide an excellent grip and a lower resistance in starting, and get a better behavior in curves. This can be achieved by equipping the wing of sliding panels that open or close the cavities, according to the request of the vehicle.

6. Conclusions and Possible Improvements

The CFD simulations indicate the effectiveness of active cavities, practiced on a formula car rear wing, in order to achieve an aerodynamic brake. Specifically, we can assert that the configuration with the best balance between downforce and drag is that with extended ducts over the entire top surface of the aerodynamic (F_l = −365.172 N, F_d = 65.88 N). Finally, by exploiting the selectivity of sliding panels, as previously explained, it can be realized different wing configurations, depending on the needs required by the race and the sensitivity of the driver.

As regards any improvements, to be made on the performance testing of the brake (the object of study of this paper), more CFD simulations could perform, using a vertical bulkhead on the terminal part of the wing, as well as 3D simulations. In this way, it would be possible to observe the effects of lift drag which should be minimized. Then the next step would be to achieve physically this device, to install it on a formula car and compare experimental data with those obtained by CFD simulations.

Acknowledgements

The research has been partially sponsored by the Italian Government on the framework of Formula Cars/Formula 4 project.

Author Contributions

Roberto Capata and Leone Martellucci contribution dealt with the tests and comparisons, while Enrico Sciubba was in charge of all aspects related to CFD simulations.

References

[1] 2014 Formula 4 Technical Regulations, FIA. http://www.fia.com/sites/default/files/basicpage/file/FIIII_plaquette_full_formula4.pdf

[2] Gamma, F., Sciubba, E., Zingaro, D. and Farello, G.E. (2002) Fluid Dynamic Behavior of Heat Exchangers with Active Cavities: A Numerical Study. *Numerical Heat Transfer Applications*, **42**, 385-400.

[3] Chandra, S., Lee, A., Gorrel, S. and Greg Jensen, C. (2011) CFD Analysis of PACE Formula 1 Car. *PACE*, **1**, 1-14. http://dx.doi.org/10.3722/cadaps.2011.PACE.1-14

[4] Prasad. A.K. and Koseff, J.R. (1989) Reynolds Number and End-Wall Effects on a Lid-Driven Cavity Flow. *Physics of Fluids A: Fluid Dynamics*, **1**, 208-218. http://dx.doi.org/10.1063/1.857491

[5] Chen, C.-L., Chung, Y.-C. and Lee, T.-F. (2012) Experimental and Numerical Studies on Periodic Convection Flow and Heat Transfer in a Lid-Driven Arc-Shape Cavity. *International Communications in Heat and Mass Transfer*, **39**, 1563-1571.

Nomenclature

A	surface [m^2]
C	coefficient, chord [m]
CFD	computational fluid dynamic [-]
DRS	drag reduction systems [-]
F	force [N]
I	turbulence intensity [%]
k	turbulent kinetic energy [J]
t	time [s]
u, U	velocity [m/s]
v	air speed [m/s]

Greek Symbol

α	wing angle of attack [°]
ε	turbolent dissipation [J]
μ	dynamic viscosity [kg/m·s]
ρ	density [kg/m^3]
τ	shear stress [Pa]
ν	kinematic viscosity [P]

Subscripts

d	drag
dw	downforce
l	lift
ε	kinetic dissipation
μ	viscous

18

Analytical Model on Steel Tanks Damaged by Corrosion

Francisco Casanova-del-Angel[1], Moisés Gaytán-López[2]

[1]National Polytechnic Institute, Mexico City, Mexico
[2]SACM SMA, Government of Mexico City, Mexico City, Mexico
Email: fcasanova@ipn.mx, mogalo_750315@hotmail.com

Abstract

The mechanical behavior of steel employed in the hull of a steel tank damaged by corrosion has been analyzed. The tank was used to filter a deep-water well for an 8-year period. Influence of porosity and dissolution of material may be introduced in the main constitutive relation adding a new damage variable *C*, which describes electrochemical damage. An elastoplastic theoretical model coupled to damage of a member, and other for damage related to thermodynamic energy are developed. This theoretical development has been used to analyze mechanical behavior of steel used in the body of a steel tank damaged by corrosion in water purifier plants, Eastern System, Mexico City, where three of every ten filters show excessive corrosion inside the steel plate filtration tanks. With samples taken from steel of the tank's hull and reinforcement of false bottom supporting filtering material inside the tank, metallography tests were carried out; localized and generalized types of corrosion were determined, as well as the type of corrosion composites generated due to anticorrosive coating used inside the tank from its manufacturing.

Keywords

Structural Corrosion, Damage Mechanics, Steel, Spectrometry, Tensile Stress Loss

1. Introduction

In presence of high magnitude overloads, structures show damage symptoms, characterized by degradation of elastoplastic properties. This is called *damaging process.* The structure, as in an uniaxial assay, goes through an elastic phase modeled by elastic behavior laws and, then, through a plastic phase with hardening (modeled by the elastoplastic behavior with hardening laws). Such hardening process is gradually lowered due to the damaging process, until the last load of the structure is reached and beginning a softening process, where damage rules over hardening, finally breaking the probe, even causing solicitation of an imposed displacement. The state law, considering the damage, is obtained taking into consideration the damaged area, as we will see throughout our theoretical development, entitled: Development of a model for damage related to thermodynamic energy.

Related to control of incrustation and corrosion when water is in rain form, it may be supposed that it is chemically balanced with the surrounding environment. When touching the soil, water dissolves certain mineral components of such, among others, calcium and bicarbonate ions. Carbon dioxide required for the reaction is continually obtained from decomposition and oxidation of organic matter of the soil due to activity of microorganisms. When water rich in calcium ions and bicarbonate is extracted from the soil and is re-exposed to air, reaction becomes inverted, carbon dioxide escapes to the air and calcium carbonate also precipitates once more.

Constitutive models define a relation between stresses and strains. Constitutive models comprised in mechanics of continuous media are presented as a promising tool to treat structural response, allowing calculation of geometrically complex structures based on finite elements codes, which allow putting together various constitutive phenomena into one formulation.

2. Mechanical Damage

The D_n scalar damage variable is defined as the ratio of the damaged area A_d to the nominal area A may be expressed as follows:

$$D_n = \frac{A_d}{A} \tag{1}$$

where D_n is the damage variable towards n direction, A is the intersection area with the plane, and A_d is the effective area damaged by corrosion, contained in A.

The value of D scalar damage variable is between 0 and 1 ($0 \leq D \leq 1$). $D = 0$, for a material without damage, and $D = 1$ for a material completely broken. In fact, failure occurs in a $D < 1$ value, through instability process.

The term *effective stress*, related to the effective surface resisting load, (A - A_d), is:

$$\sigma^{\sim} = \frac{F}{A - A_d} \tag{2}$$

Introducing in (2) the damage variable

$$D = \frac{A_d}{A} \tag{3}$$

Thus obtaining Equation (4):

$$\sigma^{\sim} = \frac{F}{(1-D)A} \tag{4}$$

Equation (4) defines effective stress in a material under tension. Finally, strain equivalent to the principle proposed by Lemaitre used in the elastic damage scalar classical form is:

$$\varepsilon_{\epsilon} = \frac{\sigma^{\sim}}{E} = \frac{\sigma}{(1-D)E} \tag{5}$$

In this model, the mechanical effect of progressive surface loss caused by corrosion and external load is described by a single internal variable, which degrades Young's model for material. The constitutive relation is:

$$\sigma_{ij} = (1-D) A_{ijkl} \varepsilon_{kl} \tag{6}$$

where σ_{ij} and ε_{kl} are the components of stress tensor and strain, respectively (i, j, k & $l \in [1, 3]$), A_{ijkl} is the starting rigidity module, and D is the damage variable defined before. Material is isotropic at the beginning and, together with E and μ, Young's model and Poisson relation are respectively considered.

3. Damage by Corrosion

Since increase in material dissolution, in turn, increases porosity, its mechanical influence has been considered to be similar to the increase of holes and micro fissures, thus producing degradation of material in the area being damaged. In light of this, it is logical to introduce variable C in the stress-strain relation, in a similar way to the treatment of mechanical damage C, also within 1 and 0: where 0 for a material without electrochemical damage and 1 for a material completely dissolved. The stress-strain relation containing the two types of damage, both mechanical and corrosion, is given in (7):

$$\sigma_{ij} = (1-D)(1-C)A_{ijkl}\epsilon_{kl} \tag{7}$$

where C is the variable for damage by corrosion obtained through an electrochemical measure system, calibrated to obtain a behavior similar to mechanical damage D. It should be noticed that a similar approach was obtained for chemical damage in [1] [2].

For calculation of damage parameter D, the damaged area value is obtained through (8), **Figure 1**, where A is the nominal area, w and t are the dimensions of the damaged area, D_n is the scalar damage variable, A_d is the damaged area and a is the thickness with corrosion damage.

$$A = w*t \text{ and } D_n = \frac{A_d}{A} \tag{8}$$

$$A_d = a*w$$

3.1. Elastoplastic Model Coupled to Damage of a Member

Generalized strains in joints in an elastic joints with damage and plasticity are expressed through equation (9) and are:

$$\phi = \{\phi^e\} + \{\phi^P\} + \{\phi^D\} = [F^0]\{M\} + \{\phi^P\} + \{\phi^D\} \tag{9}$$

where ϕ^e is the strain of elastic beam, ϕ^P is the plastic strain, ϕ^D is the strain due to damage, F^0 is the flexibility matrix of an elastic beam column in starting state without damage and $\{M\}$ the matrix of the main directions of damage.

Supposing there is an axial generalized stress N without buckling, the $\sigma = (1-D)E(\varepsilon - \varepsilon_P)$ relation allows obtaining a generalized stress-strain relation. With a uniform damage condition, the following is obtained:

$$\delta - \delta^P = \frac{N}{(1-D)K_{33}^0} \tag{10}$$

where $K_{33}^0 = AE/l$.

On the other hand, in accordance with the concentrated elasticity model, generalized strains of the member are expressed through the following relation:

$$\delta = \frac{N}{K_{33}^0} + \delta^P + \delta^D \tag{11}$$

Deleting the $\delta - \delta^P$ term between Equations (10) and (11) we obtain:

$$\delta^D = \frac{D}{(1-D)K_{33}^0} N \tag{12}$$

Equation (12) means that, in order to obtain a concentrated an elasticity model equivalent to that obtained through damage of continuous media theory, it is necessary to define axial generalized strain. It may be ob-

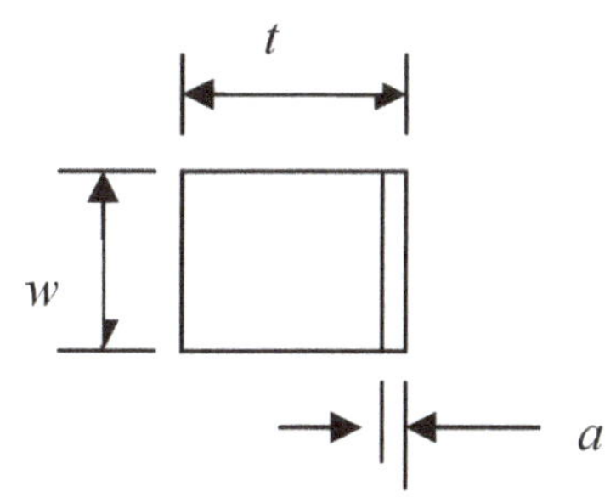

Figure 1. Mechanical damage.

served that, when damage has a zero value and joints have null flexibility (or an infinite rigidity), they behave as rigid-plastic connecting rods. If damage is equal to one, flexibility is infinite. In this case, it is equivalent to imagine that joints and beam column are disconnected and, therefore, the system may not transmit axial load.

Theoretical developments tells us that, when there are buckling effects, behavior is too complex to obtain explicit analytic expressions similar to Equation (12) valid for the general case, even using extremely simple damage evolution laws. Therefore, the existence of a set of damage internal variables $\{D\}^t = (d_i, d_j, d_a)$ is presented which may, as the damage of continuous media variable, have values within the [0, 1] interval, such that behavior of an elastic joints may be expressed as:

$$\{\phi^D\} = [C(D)]\{M\} \tag{13}$$

where $[C(D)]$ is a diagonal matrix which non-null terms are the elements of its diagonal:

$$[C(D)] = \begin{bmatrix} \dfrac{d_i}{(1-d_i)K_{11}^0} & 0 & 0 \\ 0 & \dfrac{d_j}{(1-d_j)K_{22}^0} & 0 \\ 0 & 0 & \dfrac{d_a}{(1-d_a)K_{33}^0} \end{bmatrix} \tag{14}$$

d_i and d_j parameters represent a measurement of buckling damage in joints *i* and *j*, respectively. Therefore, the state law of a degradable elastoplastic member is obtained substituting equation (14) in Equation (13):

$$\{\phi - \phi^P\} = [F^D(M,D)]\{M\} \tag{15}$$

where $[F^D(M,D)] = [C(D)] + [F^0(M)]$ is the flexibility matrix of a degradable member and in the specific case of small strains:

$$[K^D(D)] = ([C(D)] + [F^0])^{-1} \tag{16}$$

With $[K^D(D)]$ the rigidity matrix of a degradable member. And for the specific case of a constant transversal section member with area *A*, inertia *I*, elasticity module *E* and length *l*, the rigidity matrix elements $[K^D(D)]$ are expressed as follows:

$$[K^D(D)] = \begin{bmatrix} \dfrac{(1-d_i)(4-d_j)4EI}{4-d_i d_j l} & \dfrac{4(1-d_i)(1-d_j)2EI}{4-d_i d_j l} & 0 \\ \dfrac{4(1-d_i)(1-d_j)2EI}{4-d_i d_j l} & \dfrac{(1-d_j)(4-d_i)4EI}{4-d_i d_j l} & 0 \\ 0 & 0 & (1-d_a)\dfrac{AE}{l} \end{bmatrix}$$

As it may be verified, in case of the existence of a completely degraded joint $(d_i = 1, d_j = 0, d_a = 0)$ the degraded rigidity matrix coincides with that of a member with an internal joint. And the d_a damage variable reduces rigidity of the member in presence of axial actions.

Complementary strain energy of U^* member may be expressed as the addition of the complementary strain energy of the elastic beam column W^* and the complementary strain energy of joints:

$$U^* = W^* + \frac{1}{2}\{M\}^t[C(D)]\{M\}$$

And small strains

$$U^* = \frac{1}{2}\{M\}^t[F^d(D)]\{M\}$$

$$U = \frac{1}{2}\{\phi - \phi^P\}\left[K^d(D)\right]\{\phi - \phi^P\}$$

where U is the strain energy.

Generalized strain or stresses are obtained from complementary strain energy or strain energy, respectively, through:

$$\{\phi - \phi^P\} = \left\{\frac{\partial U^*}{\partial M}\right\} \text{ and } \{M\} = \left\{\frac{\partial U}{\partial \phi}\right\}$$

3.2. Damage According to Thermodynamic Energy

Based on damage mechanics, thermodynamic strength related to damage G_a is determined, provided by (17), as derived from complementary strain energy, U, related to damage d_a:

$$G_a = -\frac{\partial U}{\partial d_a} = -\frac{\partial U^*}{\partial d_a} = \frac{1}{2}\frac{1}{K_{33}^0}\left(\frac{N}{1-d_a}\right)^2 \tag{17}$$

where U^* is the complementary strain energy of member $K_{33}^0 = \frac{AE}{l}$ with A as total transverse area of the element, E as elasticity module, and l the length of the element. N is the axial generalized stress.

The concept of thermodynamic strength related to damage comes from classical theories of degradation and fracture and may be interpreted as the *strength causing damage.* Deriving strain energy related to $\{\phi^P\}$, thermodynamic stresses related to plastic generalized strains coincide with generalized strains, which justifies to have considered fluency functions as depending on such.

4. Introduction to Application

In 1869, James P. Kirkwood, Chief Engineer of the Water Department, St. Louis, USA, described European water plants in a report, which served the American Water Works Association, AWWA as a guide throughout many years to manufacture filters. Many of the first experimental works on slow-action sand filters were carried out at the Lawrence experimental station, Sanity Board, Massachusetts State, USA, which started to operate in November, 1887, and was under supervision of Allen Hazen from summer, 1888, throughout March 1893 [3]. From 1949, Jacobsen researched the dynamic effect of the fluid in aircraft containers, in 1952 Graham and Rodríguez considered that dynamic pressures of the fluid on the walls may be divided in impulsive and convective pressure. In 1957, Hosner picked up Graham and Rodríguez' approach and proposed an analysis procedure based on a simple mass-spring model. In practice, seismic analysis of storage tanks is based on such methodology.

Mexico City has a population of more than 8,851,080 inhabitants. In 1951, the City required 4 m^3/s of potable water; and in 1976 it required 9 m^3/s: this increment was achieved by drilling wells in Lerma Valley [4]-[6]. As water demand increased, water was obtained from a water collection system called *Sistema Cutzamala.* However, such source has been insufficient. Therefore, the extraction of underground water has become necessary, taking advantage of the fact that Mexico City is seated on a natural basin. Mexico City is subdivided in 16 boroughs. Iztapalapa is the most populated boroughs in the City, with 15,789 inhabitants/km^2 within a 115 km^2 territory, with an inhabitant population projection of 1,850,721 by the year 2015, and a foreseen demand of 8066.42 l/s. Given that it receives 3961.32 l/s, it has an insufficient outflow for the demand.

Related to the influence of well water components in the corrosion behavior of filtering tanks, **Table 1** shows two heavy metals, iron and manganese, with high influent values: 0.428 mg/l for iron and 0.272 mg/l for manganese. Regarding effluent, the 0.02 mg/l value is lower than the limit, for iron, and 0.01 mg/l higher for manganese, which defines the standard for both cases. In addition, this shows that industrial activity exists in the area where the well is located. Once purification processes and outflow to be potabilized are determined, reinforced concrete, steel or other material is used to construct tanks, sumps, containers, etc. **Figure 2** shows location of 16 water treatment plants in the eastern and southern areas in Mexico City, the operation area, as well as the corresponding boroughs.

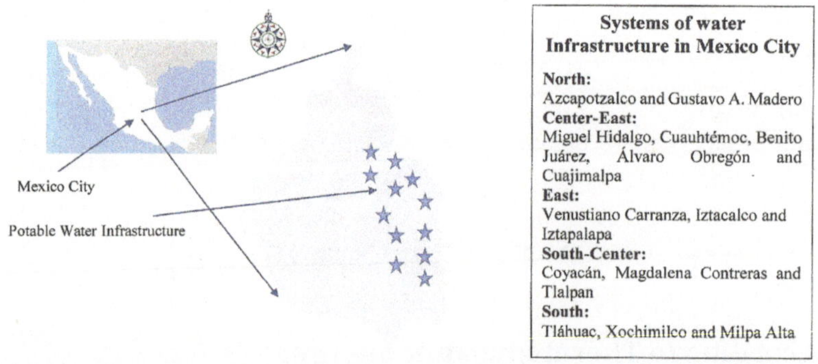

Figure 2. Areas in the Federal District for operation of potable water infrastructure. Taken from the Master Potable Water Plan for the Federal District, 1997-2010.

Table 1. Energy U_T from tensile test and ductility value of steels analyzed.

Steel	Energy in elastic area up to ε_y	Energy (U_t) in elastoplastic area between ε_y and $\varepsilon_{máx}$	Energy in damaged area between $\varepsilon_{máx}$ and ε_y limits	*Strain* ε_y	*Strain* $\varepsilon_{máx}$	*Strain* ε_u	*Ductility* $\frac{\varepsilon_{max}}{\varepsilon_y}$
Not corroded ***A284 Grade C steel***	0.3443	88.49	29.10	0.003	0.2148	0.295	53.70
Corroded ***A284 Grade C steel***	0.25	59.60	23.27	0.0025	0.185	0.263	46.25
Not corroded ***A36 steel***	0.46	37.71	22.98	0.0025	0.078	0.124	31.2
Corroded ***A36 steel***	0.35	20.97	15.04	0.0022	0.050	0.083	25.0

After the first 8 years of operation of the eastern system water treatment plants in Mexico city, 3 out of 10 filters showed exceedingly high corrosion inside the filtering tanks with A284 Grade C steel plates and $e = 9.8$ mm wall thickness: indeed the corrosion rendered the filters useless. This problem made evident the need to experiment the evolution of corrosion and the structural behavior of the tank damaged by corrosion within the laboratory. Such damage by corrosion generates a local failure in the joint of the false bottom, supporting filtering material, with the wall of the auto-supported tank. Such local failure causes total and/or partial suspension of potabilized water supply to Mexico City. Due to the functioning of such pressurized steel tanks, it is not possible to periodically check the evolution of corrosion inside them: the operation filtering tank would have be suspended, thus affecting potable water supply for up to forty days, *i.e.* the period required to remove filtering material and properly check false bottom and the wall of the tank.

We move now to look at the development of the experiment, in order to obtain data related to corrosion in the wall of the damaged filtering tank within a determined period of time; damage analysis, both mechanical and by corrosion; corrosion speed assessment; and discussion of results.

4.1. Experimental Development

Metallographic analysis was carried out to confirm steel used, the corrosion compounds, as well as to determine of the type of damage by corrosion; in A284 Grade C steel with $e = 9.8$ mm plate thickness used in the hull of the tank and A36 steel with $e = 6.35$ mm plate thickness used in the reinforcement of the false bottom.

Six samples of two different thicknesses intervening in the structure of the tank were taken, hull of the tank with 9.8 mm thickness, and a sample of the beams supporting the false bottom inside the tank, 6.35 mm thick, **Figure 3(a)** and **Figure 3(b)**, based on [7]. Dimensions of the sample were 30 × 20 × 9.5 mm and 30 × 20 × 6.35 mm.

4.2. Results of Metallographic Analysis

Figure 4(a) shows the resulting microstructure of "Small Sample" steel pieces (A36 steel, dimensions: 30 × 20 × 6.35 mm) and "Small Sample" (A284 Grade C steel, dimensions: 30 × 20 × 9.52 mm). In the Small Sample

(a) (b)

Figure 3. Steel tank damaged by corrosion. (a) Interior of a steel tank under analysis; (b) reinforcement of the false bottom plate. View from below the false bottom.

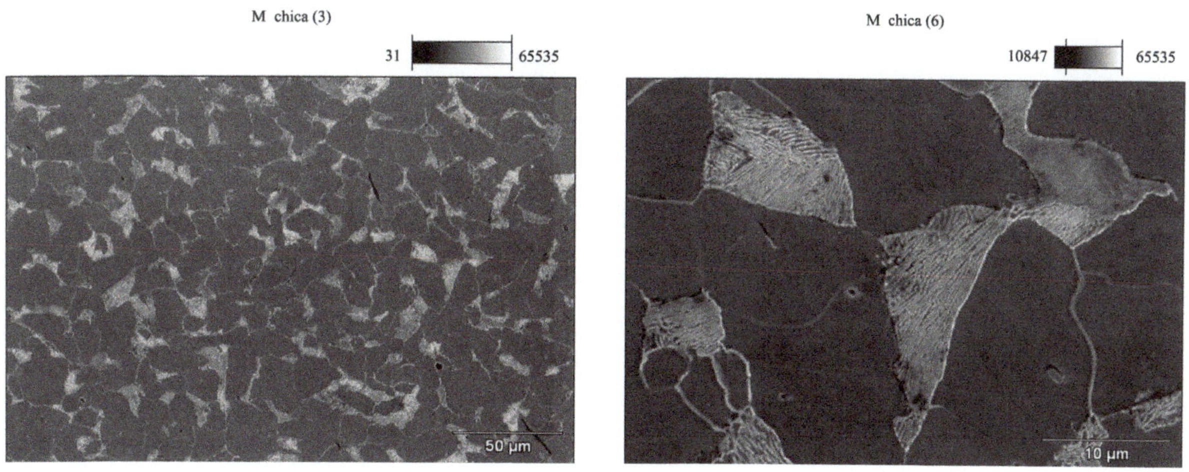

(a) (b)

Figure 4. Microstructure of A36 steel piece. (a) Small sample steel pieces. A36 steel with dimensions: 30 × 20 × 6.35 mm; (b) small sample steel pieces. A284 Grade C steel with dimensions: 30 × 20 × 9.52 mm.

shown in **Figure 4(b)** as *M_chica* (6) a high quantity of inclusions may be observed, relatively higher than those in sample *M_grande* (3). *M_chica* (3) macrograph, obtained by means of secondary electron technique, reveals microstructure of the sample, which includes perlite grains approximately 50 µm - 70 µm long, surrounded by ferrite. This micrography shows that the presence of such inclusions is still evident. Sample *M_grande* (3) shows a uniform distribution of inclusions in a lower proportion, and less when compared with the Small Sample, which has an approximate interval of 3 µm - 5 µm. Micrography of figure *M_grande* (3), obtained by means of compositional model technique (retro-dispersed electrons), shows a microstructure with perlite grains 80 µm and 100 µm long, with inclusions.

Spectrum of **Figure 5** shows, separately, elements which, in the corresponding environment, react creating compounds which generate the development of corrosion in the metal. **Figure 6** shows the profile of the A284 Grade C steel sample, where damage by corrosion may be observed.

4.3. Assessment of Corrosion Speed

In accordance with results obtained by metallographic analysis, an acceptable 0.2 mm/year corrosion speed is determined for A284 Grade C steel probe, which allows us to consider that, in another 8-year operation period, the structure will also maintain an acceptable corrosion speed, since corrosion compounds generate a passivation

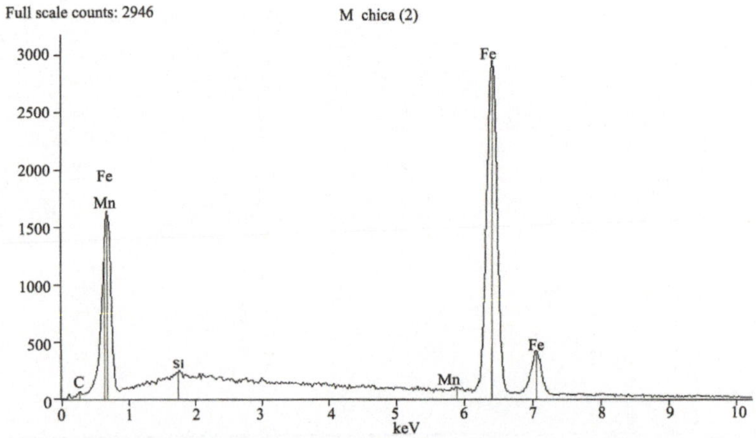

Figure 5. Distribution of chemical elements frequencies conforming metal and corrosion elements (A36 steel).

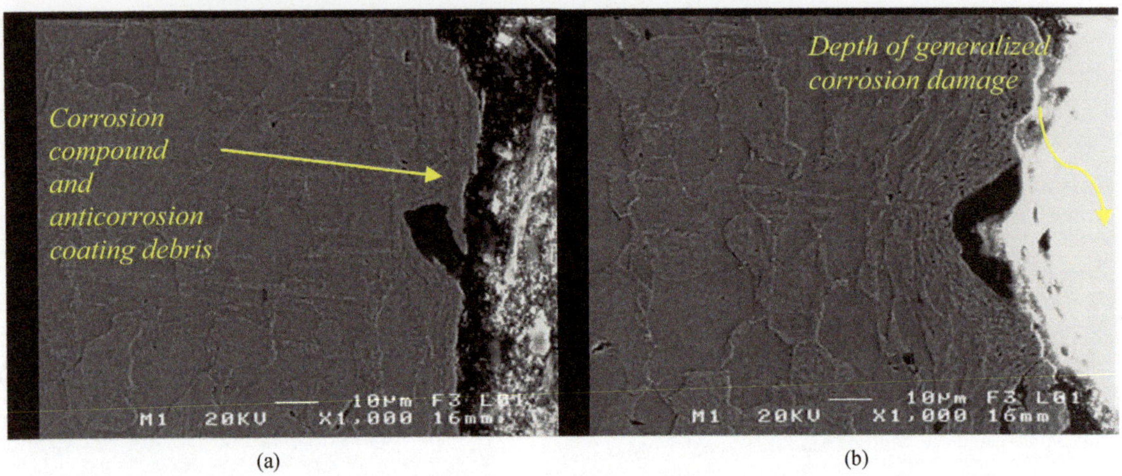

Figure 6. Microstructure (profile) of A284 Grade C steel named *M_grande* (3). (a) Corrosion compound and anticorrosion coating debris; (b) depth of generalized corrosion damage.

layer controlling and inhibiting corrosion, which allows us to expect proper behavior of structure, even in case of losing the whole coating due to corrosion, as provided by the Rules for Construction in the Federal District, which is 1/6 of calculated plate thickness. In this case, there was a 3 mm increase [8].

4.4. Tensile Strength of Analyzed Steel

Tensile strength of material is obtained under the stress-strain curve of the tensile test integrated between yield strength and maximum stress (ε_y and $\varepsilon_{\text{máx}}$). Adjusted curve for A284 Grade C not corroded steel provides us with a Degree 5 polynomial, which integration between points $\varepsilon_y = 0.0025$ and $\varepsilon_{\text{máx}} = 0.2148$ shows a 88.495 MPa tensile strength of material, **Figure** 7, blue curve. For damaged area between $\varepsilon_{\text{máx}} = 0.2148$ and $\varepsilon_u = 0.295$, energy in the damaged area in the reference probe is: 29.10 MPa, **Figure** 7, blue curve. For not corroded A284 Grade C steel, tensile strength U_t of not corroded A284 Grade C steel is: 59.60 MPa, **Figure** 7, red curve. For the damaged area, there is a Degree 4 polynomial, and the value of energy in the damaged area is 28.273 MPa, **Figure** 7, red curve.

4.5. Not Corroded A36 Steel

For not corroded A36 steel, there is a Degree 6 polynomial with limits: $\varepsilon_y = 0.0030$, $\varepsilon_{\text{máx}} = 0.078$ and 37.26 MPa tensile strength U_t, **Figure 8**, blue curve. Energy (equal to the area under the stress-strain curve for not corroded

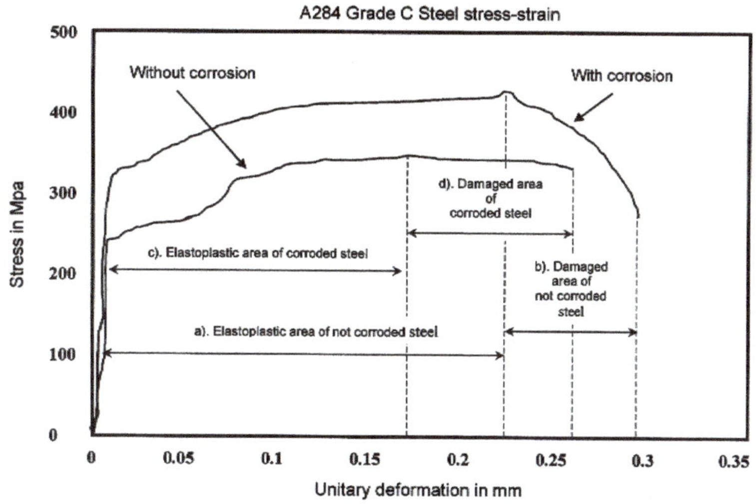

Figure 7. A284 Grade C Steel. (a) Tensile strength calculation for reference not corroded probe; (b) energy calculation in damaged area for reference not corroded probe; (c) tensile strength calculation for corroded probe; and (d) energy calculation in damaged area for corroded probe.

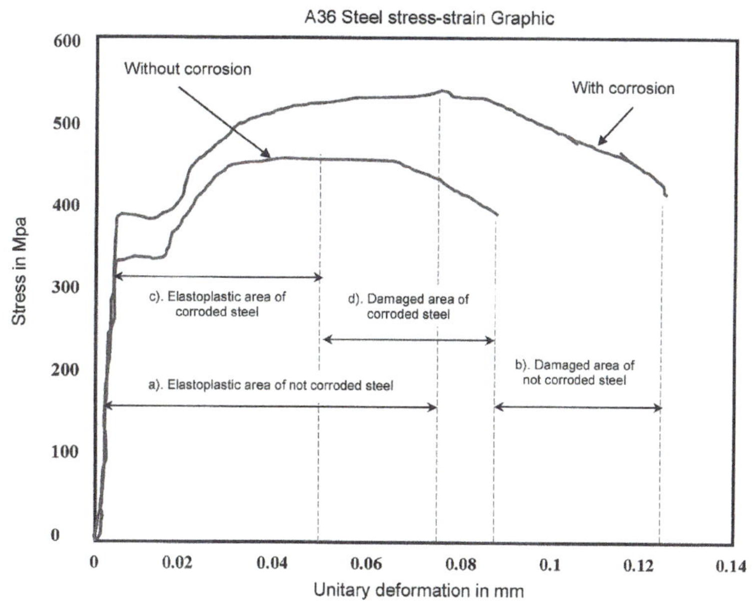

Figure 8. A36 Steel. (a) tensile strength calculation for reference not corroded probe; (b) energy calculation in damaged area for reference not corroded probe; (c) tensile strength calculation for corroded probe; and (d) energy calculation in damaged area for corroded probe.

A36 steel damaged area) is obtained integrating a Degree 4 polynomial to data between $\varepsilon_{\text{máx}} = 0.080$ and $\varepsilon_u = 0.124$ limits, **Figure 8**, blue curve, providing us with a 22.98 MPa value.

4.6. Corroded A36 Steel

Tensile strength of corroded A36 steel, based on a Degree 5 polynomial, is 20.97 MPa, **Figure 8**, in green. Energy in damaged area for corroded A36 steel is: 15.045 MPa, shown in **Figure 8**, in green. **Table 1** shows data for energy U_T from the tensile test, as well as ductility value of analyzed steels in the three areas for both steels.

For A284 Grade C steel, a 925.8 (N·m) maximum value has been observed for a 0.9 damage, which shows higher ductility of A284 Grade C steel, while for corroded probe, a 531.93 (N·m) maximum value has been obtained for a 0.9 damage, 57% energy of not corroded probe in accord to (17). For A36 steel, a 358 (N·m) the maximum value has been observed for a 0.9 damage, and a 284.067 (N·m) value for a 0.9 damage in the corroded probe, representing 79% energy of not corroded probe.

5. Discussion of Results

Heavy metal concentrations are indicators of the damaging and corrosion process already present in the tank, and the gradual detachment of metals in the inside. For influent, such concentrations may be due to damaging and corrosion process of pipes feeding the well. This shows that, in effect, the corrosion process occurs in 5 phases. In our case, the first phase is related to the *effluent effect*; the second one is *transition of humidity excess effect*; the third one is *first phase of corrosion effect*; the fourth one is *transition of corrosion effect*; and the fifth and last one is *stabilization of corrosion effect* [9].

Yield strength σ_y determined for A284 Grade C steel in the reference probe, was 295 MPa (3000 kg/cm^2), very close to the theoretical 290 MPa (2950 kg/cm^2) value. The stress value for corroded steel was 237 MPa (2416 kg/cm^2). It can be considered that A284 Grade C steel underwent a 19.5% yield strength loss. As **Figure 9** shows, such difference has low influence on structural behavior of steel, since a maximum allowable stress equal to 2/3 of yield strength σ_y, 193 MPa (1966 kg/cm^2), was taken into consideration in the design of structural elements, in this case, of the hull of the tank. Acting stresses are even lower, which is confirmed in view of radial displacements of hull of the tank, taking into consideration loss of thickness due to corrosion.

In the test carried out on A36 steel, yield strength value for non-corroded steel was determined as 368 MPa (3750 kg/cm^2), the theoretical value for such stress is 345 MPa (3500 kg/cm^2) and the value determined in test for corroded steel is 312 MPa (3181 kg/cm^2). It was determined that steel lost 15.5% stress yield strength, as shown in **Figure 10**. The structural element designed with such steel is the false bottom support for filtering material inside the steel tank. Such element is an "I" profile with plate, including the false bottom plate as higher skid. Such type of profile is more sensible to plate thickness loss due to corrosion.

A284 Grade C steel shows a 32% decrease of energy in elastoplastic area, while A 36 steel shows a 35% decrease of energy in this area, **Table 1**. This is due to higher ductility of A284 Grade C steel, which has a 13.88% ductility loss due to corrosion after 8 years of operation. A36 steel has a 17.5% ductility loss, in the same period, that is, it loses more ductility due to the composition of steel.

An ultimate 925.8 (N·m) value for 0.9 damage has been observed for A284 Grade C steel. This shows higher ductility of A284 Grade C steel. An ultimate 358 (N·m) value for 0.9 damage has been observed for A36 steel, due to lower ductility of such steel.

Upon tank examination, determination of minimal thickness of tank was found to be adequate, where internal pressure determines thickness and, since the formula does not take into consideration coating to prevent corro-

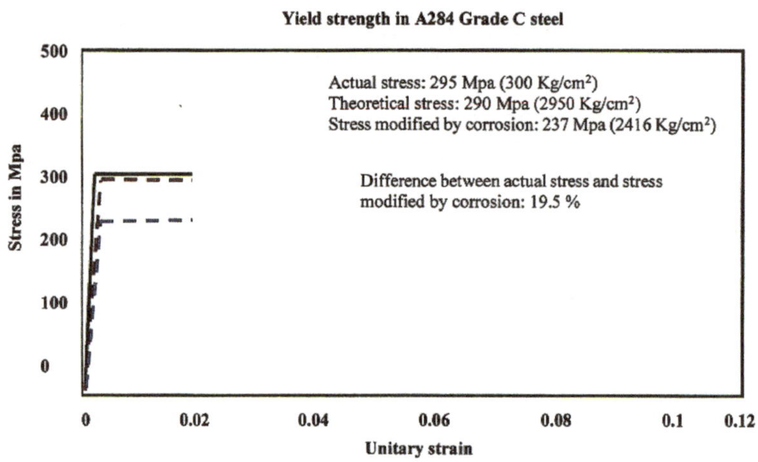

Figure 9. Yield strength stress comparison between corroded and not corroded A284 Grade C steel.

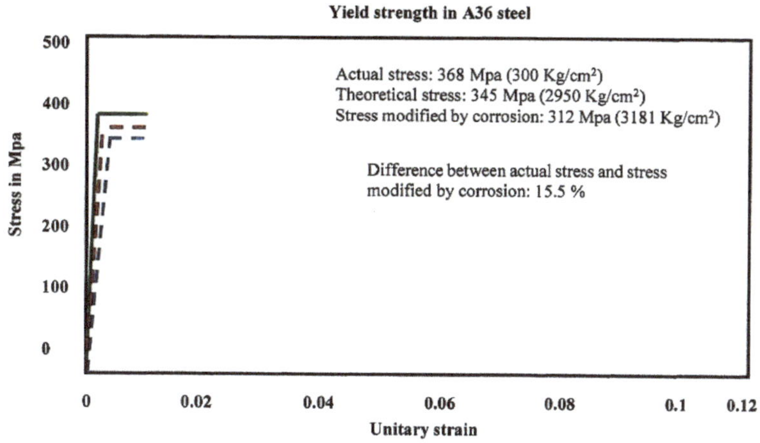

Figure 10. Yield strength stress comparison between corroded and not corroded A36 steel.

sion, the corresponding standard [10] [11], requires the addition of 1/6 of determined thickness, thus obtaining a proper safety level. This has been verified since, after 8 years of operation, steel showed generalized corrosion with a corrosion speed classified as good, being between 0.1 and 0.5 mm/year.

Stresses on the tangent line below the hull of the tank under analysis are 365.25 kg/cm^2. With stress combination due to a seismic event, the most adverse one, such stress increases up to 574.31 kg/cm^2, this being half the maximum allowable 1165 kg/m^2 stress, and corrosion does not imply an increase in stress, since the thickness increase to prevent corrosion is three times higher due to the corrosion loss shown in such period of time. In the design of false bottom and "I" profile supporting such, it was found that such profile is more sensible to variation of thickness in such section. Therefore, it shows an increase of stress where, due to thickness loss caused by corrosion, cutting stress increases from f_v = 1042 kg/cm^2 up to f_v = 1403 kg/cm^2, thus going beyond the allowable *FV* stress, even with reinforcement in both directions.

6. Conclusions

The theoretical development of damage analysis presented allows understanding the loss of strength phenomenon in metals caused by combined mechanical and electro-chemical corrosion damage effects. A-284 Grade C steel had a higher strength loss measured as fluency stress compared to loss of A-36 steel. The carbon content in steel composition influences variation of strength loss. Steel with higher carbon content has a lower fluency stress and shall also have a higher strength or fluency stress loss due to its higher ductility.

Local failure of false bottom in the filtration tank is due to cracks in tubular connections, since such are areas of high concentration of stresses and high residual stresses. In addition, in those areas the presence of defects in welding is very common. Cathode protection must be considered in the design of such filtration tanks, placing anodes in sections where corrosion is more severe. In our case, it is the joint of the hull with the false bottom.

Carbon steels show, in general, a micro-structure composed by ferrite and perlite, in normal state, as they leave the plant. Grains of material are oriented towards the lamination direction. Among these grains, a flat dislocations distribution prevails. Such dislocations grow as carbon content increases. The metallographic study of the two steels analyzed determined that there are no dislocations of material in their composition which may be considered as a factor of size increasing continuous damage.

The passivation phenomenon in analyzed steel may be considered, since creation of a protecting surface layer of corrosion products such as iron oxide and other ferrous composites derived from anticorrosion coating used in the manufacture of the tank, inhibits dissolution reactions of the metal. This could be observed with the scanning electron microscope.

The benefits of adopting a predictive maintenance strategy would benefit in knowing integrity degree of the inside of the tank, and the joint of the false bottom with the wall of the tank, which may be achieved only draining the tanks and coming into such. Some additional benefits would be extending the useful life of the tank, increasing safety and reducing maintenance costs.

Acknowledgements

This document was developed with part of the time devoted to the IPN-SIP 20120585 research project.

References

[1] Bolotin, V.V. and Shipkov, A.A. (2001) Mechanical Aspects of Corrosion Fatigue and Stress Corrosion Cracking. *International Journal of Solids and Structures*, **38**, 7297-7318.

[2] Gérard, B., Pijaudier-Cabot, G. and La Borderie, C. (1998) Coupled Diffusion-Damage Modelling and the Implications on Failure Due to Strain Localisation. *International Journal of Solids and Structures*, **35**, 4107-4120.

[3] Fair, G.M., Geyer, J.C. and Okun, D.A. (1979) Purificación de agua y tratamiento y remoción de aguas residuals. Volumen II. Editorial LIMUSA. México.

[4] DGCOH (1997-1) Elaboración de anteproyectos para solucionar los problemas de suministro de agua en diversas colonias de la Delegación Iztapalapa. Consultoría en Ingeniería Itzana S.A. de C.V. Dirección Técnica, Subdirección de Programación. U.D. Planes Hidráulicos Delegacionales. Dirección General de Construcción y Operación Hidráulica. Secretaría de Obras y Servicios. GDF. México.

[5] DGCOH (1997-2) Diagnóstico de la problemática de suministro de agua potable y drenaje en la Delegación Iztapalapa. Automatizaciones y proyectos de Ingeniería S.A. de C.V. Dirección General de Construcción y Operación Hidráulica. Secretaría de Obras y Servicios. GDF. México.

[6] DGCOH (1995) Criterio para sancionar la calidad de agua potable. Dirección General de Construcción y Operación Hidráulica. Secretaría de Obras y Servicios. DDF. México.

[7] ASTM (American Society of Testing Material) Designation: E8-01 (2002) Standard Test Methods for Tension Testing of Metallic Materials.

[8] RCDF (2004) Reglamento de Construcciones para el Distrito Federal. GDF. México.

[9] Casanova-del-Angel, F. and Toquiantzi Butrón, R. (2008) Corrosion Phases of Structural Shapes Exposed to the Atmosphere. *Corrosion Science*, **50**, 2288-2295. http://dx.doi.org/10.1016/j.corsci.2008.05.015

[10] Standard API 620 (1990) Para tanques de almacenamientos sometidos a presiones internas cercanas a 1 kg/cm^2.

[11] Standard API 650 (1998) Aplicable a grandes tanques horizontales o verticales soldados en campo y presión de operación menores de 1 Kg/cm^2.

19

Inverse Problem of Astrodynamics

Yuri Menshikov

Department of Mechanics and Mathematics, Dnepropetrovsk University, Dnepropetrovsk, Ukraine
Email: Menshikov2003@list.ru

Abstract

We consider the problem of determining the center of mass of an unknown gravitational body, using the disturbances in the motion of observed celestial bodies. In this paper an universal approach to obtain the approximate and stable estimate of problem solution is suggested. This approach can be used in other fields of Science. For example, it can be applied for investigation of interactions between fields of forces and elementary particles using known trajectories of elementary particles motions.

Keywords

Astrodynamics, Le Verrier Problem, Regularization, Main Hypothesis

1. Introduction

In 1843-1845 famous astronomers and mathematicians Urbain Jean Le Verrier (1811-1877) and John Couch Adams (1819-1892) independently of one another performed the mathematical research and came to the conclusion that the Solar system includes a celestial body (at least one) which has not observed earlier.

In fact not only the existence of a previously unknown planet has been proven, but also its orbit has been determined with an accuracy, which was sufficient for its detection and surveillance. The planet Neptune has been discovered as result. Mathematically these problems belong to the category of inverse problems of mathematical physics, *i.e.*, to ill-posed problems. The solution of this problem was executed by the method of least squares using some hypothesizes. After discovering Neptune, Le Verrier started the recalculation of theory motion of Uranus by taking into account the motion of Neptune. After finishing his investigation, Le Verrier was able to achieve results with high accuracy, which unfortunately still disagreed with results obtained by observation. This difference was not due to an error in theory or observation [1]-[3]. Theoretical success of Adams and Le Verrier was attracted thousands of enthusiasts and professional astronomers and mathematicians. Hundreds of scientific calculations were published based on the work of Le Verrier and Adams, but the results yielded nothing. There were studies, which simultaneously proved the existence of up to two dozen new celestial bodies [3].

Much later it was discovered a natural property of inverse problems their instability.

Of course, our knowledge of the Solar system has not been and will never be final and the level of our knowledge is entirely determined by the level of theoretical and observational studies. However, a theoretical analysis of the constructed motion of large (and primarily external) planets indicates, that there are yet unexplained discrepancies between theory and observation. Despite the fact that the theoretical parameters of motion were refined with results made from observation, which were made over a long period of time. For example, there are latitudinal variations in the motion of Uranus and Neptune and the deviation in the movement perihelion of Halley's comet, that cannot be explained by gravitational forces of known solar system bodies. These circumstances have led to the fact that in the 60s of the last century a hypothesis for the existence of a tenth planet emerged. This 10th planet should have a mass equal to the mass of Jupiter, with an approximate distance to the sun of 60 AU and an orbital tilt of $i = 120^\circ$. The joint solution of the equations of motion of known planets and of a hypothetical planet and subsequent thorough review of photographic plates of "suspicious" parts of the sky, have not yielded any positive results. Although, according to preliminary estimates, the hypothetical planet was supposed to be 13 - 14th magnitude, and on photographic plates were considered objects to 16.5 magnitude, but the tenth planet was not found.

Analysis of solution methods of Le Verrier and Adams shown these methods did not take into account the inaccuracy of mathematical model of planets motion. The success of their investigations was guaranteed with help of right hypothesis about tilt of unknown planet orbit to plane of the ecliptic and orbit eccentricity.

This fact is explained failure of big numbers of investigations after Le Verrier for searches of planet Pluto.

Thus the development of stable methods of approximate solutions of the inverse problem astrodynamics in more general statement remains relevant.

2. Statement of the Inverse Problem Astrodynamics

We consider n interacting masses moving under the forces of mutual attraction in an inertial coordinate system. Masses m_i are given the index $i\left(i = \overline{1,n}\right)$, $\overline{r}_{ik}$ denotes the vector connecting the mass m_i and the mass m_k. According to Newton's law the resulting force $\overline{F}_j$ acting to the mass m_j is equal [4]

$$\overline{F}_j = G\sum_{\substack{i=1 \\ i\neq j}}^{n} \frac{m_i \cdot m_j}{\left|\overline{r}_{ij}\right|^3}\overline{r}_{ij}, \tag{1}$$

where G is the gravitational constant.

The mass m_j under the influence of this force executes the motion which satisfies the differential equation

$$\frac{\mathrm{d}^2\overline{r}_{oj}(t)}{\mathrm{d}t^2} = \frac{1}{m_j}\overline{F}_j, \tag{2}$$

where $\boldsymbol{r}_{oj}$ is the radius vector connecting the origin of inertial coordinate system with the mass m_j.

Let us make the transition of the variables $\overline{r}_{ij}, j = 1,2,\cdots,n$ to the variables $\overline{r}_{0j}, j = 1,2,\cdots,n$ in Equation (2)

$$\frac{\mathrm{d}^2\overline{r}_{oj}(t)}{G\mathrm{d}t^2} = \frac{1}{Gm_j}\overline{F}_j = \sum_{\substack{i=1 \\ i\neq j}}^{n} \frac{m_i}{\left|\overline{r}_{oi} - \overline{r}_{oj}\right|^3}\left(\overline{r}_{oi} - \overline{r}_{oj}\right). \tag{3}$$

It is assumed that among n gravitational masses the location of only mass m_n is unknown. The last term in the sum on the right of Equation (3) is uncertain.

Equation (3) takes the form

$$\frac{\mathrm{d}^2\overline{r}_{oj}(t)}{G\mathrm{d}t^2} = \sum_{\substack{i=1 \\ i\neq j}}^{n-1} \frac{m_i}{\left|\overline{r}_{oi} - \overline{r}_{oj}\right|^3}\left(\overline{r}_{oi} - \overline{r}_{oj}\right) + \overline{f}_j(t),\ j \neq n, \tag{4}$$

where $\overline{f}_j(t) = \frac{m_n}{\left|\overline{r}_{oi} - \overline{r}_{oj}\right|^3}\left(\overline{r}_{on} - \overline{r}_{oj}\right)$ is the function to be determined.

In terms of projections on the axis of the inertial coordinate system the Equation (4) can be written in the form:

$$\frac{\mathrm{d}^2 x_{oj}(t)}{G\mathrm{d}t^2}=\sum_{\substack{i=1\\ i\neq j}}^{n-1}\frac{m_i}{\left|\bar{r}_{oi}-\bar{r}_{oj}\right|^3}\left(\bar{r}_{oi}-\bar{r}_{oj}\right)_x+\bar{f}_{jx}(t),\ j\neq n, \tag{5}$$

$$\frac{\mathrm{d}^2 y_{oj}(t)}{G\mathrm{d}t^2}=\sum_{\substack{i=1\\ i\neq j}}^{n-1}\frac{m_i}{\left|\bar{r}_{oi}-\bar{r}_{oj}\right|^3}\left(\bar{r}_{oi}-\bar{r}_{oj}\right)_y+\bar{f}_{jy}(t),\ j\neq n, \tag{6}$$

$$\frac{\mathrm{d}^2 z_{oj}(t)}{G\mathrm{d}t^2}=\sum_{\substack{i=1\\ i\neq j}}^{n-1}\frac{m_i}{\left|\bar{r}_{oi}-\bar{r}_{oj}\right|^3}\left(\bar{r}_{oi}-\bar{r}_{oj}\right)_z+\bar{f}_{jz}(t),\ j\neq n, \tag{7}$$

where $\bar{r}_{oj}=\bar{i}\cdot x_{oj}+\bar{j}\cdot y_{oj}+\bar{k}\cdot z_{oj}$; f_{jx}, f_{jy}, f_{jz} are the projections of the vector of $\bar{f}_j(t)$ on the corresponding axes of the inertial system coordinates, $\left(\bar{r}_{oi}-\bar{r}_{oj}\right)_x$, $\left(\bar{r}_{oi}-\bar{r}_{oj}\right)_y$, $\left(\bar{r}_{oi}-\bar{r}_{oj}\right)_z$ are the similar projections of the vector $\left(\bar{r}_{oi}-\bar{r}_{oj}\right)$.

Let us integrate Equations (5)-(7) twice from t_0 to t:

$$\begin{aligned}
\frac{x_{oj}(t)}{G}&=\int_{t_0}^{t}\mu_1(\tau)(t-\tau)\mathrm{d}\tau+\int_{t_0}^{t}f_{jx}(\tau)(t-\tau)\mathrm{d}\tau+\frac{\dot{x}_{oj}(t_0)}{G}(t-t_0)+\frac{x_{oj}(t_0)}{G},\\
\frac{y_{oj}(t)}{G}&=\int_{t_0}^{t}\mu_2(\tau)(t-\tau)\mathrm{d}\tau+\int_{t_0}^{t}f_{jy}(\tau)(t-\tau)\mathrm{d}\tau+\frac{\dot{y}_{oj}(t_0)}{G}(t-t_0)+\frac{y_{oj}(t_0)}{G},\\
\frac{z_{oj}(t)}{G}&=\int_{t_0}^{t}\mu_3(\tau)(t-\tau)\mathrm{d}\tau+\int_{t_0}^{t}f_{jz}(\tau)(t-\tau)\mathrm{d}\tau+\frac{\dot{z}_{oj}(t_0)}{G}(t-t_0)+\frac{z_{oj}(t_0)}{G},
\end{aligned} \tag{8}$$

where $\mu_1(t)=\sum_{\substack{i=1\\ i\neq j}}^{n-1}\frac{m_i}{\left|\bar{r}_{oi}-\bar{r}_{oj}\right|^3}\left(\bar{r}_{oi}-\bar{r}_{oj}\right)_x$, $\mu_2(t)=\sum_{\substack{i=1\\ i\neq j}}^{n-1}\frac{m_i}{\left|\bar{r}_{oi}-\bar{r}_{oj}\right|^3}\left(\bar{r}_{oi}-\bar{r}_{oj}\right)_y$, $\mu_3(t)=\sum_{\substack{i=1\\ i\neq j}}^{n-1}\frac{m_i}{\left|\bar{r}_{oi}-\bar{r}_{oj}\right|^3}\left(\bar{r}_{oi}-\bar{r}_{oj}\right)_z$.

Each equation of the system Equation (8) can be presented in the form

$$\int_{t_0}^{t}(t-\tau)f_{jk}(\tau)\mathrm{d}\tau=u_{jk}(t),\ k=\overline{1,3}, \tag{9}$$

where

$$u_{j1}=\frac{x_{oj}(t)}{G}-\int_{t_0}^{t}\mu_1(\tau)(t-\tau)\mathrm{d}\tau-\frac{\dot{x}_{oj}(t_0)}{G}(t-t_0)-\frac{x_{oj}(t_0)}{G},$$

$$u_{j2}=\frac{y_{oj}(t)}{G}-\int_{t_0}^{t}\mu_2(\tau)(t-\tau)\mathrm{d}\tau-\frac{\dot{y}_{oj}(t_0)}{G}(t-t_0)-\frac{y_{oj}(t_0)}{G},$$

$$u_{j3}=\frac{z_{oj}(t)}{G}-\int_{t_0}^{t}\mu_3(\tau)(t-\tau)\mathrm{d}\tau-\frac{\dot{z}_{oj}(t_0)}{G}(t-t_0)-\frac{z_{oj}(t_0)}{G}.$$

Equations (9) are known as Volterra integral equations of the first kind with respect to the unknown functions $f_{jk}(t), k=\overline{1,3}$ [5].

By finding the solutions of the Equations (9) $f_{jk}(t), k=\overline{1,3}$ you can restore the force vector $\boldsymbol{f}_j(t)$, exerted by the mass m_n on the mass m_j up to a constant factor.

Performing similar calculations and solving equations of the type Equation (9) for the mass m_l, we can determine (up to a constant factor) the force $\boldsymbol{f}_l(t)$, which is acting on the mass m_l from the mass m_n. The intersection of the lines of action of the vectors $\boldsymbol{f}_j(t)$ and $\boldsymbol{f}_l(t)$ gives the position of the mass m_n in space (in a chosen inertial system).

As is easily seen, $u_{jk}(t)$ are defined by the functions $\boldsymbol{r}_{0j}(t), j=1,2,\cdots,(n-1)$, which are assumed to be known from astronomical observations of the motions of the masses $m_j(t), j=1,2,\cdots,(n-1)$, which do contain some errors. It is assumed that functions $u_{jk}(t)$ belong to functional space $L_2(t_0,T)$.

Solution of the Equation (9) in the physical sense must also belong to $C[t_0,T]$, *i.e.*, $f_{jk}(t)\in C[t_0,T]$. Under these conditions Equations (9) are an ill-posed problem [5].

In the Equations (5)-(7) of motion the coefficients $m_j(t), j=1,2,\cdots,(n-1)$, G are determined from astronomical observations and experimental investigations and so these values are known only approximately. Thus it is assumed that each coefficient in Equations (5)-(7) is in some interval:

$$0<m_i^0\le m_i\le m_i^{up},\ i=\overline{1,n-1},\ i\ne j;\ 0<G^0\le G\le G^{up}. \tag{10}$$

Introduce into consideration the following notations

$$\bar{p}=(b_1,b_2,\cdots,b_{n-1})^*,\ \bar{R}(t)=(r_{o,1}(t),r_{o,2}(t),\cdots,r_{o,n-1}(t))^*,$$

where $b_1=m_1,\cdots,b_{j-1}=m_{j-1},b_j=m_{j+1},\cdots,b_{n-2}=m_{n-1},b_{n-1}=1/G$; $(\cdot)^*$ is the sign of transposition.

The Inequalities (10) define a closed region of $\bar{D}$ in the $(n-1)$-dimensional Euclidean space $R^{(n-1)}$. The set of vector functions $\boldsymbol{R}(t)$ forms a linear function space $C_{n-1}[t_0,T]$, which can be equipped with following norm [6]:

$$\left\|\bar{R}_1-\bar{R}_2\right\|_{C_{n-1}[t_0,T]}=\max_{1\le i\le n-1}\left|r_i^k-r_i^k\right|,$$

where $r_i^k=\left\|r_i^k(t)\right\|_{C[t_0,T]}=\max_{t\in[t_0,T]}\left|r_{o,i}^k(t)\right|$, $\bar{R}_k(t)=(r_{o,1}^k,\cdots,r_{o,n-1}^k)^*$, $k=1,2$.

The Equation (9) can be written in the form

$$\tilde{A}f=u=B_p\bar{R}, \tag{11}$$

where the operator $\tilde{A}$ is a compact operator, $\tilde{A}:F\to U$, B_p is a linear restricted operator, which transforms elements of the space $C_{n-1}[t_0,T]$ into the space $L_2(t_0,T)$. Let us assumed that the functional space $F=C[t_0,T]$ as according of the physical sense the searching functions must be the continuous functions.

The operator $\tilde{A}$ in Equation (11) doesn't depend on parameters of the mathematical model of the motion of body m_n. It is easy to see that the operator $\tilde{A}$ in Equation (11) is a completely continuous operator [6]. The operator B_p depends on the specific values of the parameters of the mathematical model of the process, which is *p*.

We denote by $A_{ex},B_{ex},\boldsymbol{R}_{ex}(t),u_{ex}$, respectively, the exact operators in Equation (11), the exact vector-valued function $\boldsymbol{R}(t)$ and the exact function of *u* in the right-hand side of Equation (11).

Suppose instead $\boldsymbol{R}_{ex}(t)$ of Equation (11) is given by the approximate function $\boldsymbol{R}_\delta(t)=(\tilde{r}_{0,1},\tilde{r}_{0,2},\cdots,\tilde{r}_{0,n-1})^*$ for which the following inequality is valid

$$\left\|\boldsymbol{R}_{ex}(t)-\boldsymbol{R}_\delta(t)\right\|_{C_{n-1}[t_0,T]}\le\delta.$$

The approximate value of $\tilde{u}$ is the right side of the Equation (11). So $\tilde{u}=B_p\boldsymbol{R}_\delta$ will match the data $\boldsymbol{R}_\delta(t)$.

Let us estimate the deviation of $\tilde{u}$ of $u_{ex}(t)$, assuming that the exact operator B_{ex} is linear:

$$\left\|\tilde{u}-u_{ex}\right\|_{L_2(t_0,T)}=\left\|B_p\boldsymbol{R}_\delta-B_{ex}\tilde{R}_{ex}\right\|_{L_2(t_0,T)}\le b_0\delta+d_1\left\|\boldsymbol{R}_\delta\right\|=\delta_o, \tag{12}$$

where $b_0=\sup_{p\in\bar{D}}\left\|B_p\right\|_{C_{n-1}[t_0,T]\to L_2(t_0,T)}$, $d_1\ge\sup_{\|\boldsymbol{R}\|\le1}\left\|B_{ex}\boldsymbol{R}-B_p\boldsymbol{R}\right\|_{C_{n-1}[t_0,T]\to L_2(t_0,T)}$.

Since real processes can be described by mathematical methods only approximately. It is assumed that the exact operator A_{ex} in the Equation (11) differs from the approximation of the operator $\tilde{A}$ (if the exact operator is linear) by a predetermined amount

$$h\ge\left\|\tilde{A}-A_{ex}\right\|_{C\to L_2}.$$

In this case it is possible to use the algorithm for solving the inverse problem with approximate operator $\tilde{A}$ proposed in [7].

However, the assumption of linearity of the exact operator A_{ex} and of information relatively of size of h in most cases does not correspond to reality.

Then the set of possible solutions $Q_{\delta_0,h}$ of Equation (11) with account of the linearity of operators $\tilde{A}, B_p$ will have the form:

$$Q_{\delta_0,h} = \left\{ f : \left\| \tilde{A}f - B_p \boldsymbol{R}_\delta \right\|_{L_2(t_0,T)} \le \delta_0 + h\|f\| \right\}. \tag{13}$$

It is easy to show that if the operator $\tilde{A}$ is a compact operator, then the set $Q_{\delta_0,h}$ is an unbounded, closed and convex set for any $\delta_0\left(\delta_0 > 0\right), h\left(h > 0\right)$. Then it is assumed that exact solution f_{ex} of Equation (11) belongs to functional space $W_2^1\left[t_0,T\right] = F_1$.

For solving ill-posed problem Equation (11) we use the Tikhonov regularization method with the stabilizing functional

$$\Omega\left[f\right] = \|f\|^2_{W_2^1[t_0,T]} = \int_{t_0}^{T} \left[q_0 f^2 + q_1 \dot{f}^2\right] \mathrm{d}\tau, \tag{14}$$

where $q_0 \ge 0, q_1 > 0$, q_0, q_1 are constants.

Thus, the problem of finding an approximate solution of Equation (11) reduces to the solution of the extreme problem:

$$\Omega\left[\tilde{f}\right] = \inf_{f \in Q_{\delta_0,h} \cap F_1} \Omega\left[f\right]. \tag{15}$$

It should be noted that there is no way to determine the size of d_1, since the exact operator B_{ex} is the unknown operator. Moreover, the exact operators A_{ex}, B_{ex} cannot be constructed, in principle, because the mathematical models, which describe the real processes, are always approximations. Of course, under certain assumptions about the exact operators A_{ex}, B_{ex} we can get some error estimates of d_1, h, but such estimates are unrealistic.

Therefore, the approximate solution of inverse problems of measurement are not of interest for practical use due to instability of the solution.

The way out of this impasse exists, if by the investigation of inverse problems of measurement restrict only some estimates of exact solutions.

3. The main Hypothesis and Results

The set of possible solutions $Q_{p,\delta}$ of Equation (11) for fixed operators $\tilde{A}, B_p$ has the form:

$$Q_{p,\delta} = \left\{ f : \left\| \tilde{A}f - B_p \boldsymbol{R}_\delta \right\|_{L_2(t_0,T)} \le \delta \left\| B_p \right\|_{C_{n-1}[t_0,T] \to L_2(t_0,T)} \right\} \tag{16}$$

Let us considered the following extreme problem

$$\Omega\left[\tilde{f}_p\right] = \inf_{f \in Q_{p,\delta} \cap F_1} \Omega\left[f\right], \tag{17}$$

The regularization parameter α can be find from condition:

$$\left\| \tilde{A}\tilde{f}_p - B_p \boldsymbol{R}_\delta \right\|_{L_2(t_0,T)} = \delta \left\| B_p \right\|_{C_{n-1}[t_0,T] \to L_2(t_0,T)}. \tag{18}$$

To obtain useful information on the exact solution of the inverse problem the use of the following hypothesis (*main hypothesis*) is suggested: the inequality is valid

$$\Omega\left[f_{ex}\right] \ge \Omega\left[\tilde{f}_p\right], \; A_{ex} f_{ex} = B_{ex} \boldsymbol{R}_{ex}, \tag{19}$$

where the function f_{ex} is an exact solution of the inverse problem of measurement Equation (11) with exact initial data, the function $\tilde{f}_p$ is the regularized solution of the inverse problem Equation (11) with the fixed operators $\tilde{A}, B_p$ (which are given the adequate description of the physical process) and stabilizing functional $\Omega\left[f\right]$.

If the exact operators A_{ex}, B_{ex} are linear, then the Inequality (19) is obvious.

Evaluation of the exact solution f_{ex}, obtained by the method of regularization on the set $Q_{p,\delta}$, for fixed operators $\tilde{A}, B_p$ is some function on vector $\boldsymbol{p} \in \bar{D}$, *i.e.*, $\tilde{f}_p$. The function $\tilde{f}_p$ can be significantly different from the exact solution f_{ex}.

The estimate $\tilde{f}_p$ allows us to conclude the existence of an unknown planet with a guarantee (in case $\tilde{f}_p \neq 0$), or its absence but without guarantee (in case $\tilde{f}_p \equiv 0$).

If $\tilde{f}_p \equiv 0$, it is also possible to have a real celestial body m_n as this does not take into account all the operators $B_p, \boldsymbol{p} \in \bar{D}$. To study the effect of the parameters of mathematical model $\boldsymbol{p} \in \bar{D}$, we need to examine all possible estimates $\tilde{f}_p$.

Note that the preparation of estimates does not use properties of the exact operators A_{ex}, B_{ex}. To assess the existence of an exact solution, it is possible to use the stabilizing functional $\Omega[f]$.

We give sufficient conditions for the existence of an element $\tilde{f}_p$.

Theorem1. Let $\Omega[f]$ be a stabilizing functional for $F = C[t_0,T], R = C_{n-1}[t_0,T], F_1 = W_2^1[t_0,T]$, the functional $\Omega[f]$ be continuous, non-negative and strictly convex on F_1. Then the solution $\tilde{f}_p$ of the extreme problem (15) exists, is unique for any $\delta > 0$ and any $\boldsymbol{p} \in \bar{D}$.

In order to study the influence of the process parameters on the estimation of the exact solution, we consider the union of the sets $Q_{p,\delta}$ for all vectors $\boldsymbol{p} \in \bar{D}$:

$$Q^{un} = \bigcup_{p \in \bar{D}} Q_{p,\delta}. \tag{20}$$

The set Q^{un} is unbounded as a union of unbounded sets.

Let f^{un} be a solution to the extreme problem:

$$\Omega\left[f^{un}\right] = \inf_{f \in Q^{un} \cap F_1} \Omega[f]. \tag{21}$$

The regularization parameter α was determined by the method of discrepancy:

$$\left\| \tilde{A} f^{un} - B_p \boldsymbol{R}_\delta \right\|_{L_2(t_0,T)} = \delta_0. \tag{22}$$

Since $Q^{un} \subset Q_{\delta_0,h}$, using Q^{un} instead of $Q_{\delta_0,h}$ in Equation (15) will yield more accurate estimates for the exact solution. This approach is based on the work of [8]. In addition, it is obvious that the inequality

$$\Omega[f_{ex}] \geq \Omega\left[f^{un}\right] \geq \Omega\left[\tilde{f}\right] \tag{23}$$

is valid.

Evaluation of the exact solution of f^{un} allows us to conclude on the existence of an unknown planet with a guarantee in case $f^{un} \neq 0$ or its absence, but without guarantee in case $f^{un} \equiv 0$.

To study the influence of the parameters p on the estimation of the exact solution of the inverse problem it is necessary to have a possibility to select an operator B_{p_0} from operators B_{p_0} which satisfies the following condition:

$$\Omega[f_1] = \Omega\left[A^{-1} B_p \boldsymbol{R}\right], \Omega[f_2] = \Omega\left[A^{-1} B_{p_o} \boldsymbol{R}\right] \tag{24}$$

implies the inequality

$$\Omega[f_1] = \|f_1\|^2_{W_2^1[t_0,T]} \geq \Omega[f_2] = \|f_2\|^2_{W_2^1[t_0,T]} \tag{25}$$

for any possible $\boldsymbol{R}$ and any $\boldsymbol{p} \in \bar{D}$; $\tilde{A}^{-1}$ is the inverse operator to $\tilde{A}$.

Subsequently, the operator B_{p_0} in the right-hand side of Equation (11) will be called as "special minimal operator" in the sense of the Inequality (25).

If the operator B_{p_0} exists and is unique, then the problem of finding the greatest lower bound of the functional $\Omega[f]$ on the set Q^{un} will have a solution, which coincides with the solution of the more simpler extreme problem [9]: find an element $f_0 \in Q_{p_0,\delta}$ for which the equality

$$\Omega[f_0] = \inf_{f \in Q_{p_0,\delta} \cap F_1} \Omega[f] \tag{26}$$

holds.

The problem (26) has a solution for every $\boldsymbol{p} \in \bar{D}$ and δ, since the conditions of the Theorem 1 are satisfied.

In this case the following inequality holds for any $\boldsymbol{p} \in \bar{D}$:

$$\Omega[f_{ex}] \geq \Omega[f_0] \geq \Omega\left[\tilde{f}_p\right] \geq \Omega\left[f^{un}\right]. \tag{27}$$

Evaluation of the exact solution f_0 allows us to conclude the existence of an unknown planet with a guarantee in case $f^{un} \neq 0$ or its absence, but without guarantee in case $f^{un} \equiv 0$.

Theorem 2. Special minimal operator B_{p_0} in the Equation (11) exists, is unique and corresponds to the vector

$$p_0 = \left(m_1^0, m_2^0, \cdots, m_{j-1}^0, m_{j+1}^0, \cdots, m_{n-1}^0, G^{up-1}\right)^*.$$

Proof. Let $\boldsymbol{R}(t)$ be a realization of an astronomical observations. Consider the problem of determining the exact lower bound of the functional $\Omega[f] = \Omega\left[A^{-1}B_p\bar{R}\right]$ in $\bar{D}$ for a fixed $\boldsymbol{R}(t)$. By the Weierstrass theorem the continuous functional $\Omega[f]$ is attained at some vector $\boldsymbol{p}_0 \in \bar{D}$.

For any $\boldsymbol{p} \in \bar{D}$ the function $\Omega[f] = \Omega\left[A^{-1}B_p\bar{R}\right]$ is strictly positive since

$\Omega[f_1] = \|f\|^2_{W_2^1[t_0,T]} > 0$ when $f^{un} \neq 0$, for all $\boldsymbol{p} \in \bar{D}$.

The function $\Omega[f]$ for a fixed $\boldsymbol{R}(t)$ can be represented as a quadratic form

$$\Omega[f] = (Cp, p) = \Omega[p]$$

where C is a real symmetric matrix $C = (c_{ik})_{k,i=1}^n$.

Coefficients of the matrix C are given by:

$$c_{ik} = \int_{t_0}^{T}\left(q_0 a_{ij}(t) a_{kj}(t) + q_1 \frac{\mathrm{d}a_{ij}}{\mathrm{d}t} \cdot \frac{\mathrm{d}a_{kj}}{\mathrm{d}t}\right)\mathrm{d}t, \quad i,k = \overline{1,(n-1)}.$$

Since $\Omega[f] > 0$ for $\boldsymbol{p} \in \bar{D}$, then the Silvester's inequalities gives us:

$$c_{11} > 0, \quad \begin{vmatrix} c_{11} & c_{12} \\ c_{21} & c_{22} \end{vmatrix} > 0, \quad \begin{vmatrix} c_{1,1} & \cdots & c_{1,n-1} \\ \vdots & \ddots & \vdots \\ c_{n-1,1} & \cdots & c_{n-1,n-1} \end{vmatrix} > 0.$$

A necessary and sufficient conditions for strong convexity of $\Omega[\boldsymbol{p}]$ to $\boldsymbol{p} \in \bar{D}$ are the following [10]:

$$\sum_{i,k=1}^{n-1} \frac{\partial^2 \Omega[p]}{\partial b_i \partial b_k} \xi_i \xi_k > 0 \tag{28}$$

for any $\xi = (\xi_1, \xi_2, \cdots, \xi_{n-1})^* \in E^{n-1}$ and any $\boldsymbol{p} \in \bar{D}$.

Quadratic form (28) is positive as

$$\bar{C} = \left(\frac{\partial^2 \Omega[p]}{\partial b_i \partial b_k} \xi_i \xi_k\right)_{i,k=1}^{n-1} = (\bar{c}_{ik})_{i,k=1}^{n-1} = (2c_{i,k})_{i,k=1}^{n-1}.$$

Therefore, $\Omega[\boldsymbol{p}]$ is strongly convex on $\bar{D}$.

Similarly, [10] we prove that $\Omega[\boldsymbol{p}]$ achieves the greatest lower bound in a single point in the domain

$$p_0 = \left(m_1^0, m_2^0, \cdots, m_{j-1}^0, m_{j+1}^0, \cdots, m_{n-1}^0, G^{up-1}\right)^* \in \bar{D}$$

for any $\boldsymbol{R}(t)$. □

Suppose that among the operators B_p there is an operator B_{p^1} such that if the condition

$$\Omega[f_1] = \Omega\left[A^{-1}B_p\boldsymbol{R}\right], \Omega[f_2] = \Omega\left[A^{-1}B_{p^1}\boldsymbol{R}\right], \tag{29}$$

then the inequality

$$\Omega[f_1] = \|f_2\|^2_{W_2^1[t_0,T]} \geq \Omega[f_1] = \|f_2\|^2_{W_2^1[t_0,T]} \tag{30}$$

for any possible $\boldsymbol{R}(t)$ and any $\boldsymbol{p} \in \bar{D}$; $\tilde{A}^{-1}$ is the inverse operator to $\tilde{A}$.

Subsequently, the operator B_{p^1} in the right-hand side of Equation (11) will be called "special maximum operator" in the sense of satisfying Inequality (30).

If the operator B_{p^1} exists and is uniquely determined then we can consider the solution of the following extreme problem: find an element $f^1 \in Q_{p^1,\delta} \cap F_1$ for which the equality

$$\Omega\left[\tilde{f}_p\right] = \sup_{p\in\bar{D}} \inf_{f\in Q_{p,\delta}\cap F_1} \Omega[f] = \inf_{f\in Q_{p^1,\delta}\cap F_1} \Omega[f] \tag{31}$$

holds.

The problem (31) has a solution for any $\boldsymbol{p} \in \bar{D}$ and δ, since the conditions of Theorem 1 are satisfied.

Thus obviously the following inequality is valid:

$$\Omega[f_{ex}] \geq \Omega\left[f^1\right] \geq \Omega\left[\tilde{f}_p\right] \geq \Omega[f_0] \geq \Omega\left[f^{un}\right] \geq \Omega\left[\tilde{f}\right]. \tag{32}$$

Theorem 3. Special maximal operator B_{p^1} in the Equation (11) exists, is unique and corresponds to the vector

$$p_0 = \left(m_1^{up}, m_2^{up}, \cdots, m_{j-1}^{up}, m_{j+1}^{up}, \cdots, m_{n-1}^{up}, G^{0-1}\right)^* \in \bar{D}$$

for any $\boldsymbol{R}(t)$.

Proof. The proof is similar to Theorem 2. □

4. Conclusion

In this paper we proposed an algorithm for finding the coordinates of an unknown gravitational mass as a result of astronomical observations. This problem was solved first by Urbain Jean Le Verrier and John Couch Adams. Here a more universal approach was suggested. The proposed hypothesis allows us to exclude the error of the approximate operator from the calculations. Also conditions for the existence of an approximate solution were obtained and several non-standard formulations of inverse problems were considered. Suggested approach can be used in other fields of Science. For example, it can be applied for investigation of interactions between fields of forces and elementary particles by help of known trajectories of elementary particles motions.

References

[1] Lykawka, P.S. and Mukai, T. (2008) An Outer Planet beyond Pluto and the Origin of the Trans-Neptunian Belt Architecture. *Astronomical Journal*, **135**, 1161-1200. http://dx.doi.org/10.1088/0004-6256/135/4/1161

[2] Luhman, K.L. (2014) The Search for a Distant Companion to the Sun with the Wide-Field Infrared Survey Explorer. *Astrophysical Journal*, **781**, 4 (7pp).

[3] Seidelmann, P.K. and Harrington, R.S. (1987) Planet X—The Current Status. *Celestial Mechanics*, **43**, 55-68. http://dx.doi.org/10.1007/BF01234554

[4] Krasovskii, N.N. (1968) Theory of Motion Control. Science, Moscow.

[5] Tikhonov, A.N. and Arsenin, V.Ya. (1979) Methods of Ill-Posed Problems Solving. Science, Moscow.

[6] Trenogin, V.A. (1980) Functional Analysis. Science, Moscow.

[7] Goncharsky, A.V., Leonov, A.S. and Yagola, A.G. (1972) Regularizing Algorithm for Ill-Posed Problems with an Approximately Given Operator. *Journal of Computational Mathematics and Mathematical Physics*, **12**, 1592-1594.

[8] Menshikov, Y.L. (1986) Regularising Algorithm for a Class of Approximate Functional Equations of the First Kind. *Journal of Differential Equations and Their Applications*, Dnepropetrovsk, 80-87.

[9] Menshikov, Y.L. (2013) Synthesis of Adequate Mathematical Description as Solution of Special Inverse Problems. *European Journal of Mathematical Sciences*, **2**, 256-271.

[10] Vasil'ev, F.P. (1980) Numerical Methods for Solving Extreme Problems. Science, Moscow.

20

Magnetic Field and Thermal Radiation Effect on Heat and Mass Transfer of Air Flow near a Moving Infinite Plate with a Constant Heat Sink

S. M. Arifuzzaman[1*], Sajal Kumar Dhali[2], Md. Abul Kalam Azad[1], Bibhuti Roy[1]

[1]Mathematics Discipline, Khulna University, Khulna, Bangladesh
[2]Computer Science and Engineering Discipline, Khulna University, Khulna, Bangladesh
Email: *arifsm42@gmail.com, sajalcseku09@gmail.com, azad0852@gmail.com, plato.roy@gmail.com

Abstract

Analytical investigation on a combined heat and mass transfer of air flow near a continuously moving infinite plate with a constant heat sink is performed in the presence of a uniform magnetic field. To observe the thermal radiation and Soret effect on the flow, thermal radiation and thermal diffusion term are added in energy and concentration equations. A flow of model is established by employing the well known boundary layer approximations. In order to obtain non-dimensional system of equations, a similarity transformation is applied on the flow model. Perturbation technique is used as main tool for the analytical approach. The numerical values of flow variables are computed by a FORTRAN program. The obtain numerical values of fluid velocity, temperature and species concentration are drawn for the different values of various parameters. To observe the effects of various parameters on the flow variables, the results are discussed in detailed with the help of graph.

Keywords

Viscous Fluid, MHD, Thermal Radiation, Heat Sink, Chemical Reaction, Soret Effects

1. Introduction

Fluid dynamics is the science which is concerned with the study of motion of fluids or that of bodies in contract of fluids. Fluids are classified as liquids and gases. The former are mot sensibly compressible expect under the

*Corresponding author.

action heavy forces whereas the latter are easily compressible and expand to fill any closed space. The heat transfer processes are of great interest in power engineering, metallurgy, astrophysics and geophysics. The processes of mass transfer have of great interest in the production of materials in order to obtain the desired properties of a substance. Chemical reactions including combustion processes are often decisively determined by the mass transfer. The combined heat and mass transfer occur due to buoyancy forces caused by temperature difference and concentration difference. A natural convective heat transfer flow of fluid was first studied by *Finston* (1956) [1]. *Sparrow* and *Gregg* (1958) computed a similar solution for laminar free convection from a non-isothermal vertical plate [2]. A Finite difference solution of transient free convective flow over an isothermal plate was obtained by *Soundalgekar* and *Ganesan* (1981) [3]. A numerical study on the natural convective cooling problem of a vertical plate was completed by *Camargo et al.* (1996) [4]. Chemical reactions including combustion processes are often decisively determined by the mass transfer. *Callahan* and *Marner* (1976) studied a free convective unsteady flow with mass transfer past a semi-infinite plate [5]. An investigation on free convective unsteady flow with mass transfer past an infinite vertical porous plate with constant suction was completed by *Soundalgekar* and *Wavre* (1977) [6]. *Soundalgekar* and *Ganesan* (1980) observed that Transient free convection flow on a semi-infinite vertical plate with mass transfer [7].

Combined heat and mass transfer problems (*Jaluria*; 1980) are of importance in many processes and have therefore received a considerable amount of attention. In many mass transfer processes, heat transfer considerations arise owing to chemical reaction and are often due to nature of the process [8]. In processes such as drying, evaporation at the surface water body, energy transfer in a wet cooling tower and flow in desert cooler, heat and mass transfer occur simultaneously. The combined heat and mass transfer flows play a special role in power engineering, metallurgy, condensation, evaporation and rectification of a fluid. In the separation processes as drying of solid materials, distillation, extraction and absorption; the combined heat and mass transfer occur due to buoyancy forces caused by temperature difference and concentration difference. *Pera* and *Gebhart* (1971) was the first author to study the natural convective heat and mass transfer problem [9]-[14]. *Hady*, *Mohamed* and *El Shehabey*, (2013) focused on the effects of heat source/sink, viscous dissipation, radiation and work done by deformation on flow and heat transfer of a viscoelastic fluid over a nonlinear stretching sheet [15]. *Viskanta* and *Grosh* (1962) considered the thermal radiation effects on the boundary layer flow and heat transfer over a wedge in an absorbing and emitting media [16]. *Bestman* (1990) observed the boundary-layer flow past a semi-infinite heated porous plate for a two-component plasma [17]. *Sattar* and *Alam* (1994) considered the thermal-diffusion as well as transpiration effects on MHD free convection and mass transfer flow past an accelerated vertical porous plate [18].

2. Analysis and Solution

2.1. Mathematical Flow

A mixed convective heat and mass transfer unsteady flow of air near an infinite vertical plate is considered here. The flow is assumed to be in the *x*-direction, which is chosen along the plate in upward direction and *y*-axis is normal to it. Initially, it is considered that the plate as well as the fluid particle is at rest at the same temperature $T\left(=T_{\infty}\right)$ and the same species concentration level $C\left(=C_{\infty}\right)$ at all points. Where, C_{∞} and T_{∞} are fluid concentration and temperature species of uniform flow respectively. It also assumed that the plate and the fluid particles outside the boundary layer move with a uniform velocity U_0 as well as a magnetic field B of uniform strength is applied normal to the flow region. To realize the model of boundary layer phenomena, we can draw the configuration of the model. The suitable physical configuration with co-ordinate systems is shown in **Figure 1**.

The unsteady two dimensional problems are governed by the following system of coupled non-linear partial differential equations.

$$\frac{\partial v}{\partial y}=0 \tag{1}$$

$$\frac{\partial u}{\partial t}+v\frac{\partial u}{\partial y}=g\beta\left(T-T_{\infty}\right)+g\beta^{*}\left(C-C_{\infty}\right)+\upsilon\left(\frac{\partial^{2}u}{\partial y^{2}}\right)-\frac{\sigma' B_{0}^{2}u}{\rho} \tag{2}$$

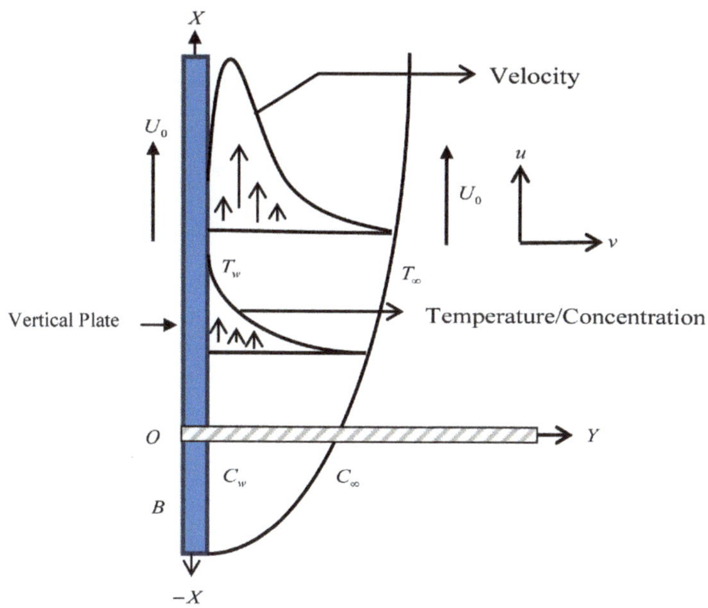

Figure 1. Physical configuration with co-ordinate system.

$$\frac{\partial T}{\partial t}+v\frac{\partial T}{\partial y}=\frac{\kappa}{\rho c_p}\frac{\partial^2 T}{\partial y^2}+\frac{\upsilon}{\rho c_p}\left(\frac{\partial u}{\partial y}\right)^2-\frac{Q_0}{\rho c_p}\left(T-T_\infty\right)-\frac{1}{\rho C_p}\frac{\partial q_r}{\partial y} \tag{3}$$

$$\frac{\partial C}{\partial t}+v\frac{\partial C}{\partial y}=D_m\left(\frac{\partial^2 C}{\partial y^2}\right)+\frac{D_m\kappa_T}{c_s c_p}\frac{\partial^2 T}{\partial y^2}-K_c\left(C-C_\infty\right) \tag{4}$$

boundary conditions,

$$u=U_0,\ T=T_w,\ C=C_w \ \text{ at } y=0$$

$$u=U_0,\ T\to T_\infty,\ C\to C_\infty \ \text{ at } y\to\infty$$

where, x, y be the Cartesian coordinate; u, v are velocity components, g is the local acceleration due to gravity, β is the is the thermal expansion coefficient, β^* is the is the concentric expansion coefficient, T_w denotes the wall temperature, C_w is the species concentration at the wall, υ is the kinematic viscosity, ρ is density, k is thermal conductivity, C_p is specific heat at constant pressure, Q_0 denotes the heat sink, B_0 is uniform magnetic field, q_r unidirectional radiative heat flux, k_c for chemical reaction, D_m is coefficient of mass diffusivity, m is mass per unit area. The radiative heat flux term by using the Rosseland approximation is given by [15].

$$q_r=-\frac{4\sigma_s}{3k_e}\frac{\partial T^4}{\partial y}$$

Where, σ_s is the Stefan-Boltzmann constant and k_e is the mean absorption coefficient, respectively. If temperature differences within the flow are sufficiently small, then the q_r can be linearized by expanding T^4 into the Taylor series about T_∞, which after neglecting higher order terms takes the form by [15]. $T^4 \cong 4T_\infty^3 T-3T_\infty^4$. Then the Equation (3) becomes,

$$\frac{\partial T}{\partial t}+v\frac{\partial T}{\partial y}=\frac{\kappa}{\rho c_p}\frac{\partial^2 T}{\partial y^2}+\frac{\upsilon}{\rho c_p}\left(\frac{\partial u}{\partial y}\right)^2-\frac{Q_0}{\rho c_p}\left(T-T_\infty\right)+\frac{16\sigma_s T_\infty^3}{3k_e\rho C_p}\frac{\partial^2 T}{\partial y^2} \tag{5}$$

2.2. Mathematical Formulation

In order to obtain similar solutions we introduce a similarity parameter σ as,

$$\sigma = \sigma(t)$$

Such that σ is the time dependent length scale. In terms of this length scale, a convenient solution of equation is considered as,

$$v = -V_0 \frac{\upsilon}{\sigma}$$

Here, the constant V_0 represents a dimensionless normal velocity at the plate which is positive for suction and negative for blowing.

We consider the following dimensionless variables,

$$\eta = \frac{y}{\sigma}$$

$$f(\eta) = \frac{u}{U_0}$$

$$\theta(\eta) = \frac{T - T_\infty}{T_w - T_\infty}$$

$$\phi(\eta) = \frac{C - C_\infty}{C_w - C_\infty}$$

Then we get,

$$f''(\eta) + f'(\eta)\left[\frac{\eta\sigma}{\upsilon}\frac{\partial\sigma}{\partial t} + V_0\right] + G_r\theta(\eta) + G_m\phi(\eta) - Mf(\eta) = 0 \tag{6}$$

$$\theta''(\eta) + P_r E_c f'^2(\eta) + P_r\theta'(\eta)\left[\frac{\sigma\eta}{\upsilon}\frac{\partial\sigma}{\partial t} + V_0\right] - P_r\beta_a\theta(\eta) - P_r R_a\theta''(\eta) = 0 \tag{7}$$

$$\phi''(\eta) + S_c\left[\frac{\sigma\eta}{\upsilon}\frac{\partial\sigma}{\partial t} + V_0\right]\phi'(\eta) + S_c S_o\theta''(\eta) - S_c C_r\phi(\eta) = 0 \tag{8}$$

where, Grashof number, $G_r = \frac{g\beta\sigma^2}{U_0\upsilon}(T_w - T_\infty)$, modified Grashof number, $G_m = \frac{g\beta^*\sigma^2}{U_0\upsilon}(C_w - C_\infty)$, Soret number, $S_o = \frac{D_m\kappa_T}{c_s c_p \upsilon}\frac{(T_w - T_\infty)}{(C_w - C_\infty)}$, magnetic force parameter, $M = \frac{\sigma' B_0^2\sigma^2}{\rho\upsilon}$, Prandtl number, $P_r = \frac{\rho\upsilon c_p}{\kappa}$, heat absorption parameter, $\beta_a = \frac{\sigma^2 Q_0}{\upsilon\rho c_p}$, Eckert number, $E_c = \frac{U_0^2}{\rho c_p (T_w - T_\infty)}$, chemical reaction parameter $= C_r$, thermal radiation parameter, $R_a = \frac{16\sigma_s T_\infty^3}{3k_e\rho C_p\upsilon}$, Schmidt number, $S_c = \frac{\upsilon}{D_m}$.

Hence following the works of *Sattar* and *Alam* (1994) [18] one can try a class of solutions of the Equations (6), (7) and (8) by assuming that,

$$\frac{\sigma}{\upsilon}\frac{\partial\sigma}{\partial t} = c \quad (c \text{ is a constant}) \tag{9}$$

Now integrating (9) we obtain

$$\sigma = \sqrt{2c\upsilon t} \tag{10}$$

where the constant of integration is determined through the condition that $\sigma = 0$ when, $t = 0$. It thus appears from (10) that, by making a realistic choice of c to be equal to 2 in (9) the length scale σ becomes equal to $\sigma = 2\sqrt{\upsilon t}$ which exactly corresponds to the usual scaling factor considered for various unsteady boundary layer flows (Schlichting, 1968). Since σ is a scaling factor as well as a similarity parameter, any value of c in (9)

would not change the nature of the solution except that the scale would be different. Finally, introducing (9) with $c = 2$ in Equations (6), (7) and (8) we respectively have the following dimensionless ordinary differential equations.
where, $f_w = 2\eta + V_0$ (Suction Parameter).

The dimensionless similar equations,

$$f''(\eta) + f'(\eta) f_w(\eta) + G_r\theta(\eta) + G_m\phi(\eta) - Mf(\eta) = 0 \tag{11}$$

$$\theta''(\eta) + P_r E_c f'^2(\eta) + P_r\theta'(\eta) f_w(\eta) - P_r\beta_a\theta(\eta) - P_r R_a\theta''(\eta) = 0 \tag{12}$$

$$\phi''(\eta) + S_c f_w(\eta)\phi'(\eta) + S_c S_o\theta''(\eta) - S_c C_r\phi(\eta) = 0 \tag{13}$$

Non-dimensional appropriate boundary conditions,

$$f(\eta) = 1,\ \theta = 1,\ \phi = 1 \text{ at } \eta = 0$$

$$f(\eta) = 1,\ \theta = 0,\ \phi = 0 \text{ at } \eta \to \infty$$

2.3. Mathematical Analysis

Since the solution is sought for the large suction farther transformation can be made as,

$$\xi = \eta f_w,\ f = f_w^2 F(\xi),$$
$$\theta = f_w^2 H(\xi) \text{ and } \phi = f_w^4 G(\xi)$$

Model with small quantity,

$$F''(\xi) + F'(\xi) + \varepsilon G_r H(\xi) + \varepsilon G_m G(\xi) - MF(\xi)\varepsilon = 0$$

$$H''(\xi) + \frac{1}{\varepsilon} A_{24} P_r E_c F'^2(\xi) + A_{24} P_r H'(\xi) - P_r\beta_a\varepsilon A_{24} H(\xi) = 0$$

where, $A_{24} = (1 - P_r R_a)^{-1}$

$$G''(\xi) + S_c G'(\xi) + \varepsilon^2 S_c S_o H''(\xi) - \varepsilon S_c C_r G(\xi) = 0$$

Appropriate boundary conditions,

$$F = \varepsilon, H = \varepsilon, G = \varepsilon^2 \text{ at } \eta = 0$$

$$F = \varepsilon, H = 0, G = 0 \text{ at } \eta \to 0$$

2.4. Solution

Now, for the large suction $(f_w > 1)$, ε will be very small. Therefore following *Bestman* (1990) [17], *F*, *H* and *G* can be expended in terms of the small perturbation quantity ε,

$$F(\xi) = \varepsilon F_1(\xi) + \varepsilon^2 F_2(\xi) + \varepsilon^3 F_3(\xi) + \cdots \tag{14}$$

$$H(\xi) = \varepsilon H_1(\xi) + \varepsilon^2 H_2(\xi) + \varepsilon^3 H_3(\xi) + \cdots \tag{15}$$

$$G(\xi) = \varepsilon G_1(\xi) + \varepsilon^2 G_2(\xi) + \varepsilon^3 G_3(\xi) + \cdots \tag{16}$$

The first order equations,

$$\left.\begin{aligned} &F'' + F_1' = 0 \\ &H_1'' + A_{24} P_r E_c F_1'^2 + A_{24} P_r H_1' = 0 \\ &G_1'' + S_c G_1' = 0 \end{aligned}\right\} \tag{17}$$

with the boundary conditions,

$$F_1 = 1,\ H = 0,\ G_1 = \varepsilon \ \text{at}\ \xi = 0$$

$$F_1 = 0,\ H = 0,\ G_1 = 0 \ \text{at}\ \xi \to \infty$$

The second order equations,

$$\left.\begin{array}{l} F_2'' + F_2' + G_r H_1 + G_m G_1 - M F_1 = 0 \\ H_2'' + 2F_1' F_2' A_{24} P_r E_c + A_{24} P_r H_2' - A_{24} P_r \beta_a H_1 = 0 \\ G_2'' + S_c G_2' - S_c C_r G_1 = 0 \end{array}\right\} \quad (18)$$

with the boundary conditions,

$$F_2 = 0,\ F_2' = 0,\ H_2 = 0,\ G_2 = 0 \ \text{at}\ \xi = 0$$

$$F_2' = 0,\ H_2 = 0,\ G_2 = 0 \ \text{at}\ \xi \to \infty$$

Also the third order equations,

$$\left.\begin{array}{l} F_3'' + F_3' + G_r H_2 + G_m G_2 + M F_2 = 0 \\ H_3'' + P_r A_{24} F_2'^2 E_c + 2A_{24} P_r F_1' F_3' + A_{24} P_r H_3' - A_{24} P_r \beta_a H_2 = 0 \\ G_3'' + S_c G_3' + S_c S_o H_1'' - S_c C_r G_2 = 0 \end{array}\right\} \quad (19)$$

with the boundary conditions,

$$F_3 = 0,\ F_3' = 0,\ H_3 = 0 \ \text{and}\ G_3 = 0 \ \text{at}\ \xi = 0$$

$$F_3 = 0,\ F_3' = 0,\ H_3 = 0 \ \text{and}\ G_3 = 0 \ \text{at}\ \xi \to \infty.$$

From Equation (17) we get first order solutions,

$$F_1 = 1,\ H_1 = \mathrm{e}^{-P_r A_{24} \xi} \ \text{and}\ G_1 = \varepsilon \mathrm{e}^{-S_c \xi}$$

From Equation (18) we get second order solutions,

$$F_2 = A_{21} \mathrm{e}^{-A_{24} P_r \xi} + A_{22} \mathrm{e}^{-S_c \xi} + M(\xi - 1),$$

$$H_2 = -\beta_a \mathrm{e}^{-A_{24} P_r \xi} \ \text{and}\ G_2 = -\varepsilon S_c C_r \mathrm{e}^{-S_c \xi}$$

From Equation (19) we get third order solutions,

$$F_3 = A_{31} \mathrm{e}^{-A_{24} P_r \xi} + A_{23} \mathrm{e}^{-S_c \xi} + A_{32} \mathrm{e}^{-S_c \xi} - 2M^2 \xi + \frac{M^2 \xi^2}{2} + 2M^2 + A_{33} \mathrm{e}^{-A_{24} P_r \xi}$$

$$H_3 = A_{40} \mathrm{e}^{-A_{24} P_r \xi} + A_{41} \mathrm{e}^{-S_c \xi} - A_{42} \mathrm{e}^{-2A_{24} P_2 \xi} - A_{43} \mathrm{e}^{-2S_c \xi} - A_{44} \mathrm{e}^{-(A_{24} P_r + S_c)\xi} - A_{24} P_r E_c M^2 (\xi - 1)$$

and

$$G_3 = A_{45} \mathrm{e}^{-A_{24} P_r \xi} + \varepsilon S_c C_r^{\,2} \mathrm{e}^{-S_c \xi}$$

From the Equation (14), (15) and (16) we have series for the solution. Putting the equations of first, second and third order solution in the Equations (14), (15) and (16), we get the values of F, H and G. Substituting the values of F, H and G to the equations $f = f_w^2 F(\xi)$, $\theta = f_w^2 H(\xi)$ and $\phi = f_w^4 G(\xi)$ we get,

The fluid velocity equation,

$$f = \mathrm{e}^{-A_{24} P_r \xi} \left(\varepsilon A_{21} + \varepsilon^2 A_{31} + \varepsilon^2 A_{33} \right) + \mathrm{e}^{-S_c \xi} \left(\varepsilon A_{22} + \varepsilon^3 A_{23} + \varepsilon^2 A_{32} \right) + 1 + \varepsilon^2 M(\xi - 1) + M^2 \left(2\varepsilon^2 + \frac{\varepsilon^2 \xi^2}{2} - 2\varepsilon^2 \xi \right)$$

The fluid temperature equation,

$$\theta = e^{-A_{24}P_r\xi}\left(1+\varepsilon^2 A_{40} - \varepsilon\beta_a\right) + \varepsilon^2 e^{-S_c\xi} A_{41} + \varepsilon^2 A_{42} e^{-2A_{24}P_r\xi} + \varepsilon^2 A_{43} e^{-2S_c\xi} - \varepsilon^2 A_{44} e^{-(A_{24}P_r+S_c)\xi} - \varepsilon^2 A_{24} P_r E_c M^2 (\xi - 1)$$

The fluid concentration equation, $\phi = e^{-S_c\xi} - \varepsilon S_c C_r e^{-S_c\xi} + \varepsilon A_{45} e^{-A_{24}P_r\xi} + \varepsilon^2 S_c C_r^2 e^{-S_c\xi}$

3. Results and Discussion

To discuss the results of the problem, the analytical solutions are obtained by using the perturbation technique. In order to analyze the physical situation of the model, we have computed the numerical values of the flow variables for different values of dimensionless normal velocity (V_o), Grash of number (G_r), modified Grashof number (G_m), Schmidt number (S_C), Eckert number (E_c), Prandtl Number (P_r). Soret number (S_o), magnetic force parameter (*M*), Chemical reaction parameter (C_r), thermal radiation parameter (R_a), heat absorption parameter (β_a). The fluid velocity, temperature and concentration versus the non-dimensional coordinate variable η are displayed in **Figures 2-12**.

In **Figure 2** velocity profiles increase with the increase of heat absorption parameter β_a. In **Figure 3**, we observe that velocity profiles increase with the increase of magnetic force parameter *M*. In **Figure 4**, velocity profiles decrease with the increase of Prandtl number P_r. In **Figure 5**, we observe that velocity profiles decrease with the increase of radiation parameter R_a. In **Figure 6** temperature profiles decrease with the increase of heat absorption parameter β_a. In **Figure 7**, temperature profiles decrease with the increase of Prandtl number P_r. In **Figure 8**, we observe that temperature profiles decrease with the increase of radiation parameter R_a. In **Figure 9**, concentration profiles decrease with the increase of Schmidt number S_c. In **Figure 10**, we observe that concentration profiles increase with the increase of Soret number S_o. In **Figure 11**, concentration profiles decrease with the increase of chemical reaction parameter C_r. In **Figure 12**, we observe that concentration profiles increase with the increase of radiation parameter R_a. The other parameters remain constant for every profile.

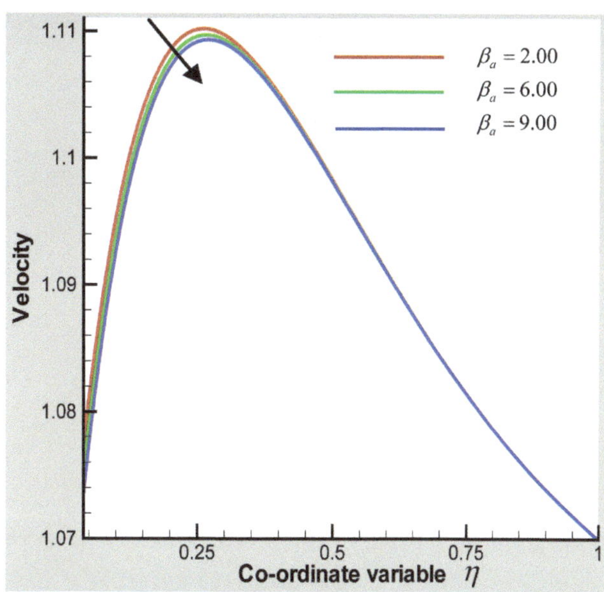

Figure 2. Velocity profiles for different values of β_a with $G_m = 0.10$, $G_r = 0.20$, $P_r = 0.71$, $V_0 = 3.80$, $E_c = 0.70$, $S_c = 0.30$, $M = 5.00$, $S_o = 0.60$, $R_a = 0.30$ and $C_r = 1.00$.

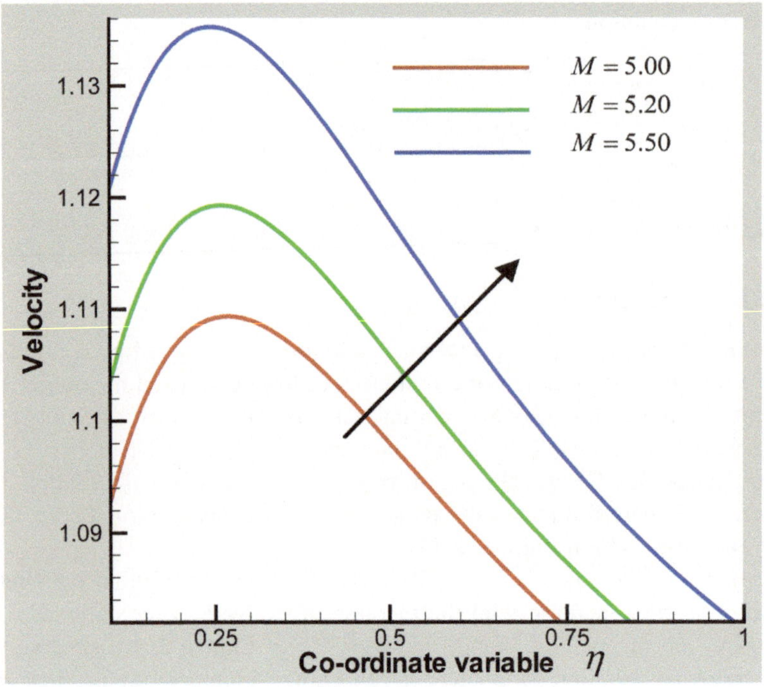

Figure 3. Velocity profiles for different values of M with $G_m = 0.10$, $G_r = 0.20$, $P_r = 0.71$, $V_0 = 3.80$, $E_c = 0.70$, $S_c = 0.30$, $\beta_a = 2.00$, $S_o = 0.60$, $R_a = 0.30$ and $C_r = 1.00$.

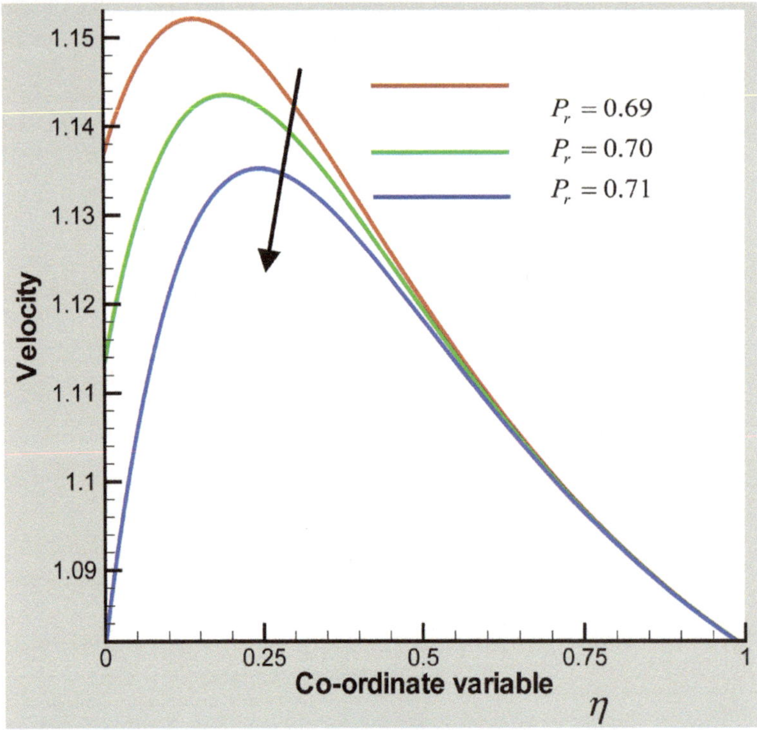

Figure 4. Velocity profiles for different values of P_r with $G_m = 0.10$, $G_r = 0.20$, $M = 5.00$, $V_0 = 3.80$, $E_c = 0.70$, $S_c = 0.30$, $\beta_a = 2.00$, $S_o = 0.60$, $R_a = 0.30$ and $C_r = 1.00$.

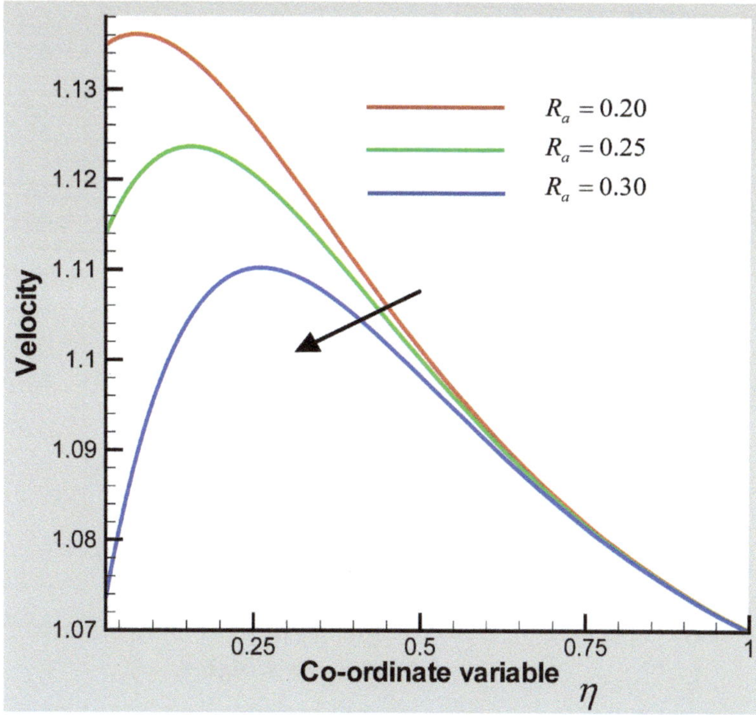

Figure 5. Velocity profiles for different values of R_a with $G_m = 0.10$, $G_r = 0.20$, $M = 5.00$, $V_0 = 3.80$, $E_c = 0.70$, $S_c = 0.30$, $\beta_a = 2.00$, $S_o = 0.60$, $P_r = 0.71$ and $C_r = 1.00$.

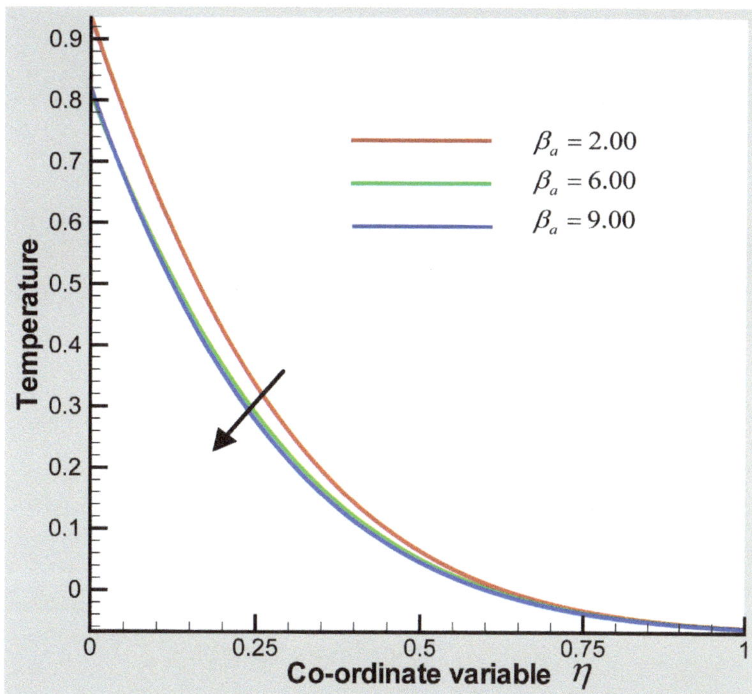

Figure 6. Temperature profiles for different values of β_a with $G_m = 0.10$, $G_r = 0.20$, $M = 5.00$, $R_a = 0.30$, $V_0 = 3.80$, $S_c = 0.30$, $E_c = 0.70$, $S_o = 0.60$, $P_r = 0.71$ and $C_r = 1.00$.

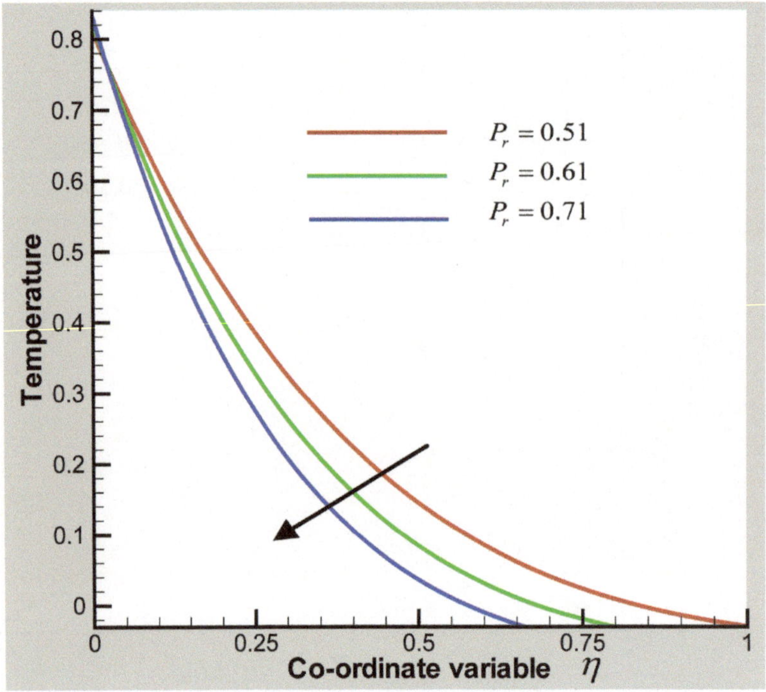

Figure 7. Temperature profiles for different values of P_r with $G_m = 0.10$, $G_r = 0.20$, $\beta_a = 2.00$, $R_a = 0.30$, $V_0 = 3.80$, $S_c = 0.30$, $E_c = 0.70$, $S_o = 0.60$, $M = 5.00$ and $C_r = 1.00$.

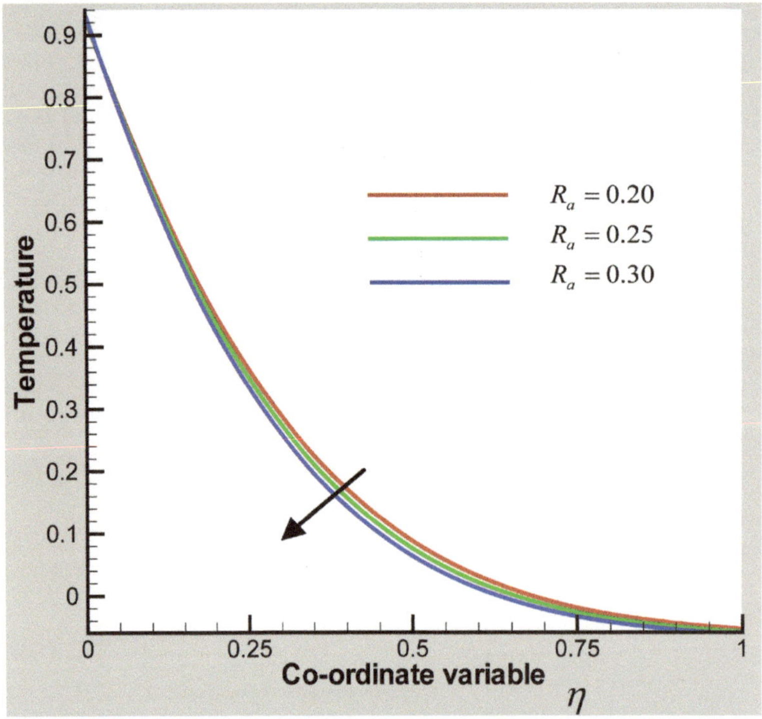

Figure 8. Temperature profiles for different values of R_a with $G_m = 0.10$, $G_r = 0.20$, $\beta_a = 2.00$, $P_r = 0.71$, $V_0 = 3.80$, $S_c = 0.30$, $E_c = 0.70$, $S_o = 0.60$, $M = 5.00$ and $C_r = 1.00$.

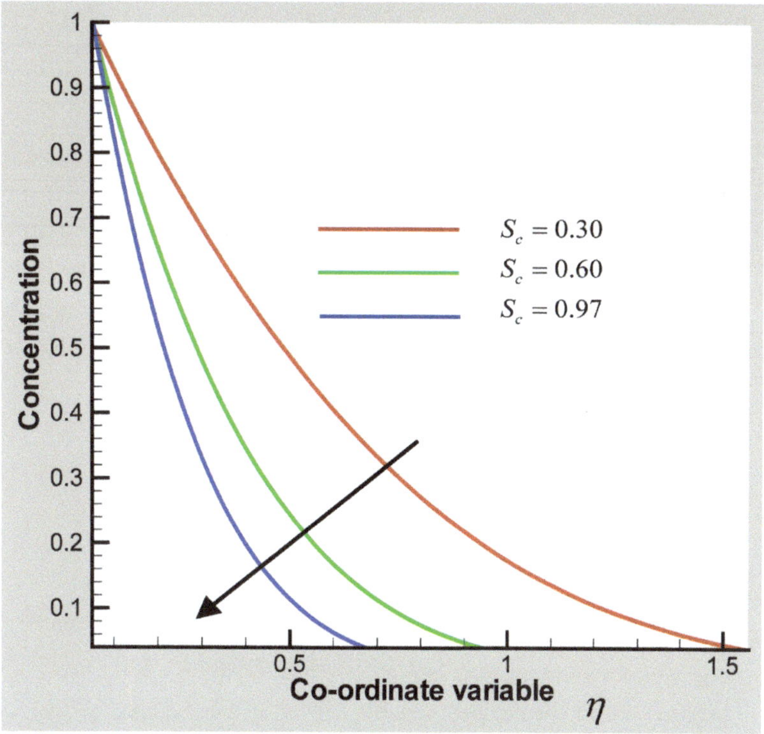

Figure 9. Concentration profiles for different values of S_c with $G_m = 0.10$, $G_r = 0.20$, $\beta_a = 2.00$, $P_r = 0.71$, $M = 5.00$, $R_a = 0.30$, $E_c = 0.70$, $S_o = 0.60$, $V_0 = 3.80$ and $C_r = 1.00$.

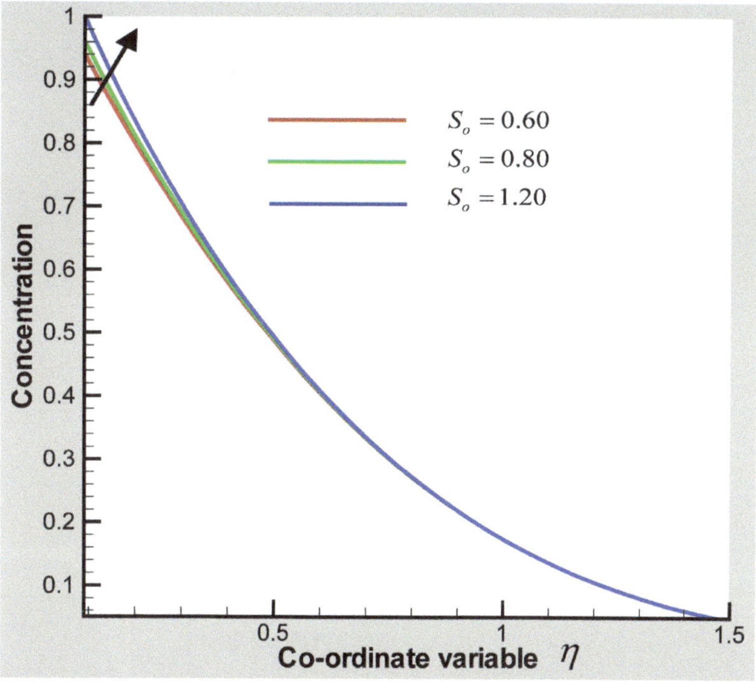

Figure 10. Concentration profiles for different values of S_o with $G_m = 0.10$, $G_r = 0.20$, $\beta_a = 2.00$, $P_r = 0.71$, $M = 5.00$, $R_a = 0.30$, $E_c = 0.70$, $S_c = 0.30$, $V_0 = 3.80$ and $C_r = 1.00$.

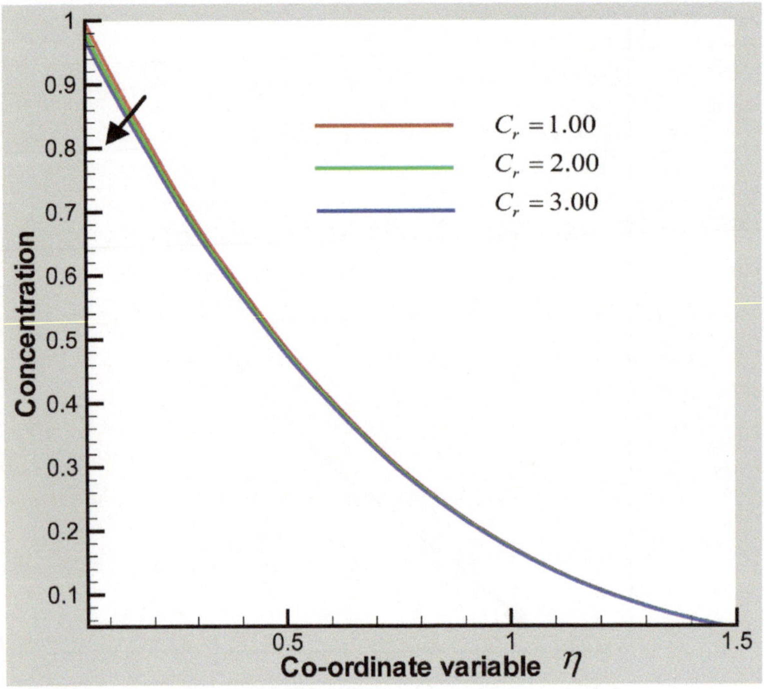

Figure 11. Concentration profiles for different values of C_r with $G_m = 0.10$, $G_r = 0.20$, $\beta_a = 2.00$, $P_r = 0.71$, $M = 5.00$, $R_a = 0.30$, $E_c = 0.70$, $S_c = 0.30$, $S_o = 0.60$ and $V_0 = 3.80$.

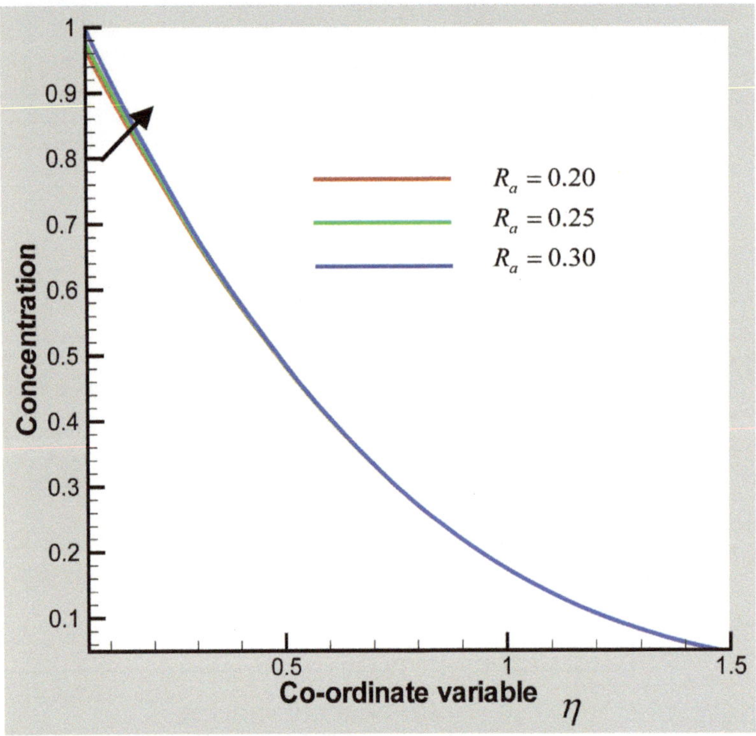

Figure 12. Concentration profiles for different values of R_a with $G_m = 0.10$, $G_r = 0.20$, $\beta_a = 2.00$, $P_r = 0.71$, $M = 5.00$, $S_o = 0.60$, $E_c = 0.70$, $S_c = 0.30$, $V_0 = 3.80$ and $C_r = 1.00$.

4. Conclusions

The analytic solutions o for MHD flow of a viscous fluid near a moving infinite plate with thermal radiation, heat sink, chemical reaction, Soret is analyzed. The results are presented graphically with various parameters. Form the graphical representation, we have the following observations:

- Velocity and temperature profiles decrease with the increase of heat absorption parameter β_a.
- Velocity profiles increase with the increase of magnetic force parameter M.
- Velocity and temperature profiles decrease with the increase of radiation parameter R_a while concentration increase near the plate and away from the plate remain unchanged for increase of radiation parameter R_a.
- Concentration profiles increase near the plate and away from the plate remain unchanged with the increase of Soret number S_o.
- Concentration profiles decrease near the plate and away from the plate remain unchanged with the increase of chemical reaction parameter C_r.

References

[1] Finston, M. (1956) Free Convection past a Vertical Plate. *Journal of Applied Mathematical Physics* (*ZAMP*), **7**, 527-529.

[2] Sparrow, E.M. and Gregg, J.L. (1958) Similar Solutions for Free Convection from a Non Isothermal Vertical Plate. *ASME Journal of Heat Transfer*, **80**, 379-386.

[3] Soundalgekar, V.M. and Ganesan, P. (1981) Finite-Differential Analysis of Transient Free Convection with Mass Transfer on an Isothermal Vertical Plate. *International Journal of Engineering Science*, **19**, 757-770. http://dx.doi.org/10.1016/0020-7225(81)90109-9

[4] Camargo, R., Luna, E. and Trevino, C. (1996) Numerical Study of the Natural Convective Cooling of a Vertical Plate. *International Journal of Heat and Mass Transfer*, **19**, 89-95. http://dx.doi.org/10.1007/s002310050096

[5] Callahan, G.D. and Marner, W.J. (1976) Transient Free Convection with Mass Transfer on an Isothermal Vertical Flat Plate. *International Journal of Heat and Mass Transfer*, **19**, 165-174. http://dx.doi.org/10.1016/0017-9310(76)90109-5

[6] Soundalgekar, V.M. and Wavre, P.D. (1977) Unsteady Free Convection Flow past an Infinite Vertical Plate with Constant Suction and Mass Transfer. *International Journal of Heat Mass Transfer*, **20**, 1363-1373. http://dx.doi.org/10.1016/0017-9310(77)90033-3

[7] Soundalgekar, V.M. and Ganesan, P. (1980) Transient Free Convective Flow past a Semi-Infinite Vertical Plate with Mass Transfer. *Journal of Energy, Heat and Mass Transfer*, **2**, 83.

[8] Jaluria, Y. (1980) Natural Convection Heat and Mass Transfer. Pergamon Press, Oxford.

[9] Gebhart, B. and Pera, L. (1971) The Nature of Vertical Natural Convection Flows Resulting from the Combined Buoyancy Effect of Thermal and Mass Diffusion. *International Journal of Heat Mass Transfer*, **14**, 2025-2050. http://dx.doi.org/10.1016/0017-9310(71)90026-3

[10] Singh, A.K., Singh, A.K. and Singh, N.P. (2003) Heat and Mass Transfer in MHD Flow of a Viscous Fluid past a Vertical Plate under Oscillatory Suction Velocity. *Indian Journal of Pure and Applied Mathematics*, **34**, 429.

[11] Chamka, A. and Khaled, A.R.A. (2001) Simultaneously Heat and Mass Transfer in Free Convection. *Industrial Engineering Chemical*, **49**, 961-968.

[12] Haque, M.M. and Sarder, U.K. (2015) Thermal Diffusion Effect on Convective Heat and Mass Transfer of High Speed MHD Flow over a Stretching Sheet. *Journal of Scientific Research and Reports*, **8**, 1-14. http://dx.doi.org/10.9734/jsrr/2015/18075

[13] Anuradha, S. and Priyadharshini P. (2014) Heat and Mass Transfer on Unsteady MHD Free Convective Flow pass a Semi-Infinite Vertical Plate with Soret Effect. *The International Journal of Science & Technoledge*, **2**, 328-337.

[14] Bhattacharyya, K. and Layek, G.C. (2011) Effects of Suction/Blowing on Steady Boundary Layer Stagnation-Point Flow and Heat Transfer towards a Shrinking Sheet with Thermal Radiation. *International Journal of Heat Mass Transfer*, **54**, 302-307. http://dx.doi.org/10.1016/j.ijheatmasstransfer.2010.09.043

[15] Hady, F.M., Mohamed, R.A. and ElShehabey, H.M. (2013) Thermal Radiation, Heat Source/Sink and Work Done by Deformation Impacts on MHD Viscoelastic Fluid over a Nonlinear Stretching Sheet. *World Journal of Mechanics*, **3**, 203-214.

[16] Viskanta, R. and Grosh, R.J. (1962) Boundary Layer in Thermal Radiation Absorbing and Emitting Media. *International Journal of Heat Mass Transfer*, **5**, 795-806. http://dx.doi.org/10.1016/0017-9310(62)90180-1

[17] Bestman, A.R. (1990) The Boundary-Layer Flow past a Semi-Infinite Heated Porous Plate for a Two-Component

Plasma. *Astrophysics and Space Science*, **173**, 93-100. http://dx.doi.org/10.1007/BF00642565

[18] Sattar, M.A. and Alam, M.M. (1994) Thermal-Diffusion as Well as Transpiration Effects on MHD Free Convection and Mass Transfer Flow past an Accelerated Vertical Porous Plate. *Indian Journal of Pure and Applied Mathematics*, **25**, 679-688.

Stability Analysis of a SDOF Mechanical Model with Distinct Critical Points: II. Catastrophe Theory Approach

Dimitrios S. Sophianopoulos, Vasiliki S. Pantazi

Department of Civil Engineering, University of Thessaly, Volos, Greece
Email: dimsof@civ.uth.gr

Abstract

In a recent publication [1], the fully nonlinear stability analysis of a Single-Degree-of Freedom (SDOF) model with distinct critical points was dealt with on the basis of bifurcation theory, and it was demonstrated that this system is associated with the butterfly singularity. The present work is the companion one, tackling the problem via the Theory of Catastrophes. After Taylor expanding the original potential energy function and introducing Padè approximants of the trigonometric expression involved, the resulting truncated potential is a universal unfolding of the original one and an extended canonical form of the butterfly catastrophe potential energy function. Results in terms of equilibrium paths, bifurcation sets and manifold hyper-surface projections fully validate the whole analysis, being in excellent agreement with the findings obtained via bifurcation theory.

Keywords

Mechanical Models, Nonlinear Stability, Critical Points, Catastrophe Theory, Butterfly Singularity

1. Introduction

Problems in Non-Linear Buckling, where various phenomena may arise, such as discontinuities, singularities and instabilities, can be efficiently treated by using Catastrophe Theory [2]-[4], and a better insight on them can be readily gained. Due to its topological character, this particular Theory allows for the establishment of universal solutions and the direct production of qualitative results, since it deals with various forms of potential energy functions incorporating the behavior of classes of complex systems.

The Theory of Catastrophes supplies the tools for comprehension and detection of numerous types of bifurcations and it has been proven highly efficient as far as potential discrete or continuous systems are concerned. Among the former type of systems, mechanical models (original ones or products of adequate discretization of the latter systems) with a few degrees of freedom are included, which, despite their geometrical and overall simplicity, may exhibit a quite complicated post-buckling response, associated with all kinds of distinct critical points. Their treatment via the foregoing Theory however is not an easy task, and this remark is fully justified from the admittedly limited relevant publications [5]-[13].

Aiming to contribute a further step towards the goal of using Catastrophe Theory in achieving universal results on the buckling and post-buckling of a simple model, the present work reconsiders the single DOF system dealt with in [1], utilizing advanced mathematical tools [14]. After Taylor expansions and Padé approximations of trigonometric functions, the universal unfolding of the system's original potential energy yields the butterfly singularity, in full accordance with the findings of the Theory of Bifurcations [1].

2. Single DOF System and Its Truncated Total Potential Energy Function

We reconsider the single DOF mechanical model studied in the companion paper [1], depicted in **Figure 1**.

The system consists of a weightless rigid bar of length ℓ, partially pinned at its base via a linear rotational spring of stiffness c; its tip is connected via an inclined linear extensional spring of stiffness k, the angle of inclination a of which is always constant, since the other end of the spring freely slides along an equally inclined support. The tip of the bar (point A) is acted upon by a gravitational force P. Hence, the single degree of freedom (active coordinate) characterizing the foregoing model is the rotation θ. This model is considered generally imperfect, by introducing an initial rotation ε. Before the action of the loading P the springs are considered unstressed. Angle α is measured (*i.e.* is positive) from the horizontal line passing through point A clockwise, while both rotations θ and ε are measured from the vertical line through the base again clockwise.

After the introduction of the dimensionless parameters $\beta = c/k\ell^2$ and $\lambda = P/k\ell$, the non-dimensional total potential energy function governing the system is given by

$$V = \frac{1}{2}\left[\sin(\alpha-\varepsilon)-\sin(\alpha-\theta)\right]^2 + \frac{1}{2}\beta(\theta-\varepsilon)^2 - \lambda(\cos\varepsilon - \cos\theta). \tag{1}$$

Aiming to explore the nature of the above potential and classify the system to one of the seven Elementary Catastrophes [4] if possible, we perform a Taylor series expansion of the function given in Expression (1) up to the 6th order and introduce in the outcome of this process Padé approximations of the trigonometric functions involved (associated only with a and ε), given in general by:

$$\cos(y) \cong \frac{1-\frac{3665y^2}{7788}+\frac{711y^4}{25960}-\frac{2923y^6}{7850304}}{1+\frac{229y^2}{7788}+\frac{y^4}{2360}+\frac{127y^6}{39251520}}, \quad \sin(y) \cong \frac{y-\frac{2363y^3}{18183}+\frac{12671y^5}{4363920}}{1+\frac{445y^2}{12122}+\frac{601y^4}{872784}+\frac{121y^6}{16662240}}$$

$$\cos^2(y) \cong \frac{1-\frac{1718y^2}{1947}+\frac{722y^4}{3245}-\frac{7244y^6}{613305}}{1+\frac{229y^2}{1947}+\frac{2y^4}{295}+\frac{127y^6}{613305}}, \quad \sin^2(y) \cong \frac{y^2-\frac{140y^4}{649}+\frac{39y^6}{3245}}{1+\frac{229y^2}{1947}+\frac{2y^4}{295}+\frac{127y^6}{613305}}. \tag{2a-d}$$

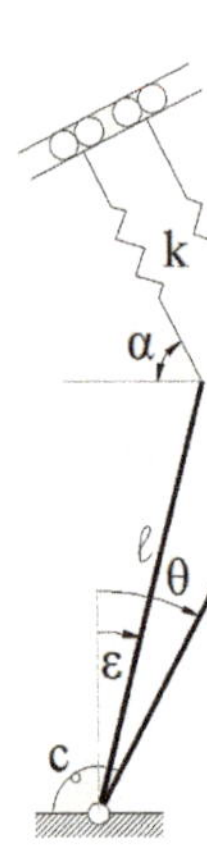

Figure 1. Single DOF mechanical model dealt with herein.

After cumbersome symbolic manipulations in *Mathemetica* [14], the truncated potential of the foregoing model takes a 6th-order polynomial form with respect to θ, given by

$$V_{tr} = A\theta^6 + H\theta^5 + B\theta^4 + \Gamma\theta^3 + \Delta\theta^2 + E\theta + Z. \tag{3}$$

where coefficients A, B, Γ, Δ, E, Z and H are strongly nonlinear functions of the four parameters a, β, λ, ε; their general form is given below:

$$\left.\begin{array}{l} A = A(\alpha,\varepsilon,\lambda),\ B = B(\alpha,\varepsilon,\lambda),\ \Gamma = \Gamma(\alpha,\varepsilon,\lambda),\ \Delta = \Delta(\alpha,\varepsilon,\beta,\lambda) \\ E = E(\alpha,\varepsilon,\beta),\ Z = Z(\alpha,\varepsilon,\beta,\lambda),\ H = H(\alpha,\varepsilon) \end{array}\right\}. \tag{4}$$

The exact expressions of these coefficients are enormously lengthy, and so they will not be given herein for reasons of brevity. Following standard procedures (monomials, k-jets and so on) described in [6], it was found that the truncated potential given above is a universal unfolding of the original one, since it has a co-dimension equal to four and a determinacy equal to six. These procedures however were extremely time-consuming and literally almost exhausted our computer resources.

Evidently, V_{tr} has one active variable and resembles the standard form of the potential associated with the Butterfly Singularity. There are however three distinguished differences/obstacles, which do not allow us to fully classify it among the seven elementary Catastrophes, namely: (a) the "constant" term Z, which has a negligible influence [4] [6], since Catastrophe Manifolds (equilibria) and Bifurcation Sets (critical states), associated with 1st and 2nd order derivatives of V_{tr} with respect to the active variable θ, are not affected by its presence, (b) the 5th order term related to coefficient H, which however does not depend on the external load, remaining constant as λ is varied, and (c) the effect of the presence of coefficient A on the truncated potential.

We initially address the latter of the aforementioned obstacles. The standard perturbation of the germ x^6, associated with the butterfly singularity, is of the form $V = x^6 + t_1x^4 + t_2x^3 + t_3x^2 + t_4x$ [5], with four unfolding parameters. The presence of A initially reveals a fifth unfolding parameter, which would not allow the sought classification. However, within accepted ranges of the parameters α, ε and λ (on which A depends), if its mean value is independent of these parameters, then one could easily consider A as a quasi-constant. Consequently, the whole analysis could in this case be continued smoothly, overcoming this particular drawback.

Indeed, after cumbersome symbolic and algebraic calculations [14], it was found than the mean value of A, for a ranging from $-\pi/2$ to $3\pi/2$, ε ranging from $-\pi/2$ to $\pi/2$ and λ ranging from -5 to 10, it was found that A varies from 0.025 to -0.10, with a mean value -0.0375 and a negliblibly small standard deviation. Consequently, the presence of A does not affect our classification effort whatsoever.

Considering at this point the second of the above obstacles, the following are valid. Within the range of values of the angular parameters α and ε, on which H is dependent, given above, one may plot the surface $H(\alpha,\varepsilon)$, which is depicted in **Figure 2**. Its absolute maximum value does not exceed 0.0695, implying that the term

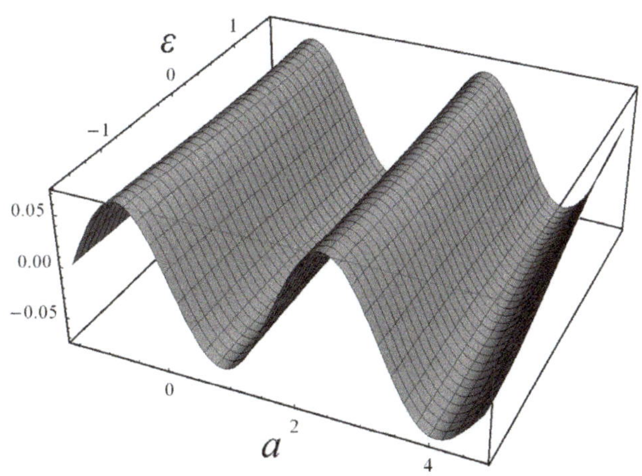

Figure 2. Variation, via 3D plot, of the value of coeffient H within accepted ranges of values of control parameters α and ε.

$H\theta^5$ becomes negligibly small for values of θ smaller than 1, while for the extreme model deformation of θ equal to $-\pi/2$ or $\pi/2$ this term remains smaller than 0.7. These preliminary findings are a strong initial indication that the above term may be omitted from the expression of the truncated potential and hence one may assume that the related Singularity is indeed of the Butterfly type. The qualitative analysis that follows will provide strong evidence and justification of the above, taking also into account the local nature of the whole approach. and the Splitting Lemma of the Theory of Catastrophes [4].

3. Qualitative Analysis

Based on the preceding findings, the qualitative analysis that follows consists of two steps. In the 1st step, a brief description of the theoretical background is presented, while the 2nd step concerns symbolic manipulations for the computation of Catastrophe Manifolds (CMs) and Bifurcation Sets (BSs).

Thereafter, verification, validation and proof of the existence of the Butterfly Singularity via a parametric comparative study of the CMs, BSs (and related projections) of the system are given in the Numerical Results and Discussion section.

3.1. Theoretical Background

For any given gradient system, governed by a potential $\boldsymbol{V}$, which is associated with one state (independent) variable–active coordinate, say $\boldsymbol{x}$, and four control parameters, defined by $\boldsymbol{r}_i$ (i = 1 – 4), *i.e.* $V = V(x; r_i)$, Nonlinear Stability and Bifurcation Theory dictates that the equilibrium configurations are the solutions of Equation (5a), while the critical states should also satisfy Equation (5b).

$$\frac{\mathrm{d}V}{\mathrm{d}x} = 0 \ (\mathrm{a}), \quad \frac{\mathrm{d}^2V}{\mathrm{d}x^2} = 0 \ (\mathrm{b}). \tag{5a,b}$$

We assume that $\boldsymbol{U}$ is a universal unfolding of $\boldsymbol{V}$, being a product of adequate Taylor expansion and truncation. We also assume that $\boldsymbol{U}$ is a 6th order polynomial of the most general form with respect to $\boldsymbol{x}$ (as the one shown in Equation (3)), the coefficients of which are known functions of $\boldsymbol{r}_i$, implying that the unfolding parameters are directly dependent on the control parameters of $\boldsymbol{V}$. Hence, instead of seeking equilibria, critical states and bifurcations (ought to infinitesimal changes of one or more of $\boldsymbol{r}_i$, individually or in an combined manner) of the original system, it is convenient to apply Catastrophe Theory on the truncated system and afterwards compare the results with those of Nonlinear Stability Theory. Aiming to present qualitative proof that the original system, governed by $\boldsymbol{V}$, is associated with the Butterfly Catastrophe, a short reminder of the salient features of this Singularity is given below:

- Depending on the parameter values, the potential function may have three, two, or one different local minima, separated by the loci of fold bifurcations. At the butterfly point, the different 3-surfaces of fold bifurcations, the 2-surfaces of cusp bifurcations, and the lines of swallowtail bifurcations all meet up and disappear, leaving a single cusp structure remaining.
- The set of critical points of the $\boldsymbol{U}$-unfolding is the so-called four-dimensional Catastrophe Manifold $\boldsymbol{CM}_U$, given by [5]

$$\boldsymbol{CM}_U = \frac{\partial \boldsymbol{U}}{\partial \boldsymbol{x}} = 0. \tag{6}$$

- The set of the singular (degenerate) critical points S_U is a four-dimensional Sub-Manifold of the previous one, satisfying [5]:

$$S_U = \boldsymbol{CM}_U \big|_{\frac{\partial^2 U}{\partial x^2}=0}. \tag{7}$$

- The projection of the above Manifold in the space of the control parameters, which forms the Bifurcation Set BS_U, should combine Equations (6) and (7) yielding:

$$BS_U = \exists \boldsymbol{x} : \left\{ \frac{\partial \boldsymbol{U}}{\partial \boldsymbol{x}} = 0, \frac{\partial^2 \boldsymbol{U}}{\partial \boldsymbol{x}^2} = 0 \right\}. \tag{8}$$

- This Bifurcation Set constitutes a hyper-surface in the four-dimensional control space, and thus its graphic representation is impossible in the three-dimensional physical space. Only 3D sections of this hyper-surface can be drawn as 3D contour plots or bifurcation diagrams, for some characteristic combinations of the control parameters, with at least one of them kept constant. Furthermore, it should be noted that pioneer publications on the subject [2] [4] [6] have revealed the immense difficulty of the sought representations, since it is valid that: (a) the BS is made of fold hyper-surfaces (lines on a 2D slice of control space) which meet on surface of cusp points which themselves meet on lines of swallowtail points, (b) these lines of swallowtail points meet together at a butterfly point and (c) at this point in control space, potential curve (germ) has only one minimum which corresponds to the collapse of three minimums and two maximums. Furthermore, the potential energy described by $\boldsymbol{U}$ will have in general an odd number of minima and an even number of maxima, and therefore, in some regions of the Catastrophe Manifold (and for some orientation of the surface) there will be up to five layers of the $\boldsymbol{CM}_U$ overlying each other.

3.2. Computation of Catastrophe Manifolds and Bifurcation Sets via Elimination Scheme

From the One-Dimensional Elementary Catastrophes (Cuspoids), the Butterfly (A_5 according to the classification of Thom [4]), possesses the highest degree of complexity and moreover it rarely appears in the Theory of Elastic Stability [5]; however, it has been found to occur in other engineering disciplines, as for instance in the mesophase stability response of external magnetic or electric fields [7].

In order to show that the one-dimensional system studied herein exhibits the Butterfly Singularity, one may either use Equations (6) and (8) or employ and extend the alternative elimination procedure of Deng [8]. The main drawback of the latter method lies on the fact that, both CM and BS, seem to depend on all the parameters involved, namely λ, β, θ, ε and a. Hence, the former scheme will be employed hereafter.

After computing the Catastrophe Manifold, the Bifurcation Set is evaluated, by symbolically eliminating favorably either λ or β (due to their apparent linear presence). In the foregoing analysis β was eliminated, based either on the original potential, given in Equation (1), or on the truncated one with $H = 0$ from Equation (3). The expression governing the BS of the original system is given in Equation (9), while the one of the truncated system is extremely lengthy and will not be presented for brevity.

$$\begin{aligned} BS :\rightarrow (\varepsilon-\theta)\left(2\cos\left[2(\alpha-\theta)\right]-\cos\left[2\alpha-\varepsilon-\theta\right]+\cos\left[\varepsilon-\theta\right]-2\lambda\cos\theta\right) \\ +2\cos\left[\alpha-\theta\right]\sin\left[\alpha-\varepsilon\right]=\sin\left[2(\alpha-\theta)\right]+2\lambda\sin\theta \end{aligned} \tag{9}$$

In the sequel, one may produce reliable results in terms of equilibrium paths, bifurcation diagrams and bifurcation sets, and qualitatively justify that the system dealt with is governed by the Butterfly Singularity, in agreement with the findings of the companion paper [1].

4. Numerical Results and Discussion

Our first goal is to compare 3D Contour Plots of the Bifurcation Set, as products (a) of the original potential, and (b) of the truncated potential, with $H = 0$. The corresponding plots, for two characteristic values of imperfection parameter ε, are illustrated in **Figure 3** and **Figure 4**.

Comparing the contents of the above figures, it is evident that the plots of the two cases are almost identical. This result verifies to a certain extent the validity of the approximate Catastrophe Theory approach adopted in this work. Further validation and evidence will be provided in what follows, by reproducing equilibrium paths and bifurcation diagrams of the system, based on its truncated potential with $H = 0$.

As far as equilibrium paths are concerned, the cases of $\beta \neq 0$, α, $\varepsilon = 0$ and β, $\varepsilon \neq 0$, $\alpha = 0$ where chosen. The corresponding plots are shown in **Figure 5** and **Figure 6** respectively.

These paths are identical with the ones presented in the companion paper, via a Nonlinear Stability analysis.

Furthermore, several Bifurcation Diagrams (imperfection sensitivity plots) are presented, as products of the present analysis. These correspond to $\alpha = 0$, shown in **Figure 7**, to $a = 0.10$, $\beta = 0$, shown in **Figure 8**, and to $a = 0.20$, $\beta = 0.20$, shown in **Figure 9**.

Once again, full accordance with the relevant findings of the companion paper is identified.

From all the above results and the preceding analysis, it is proven that the model exhibits the Butterfly Singularity.

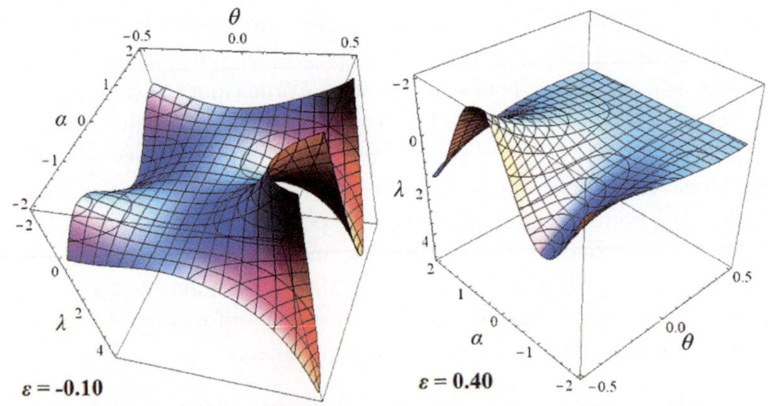

Figure 3. Bifurcation Sets based on the original potential.

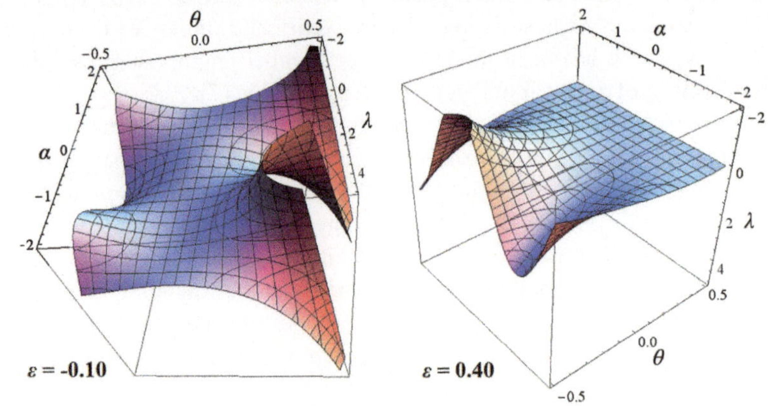

Figure 4. Bifurcation Sets based on the truncated potential with $H = 0$.

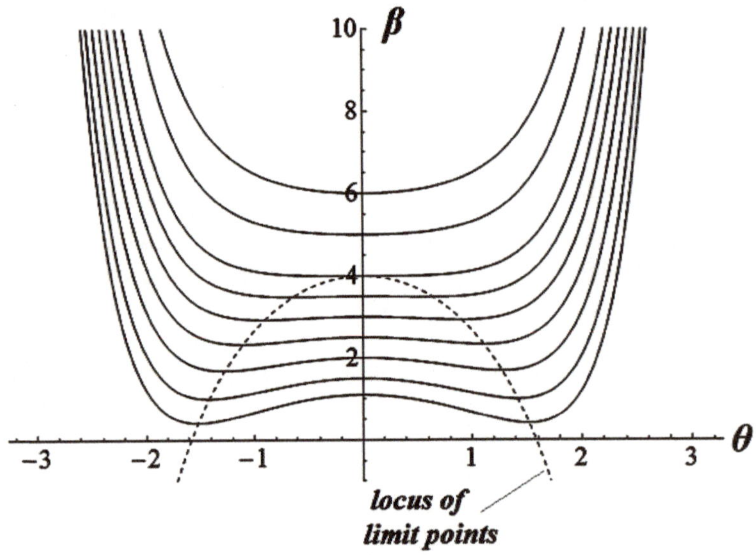

Figure 5. Equilibrium paths for $\beta \neq 0$, α, $\varepsilon = 0$, based on the truncated potential with $H = 0$.

5. Concluding Remarks

The single degree-of-freedom model analyzed herein, although quite simple, is associated with all kinds of dis-

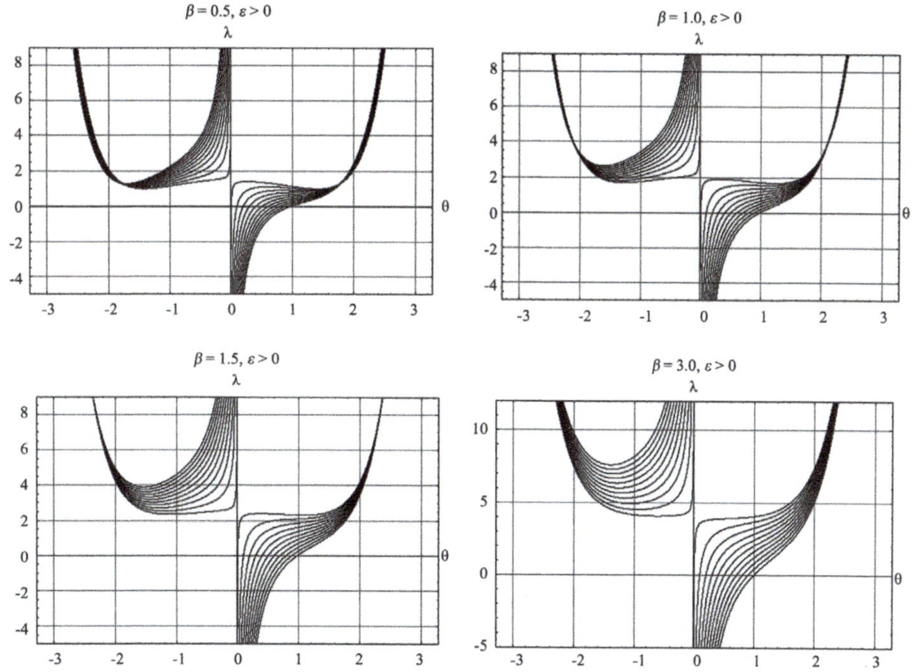

Figure 6. Equilibrium paths for β, $\varepsilon \neq 0$, $\alpha = 0$, based on the truncated potential with $H = 0$.

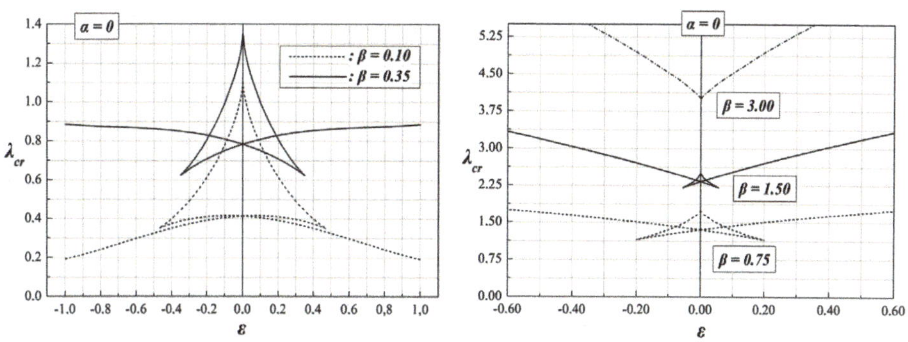

Figure 7. Bifurcation diagrams for $\alpha = 0$, based on the truncated potential with $H = 0$.

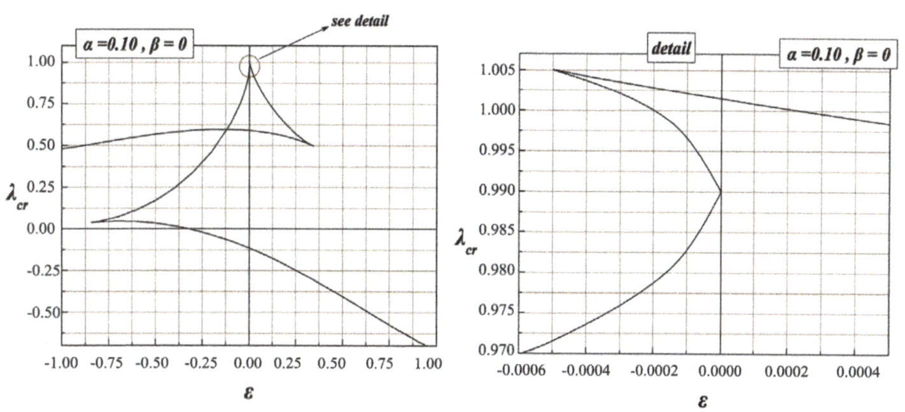

Figure 8. Bifurcation diagrams for $\alpha = 0.10$, $\beta = 0$, based on the truncated potential with $H = 0$.

tinct critical points and related to the Butterfly Catastrophe, a situation scarcely encountered in Structural Mechanics. The use of the truncated potential, within reasonable parameter ranges, lead to results in excellent

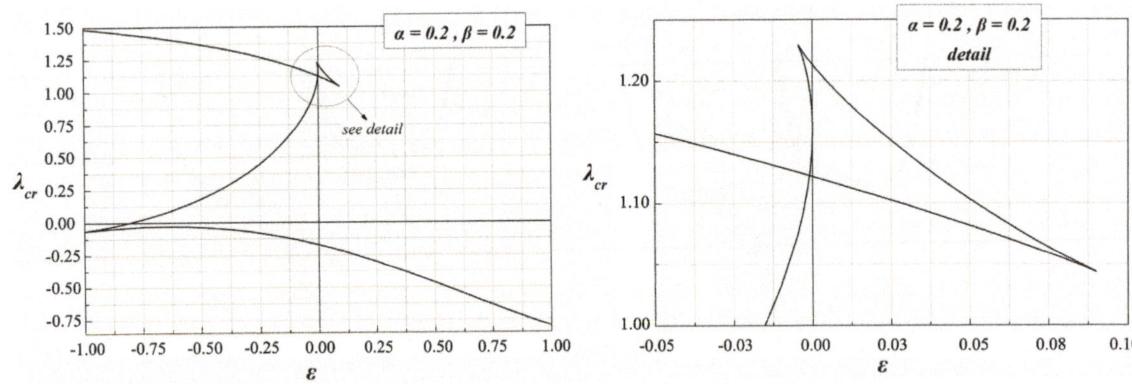

Figure 9. Bifurcation diagrams for $a = 0.20$, $\beta = 0.20$, based on the truncated potential with $H = 0$.

agreement with ones obtained from the original potential in terms of equilibria, imperfection sensitivity and Bifurcation Sets, as also reported in the companion paper.

References

[1] Sophianopoulos, D.S. (2013) Stability Analysis of a Single-Degree-of-Freedom Mechanical Model with Distinct Critical Points: I. Bifurcation Theory Approach. *World Journal of Mechanics*, **3**, 62-81. http://dx.doi.org/10.4236/wjm.2013.31005

[2] Woodcock, A.E.R. and Poston, T. (1974) A Geometrical Study of the Elementary Catastrophes. Springer, New York. http://dx.doi.org/10.1007/BFb0068967

[3] Stewart, I.A. (1980) Catastrophe Theory and Equations of State: Conditions for a Butterfly Singularity. *Mathematical Proceedings of the Cambridge Philosophical Society*, **88**, 429-449. http://dx.doi.org/10.1017/S0305004100057789

[4] Gilmore, R. (1981) Catastrophe Theory for Scientists and Engineers. Dover, New York.

[5] Theocaris, P.S. (1984) Instability of Cantilever Beams with Non-Linear Elements: Butterfly Catastrophe. *International Journal of Mechanical Sciences*, **26**, 265-275. http://dx.doi.org/10.1016/0020-7403(84)90047-X

[6] Steindl, H. and Troger, A. (1991) Nonlinear Stability and Bifurcation Theory. Springer, New York.

[7] Pershin, K.Vl. (1992) The Butterfly Catastrophe and Mesophase Stability in an External Field. *Physics Letters A*, **161**, 510-516. http://dx.doi.org/10.1016/0375-9601(92)91083-4

[8] Deng, C.-G. (1994) Equations of Bifurcation Sets of Three-Parameter Catastrophes and Application in Imperfection Sensitivity Analysis. *International Journal of Engineering Science*, **32**, 1811-1822. http://dx.doi.org/10.1016/0020-7225(94)90110-4

[9] Lignos, X.A., Parke, G.A.R., Harding, J.E. and Kounadis, A.N. (2002) A Comprehensive Catastrophe Theory for Non-Linear Buckling of Simple Systems Exhibiting Fold and Cusp Catastrophes. *International Journal of Numerical Methods in Engineering*, **54**, 175-193. http://dx.doi.org/10.1002/nme.416

[10] Lignos, X., Ioannidis, G. and Kounadis, A.N. (2003) Non-Linear Buckling of Simple Models with Tilted Cusp Catastrophe. *International Journal of Nonlinear Mechanics*, **38**, 1163-1172. http://dx.doi.org/10.1016/S0020-7462(02)00061-6

[11] Lengyel, A. and You, Z. (2004) Bifurcations of SDOF Mechanisms Using Catastrophe Theory. *International Journal of Solids and Structures*, **41**, 559-568. http://dx.doi.org/10.1016/j.ijsolstr.2003.09.024

[12] Raftoyiannis, I.G., Constantakopoulos, T.G., Michaltsos, G.T. and Kounadis, A.N. (2006) Dynamic Buckling of a Simple Geometrically Imperfect Frame Using Catastrophe Theory. *International Journal of Mechanical Sciences*, **48**, 1021-1030. http://dx.doi.org/10.1016/j.ijmecsci.2006.05.010

[13] Sophianopoulos, D.S. (2007) Bifurcations and Catastrophes of a Two-Degrees-of-Freedom Nonlinear Model Simulating the Buckling and Postbuckling of Rectangular Plates. *Journal of the Franklin Institute*, **344**, 463-488. http://dx.doi.org/10.1016/j.jfranklin.2006.02.012

[14] Wagon, S. (2010) Mathematica® in Action: Problem Solving Through Visualization and Computation. Springer, New York. http://dx.doi.org/10.1007/978-0-387-75477-2

22

Effects of Radiations and Heat Source/Sink on a Casson Fluid Flow over Nonlinear Stretching Sheet

Chenna Sumalatha, Shankar Bandari

Department of Mathematics, Osmania University, Hyderabad, India

Email: sumalatha.chenna@gmail.com

Abstract

The present study deals with the flow over a nonlinearly stretching sheet of Casson fluid with the effects of radiation and heat source/sink. The Casson fluid model is used to characterize the non-Newtonian fluid behaviour. With the help of justified similarity transformations the governing equations were reduced to couple nonlinear ordinary differential equations. The effective numerical technique Keller Box method is used to solve these equations. The variations in velocity, temperature profiles were presented with the various values of nonlinear stretching parameter n and Casson parameter β. The nature of Skinfriction and Local nusselt number has presented. Effects of radiation and heat source/sink on temperature profiles have been discussed.

Keywords

Nonlinear Stretching Sheet, Casson Fluid, Keller Box Method, Radiation, Heat Source/Sink

1. Introduction

In recent decades the scientists and engineers are interested to study the flow caused by steady or unsteady stretching sheets because of its rigorous applications in various engineering fields for eg: Polymer sheet extruded continuously from a dye, cooling of metallic plate in a cooling bath and heat-treated materials that travel between feed and wind-up rolls or on a conveyor belt, etc. The exact solution for the flow due to stretching of flat surface was first obtained by Crane [1]. He was investigated the steady flow of an incompressible viscous fluid over an elastic sheet, where as the flow was caused by the stretching of the sheet in its own plane with a velocity varying linearly with the distance from a fixed point. The solution of the three-dimensional fluid motion caused by the stretching boundary was studied by Wang C.Y. [2]. H. I. Andersson [3] has explained an exact solution of the flow of a viscoelastic fluid past a stretching sheet in the presence of a transverse magnetic field is considered.

All the above studies are considered only for a linear stretching sheet, but it is true that not in all cases the stretching sheet is linear. It was first identified by Gupta and Gupta [4]. The study of heat transfer in a viscous fluid over a non-linearly stretching sheet with viscous dissipation was done by Vajravelu [5]. The flow of an incompressible viscous fluid over a non-linear stretching sheet in the presence of chemical reaction and magnetic field was investigated by Raptis and Perdiks [6]. Cortell [7] explained a numerical analysis for flow and heat transfer in a viscous fluid over a non-linear stretching sheet. The Homotopy perturbation method [HPM] and the pade approximation are employed to investigate the MHD boundary layer flow over a non-linear stretching sheet can be found in refs. [8] [9]. Some important aspects of the flow of various non-Newtonian fluids over a non-linear stretching sheet can be found in article [10]. Very recently Pramanic S. [11] has obtained the solution for Casson fluid flow and heat transfer past an exponentially porous stretching surface in presence of thermal radiation and he found that the effect of increasing values of the Casson parameter is seen to suppress the velocity field. But the temperature is enhanced with increasing Casson parameter. Whereas Krishendu Battacharya [12] worked on MHD Stagnation-Point Flow of Casson Fluid and Heat Transfer over a Stretching Sheet with Thermal Radiation. The unsteady boundary layer flow and heat transfer of a Casson fluid over a moving plat plate with a parallel free stream were studied by Mustafa *et al.* [13] and they solved the problem an alytically using homotopy analysis method (HAM).Different from stretching sheet the Nadeem S. [14] has obtained the numerical solution for MHD flow of a Casson fluid over an exponentially shrinking sheet. The study on Casson fluid flow on both stretching/ shrinking sheet was done by Bhattacharya [15]. The extension of study on the MHD boundary layer flow with heat and mass transfer of a nanofluid over non-linear stretching sheet can be studied by Mabood [16]. Raftari [17] worked an approximate solution of magnetohydrodyn amic [MHD] boundary layer flow.

Casson fluid is one kind of Non-Newtonian fluid. This behaves like a solid elastic and the yield shear stress exists in the consecutive equation for this fluid. Fredrickson [18] investigated the steady flow of a casson fluid in tube. The analytical solution of the unsteady boundary layer flow and heat transfer of a casson fluid over a moving flat plate with a parallel free stream was studied by Mustafa [19]. The analytical solution of the boundary layer flow of an incompressible viscous fluid over a non-linear stretching sheet was studied by Hayat T. [20]. Fekry M. Hady [21] studied the characteristics of a viscous nanofluid over a non-linearly stretching sheet in the presence of thermal radiation and heat transfer. The effect of transpiration on casson fluid heat transfer toward a linearly stretching surface is reported by S. Mukhopadhyay [22]. Hayat [23] determined the effect of thermal radiation on the flow of a second grade fluid.

Since most of the Fluid flow problems are nonlinear so the methods of solving the nonlinear equations are very important. Till today there are different types of Exact methods. In tradition, the perturbation technique is used to find out the approximations to the nonlinear equation similarly Adomin Methods [24]. Even though there are different kinds of analytical techniques available, there is an importance for Numerical techniques. Ibrahim [25] obtained the numerical solution of the boundary layer flow and heat transfer over non-isothermal stretching sheet in a nano fluid with effect of magnetic field. Mahantesh [26] investigated the effect of thermal radiation and non-uniform heat source/sink as heat transfer of MHD visco elastic boundary layer flow over a stretching surface. Bhattacharyya [27] studied the effect of radiation and heat source/sink on the unsteady boundary layer flow and heat transfer past a shrinking sheet with suction/injection. The effect of radiation and heat transfer over unsteady stretching surface is analysed by Mahamed [28]. The effect of radiation on free convection from a porous vertical plate studied by Hossain [29]. The Keller-Box method introduced by Keller [30] is one of the best numerical method basically it's a mixed finite volume method which consists in taking the average of a conservation law and of the associated constitutive law at the level of the same mesh cell. Sarif [31] worked the numerical solution of the steady boundary layer flow and heat transfer over a stretching sheet with Newtonian heating by using Keller box method. Finally the objective of this paper is to study the effects of radiation and heat source/sink of Casson fluid flow over a nonlinearly stretching sheet. The governing partial differential equations are converted into nonlinear differential equations by using suitable similarity transformations and those equations are then solved by using Keller box method.

2. Mathematical Formulation

The steady two-dimenional incompressible flow of a Casson fluid bounded by a stretching sheet at $y = 0$ has considered. The flow is confined to $y > 0$. Here x-, y-axes are the directions of the plate and normal to the plate respectively. The rheological equation of state for an isotropic and incompressible flow of the Casson fluid is given by

$$\tau_{ij} = \begin{cases} 2\left(\mu_B + p_y/\sqrt{2\pi}\right)e_{ij}, & \pi > \pi_c, \\ 2\left(\mu_B + p_y/\sqrt{2\pi_c}\right)e_{ij}, & \pi < \pi_c, \end{cases}$$

where μ_B and p_y the plastic dynamic viscosity, yield are stress respectively. Similarly π is the product of the component of deformation rate with itself, $\pi = e_{ij}e_{ij}$, e_{ij} is the (i, j)-th component of the deformation rate and π is a critical value of this product based on the non-Newtonian model. Finally the resultant boundary layer equations are

$$\frac{\partial u}{\partial x} + \frac{\partial v}{\partial y} = 0 \tag{1}$$

$$u\frac{\partial u}{\partial x} + v\frac{\partial v}{\partial y} = \nu\left(1 + \frac{1}{\beta}\right)\frac{\partial^2 u}{\partial y^2} \tag{2}$$

$$u\frac{\partial T}{\partial x} + v\frac{\partial T}{\partial y} = k\frac{\partial^2 T}{\partial y^2} - \frac{1}{\rho C_p}\frac{\partial q_r}{\partial y} + \frac{Q_0}{\rho C_p}\left(T - T_\infty\right) \tag{3}$$

where u, v are the velocity components in x, y directions respectively, v is the kinematic viscosity, ρ is the density, $\beta = \mu_B\sqrt{2\pi_c}/p_y$ is the Casson fluid parameter and κ is the thermal diffusivity, C_p is the specific heat.

The suitable boundary conditions are given by

$$\begin{aligned} &u = U = cx^n, v = 0, T = T_w \text{ at } y = 0, \\ &u \to 0, T \to T_\infty \text{ at } y \to \infty. \end{aligned} \tag{4}$$

Using the Rosseland approximation $q_r = -\left(\frac{4\sigma^*}{3k^*}\right)\frac{\partial T^4}{\partial y}$ is obtained where σ^* is the Stefan-Boltzmann constant and k^* is the absorption coefficient. We assume that the temperature variation within the flow is such that T^4 may be expanded in a Taylor's series. Expanding T^4 about T_∞ and neglecting higher-order terms we get

$$T^4 = 4T_\infty^3 T - 3T_\infty^4 \tag{5}$$

Here, $c(c > 0)$ is the surface stretching sheet related parameter. T_w, T_∞ are the uniform temperature at the sheet, free stream temperature respectively and n is the power index related to the surface stretching speed.

With the help of following similarity transformations

$$\begin{aligned} &u = cx^n f'(\eta),\ v = \sqrt{cv\frac{(n+1)}{2}}x^{\frac{n-1}{2}}\left[f(\eta) + \frac{n-1}{n+1}\eta f'(\eta)\right], \\ &\theta(\eta) = \frac{T - T_\infty}{T_W - T_\infty},\ \eta = y\sqrt{\frac{c(n+1)}{2v}}x^{\frac{n-1}{2}} \end{aligned} \tag{6}$$

The Equations (1), (2) and (3) are transformed in to coupled non linear ordinary differential equations as follows.

$$\left(1 + \frac{1}{\beta}\right)f''' + ff'' - \frac{2n}{n+1}f'^2 = 0 \tag{7}$$

$$\left(1 + \frac{4}{3}R_d\right)\theta'' + Pr\left(f\theta' + \frac{2}{n+1}Q\theta\right) = 0 \tag{8}$$

and the boundary conditions are transformed into

$$f'(\eta)=1, f(\eta)=0, \theta(\eta)=1 \text{ at } \eta=0$$
$$f'(\eta)\to 0, \theta(\eta)\to 0 \text{ at } \eta=\infty \tag{9}$$

where $Pr=\dfrac{\mu c_p}{\kappa}$ is the Prandtl Number, $Q=\dfrac{Q_0 x}{u_w \rho c_p}$ is the Heat source for $Q>0$, and for sink $Q<0$ and $R_d=\dfrac{4\sigma^* T_\infty^3}{kk^*}$ is the Radiation parameter.

Hence the dimensionless form of Skin friction C_f and the Local Nusselt number Nu_x are given by

$$C_f\left(Re_x\right)^{1/2}=\left(1+\frac{1}{\beta}\right)f''(0),\ \frac{Nu_x}{\left(Re_x\right)^{1/2}}=-\theta'(0) \tag{10}$$

where $Re_x=\dfrac{U_w x}{\nu}$ is the local Reynolds number.

3. Numerical Discussion

The effect of Radiation parameter (R_d), Casson parameter (β), Heat source (Q), Prandtl number (Pr) and the nonlinear stretching parameter on flow of a Casson fluid over a Nonlinear stretching sheet has been discussed in this section. The Keller Box method is used to solve the ordinary differential equations.

The results obtained with this method agreed with Vajravelu results for Skin friction coefficient $f''(0)$ and Local Nusselt number $\theta'(0)$ of the viscouss incompressible fluid. Which are mentioned in **Table 1**.

Figure 1 depicts the variations in Skin friction coefficient with Casson Parameter β for two different values

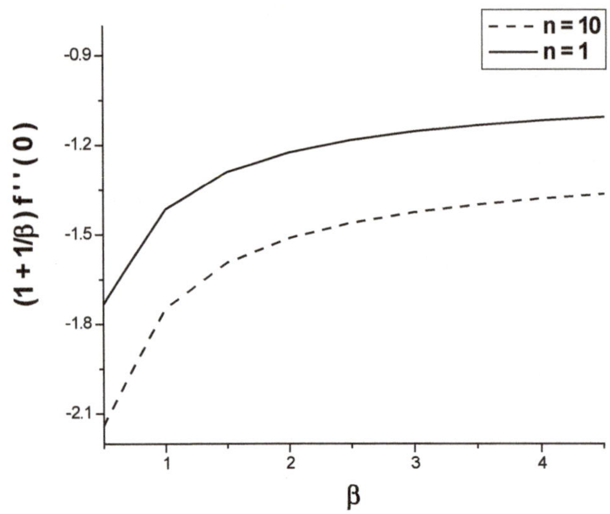

Figure 1. Skin friction coefficient versus Casson parameter β.

Table 1. Variations in $f''(0)$, $\theta'(0)$ when $R_d = 0$, $Q = 0$.

n	Vajravelu			Present Study		
		$f''(0)$	$\theta'(0)$		$f''(0)$	$\theta'(0)$
	Pr	0.72	7.0	Pr	0.72	7.0
1	−1.0000	−0.4590	−1.8953	−1.0000	−0.4590	−1.8954
5	−0.1945	−0.4394	−1.8610	−1.1944	−0.4396	−1.8616
10	−1.2348	−0.4357	−1.8541	−1.2348	−0.4356	−1.8547

of nonlinear stretching parameters n. It is noticed that the skin friction coefficient is increased with the increase of β for both values of non linearly stretching parameter *i.e.*, for $n = 1$, $n = 10$. It is also cleared that the skin friction coefficient is suppressed for $n = 10$ when compared to $n = 1$ and the negative sign in the graph indicates that there is a drag force on fluid at the wall. **Figure 2** indicates the variations of heat source $\theta'(0)$ with β. From this figure we noticed that there is an increase in heat source with the increase of β which indicates more heat is transferred from the sheet. Negative values indicates that heat is transferred from the sheet to the ambient fluid.

The nature of velocity profile with variation in non linearly stretching parameter n has been discussed in **Figure 3**. The velocity of the fluid is found to be decreased with the increase of n. But the decrease of the velocity profile is negligible for large values of n since the coefficient $\frac{2n}{n+1}$ approaches to 2 when $n \to \infty$ [13]. On the other hand the temperature increases with increase of n this result is exhibited in **Figure 4**.

The effect of Casson parameter β on velocity and temperature profiles has been discussed in the **Figure 5** and **Figure 6** with nonlinear stretching parameter $n = 10$. It is noticed that the momentum boundary layer thickness decreases and thermal boundary layer thickness increases with increase of β.

The effect of the Prandtl number on the velocity profile has been depicted in **Figure 7**. The temperature and the thermal boundary layer thickness decrease with the increase of Pr. Physically it signifies that an increase in

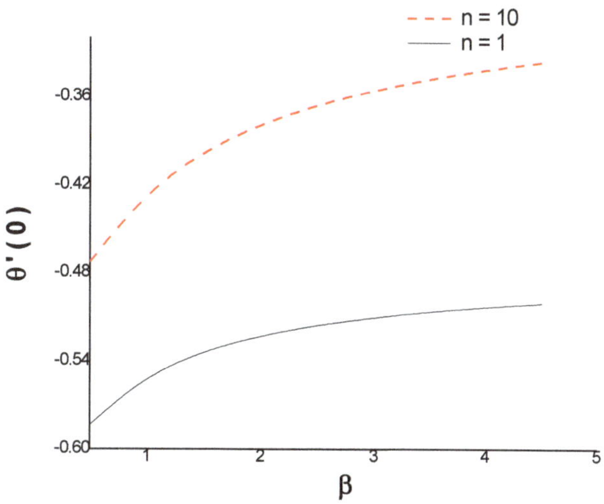

Figure 2. Wall temperature versus Casson parameter β.

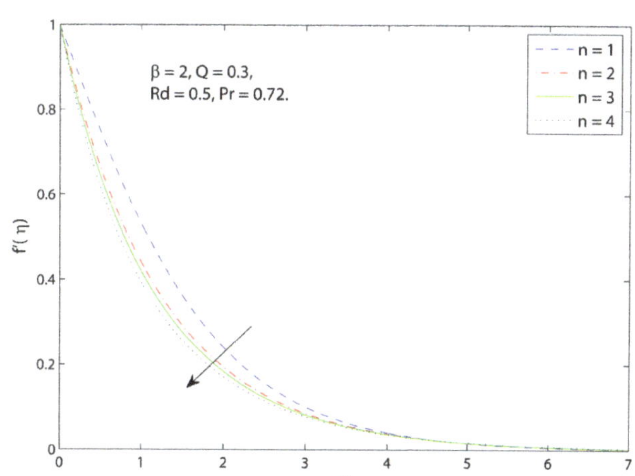

Figure 3. Velocity profile with variation in n.

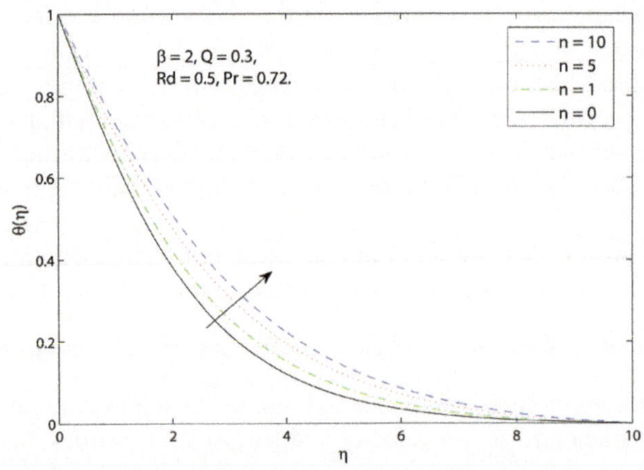

Figure 4. Temperature profile with variation in *n*.

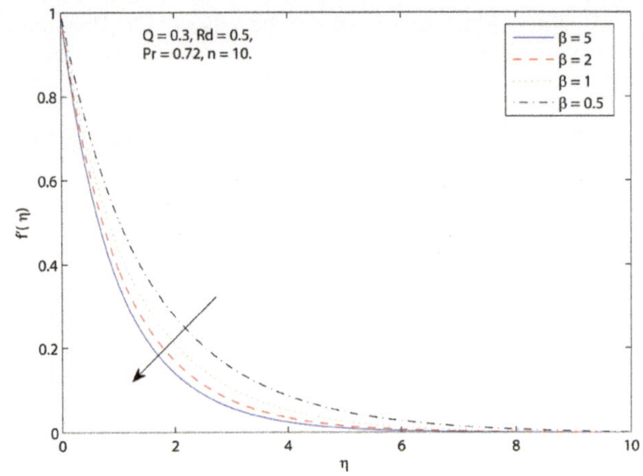

Figure 5. Velocity profile with variation in β.

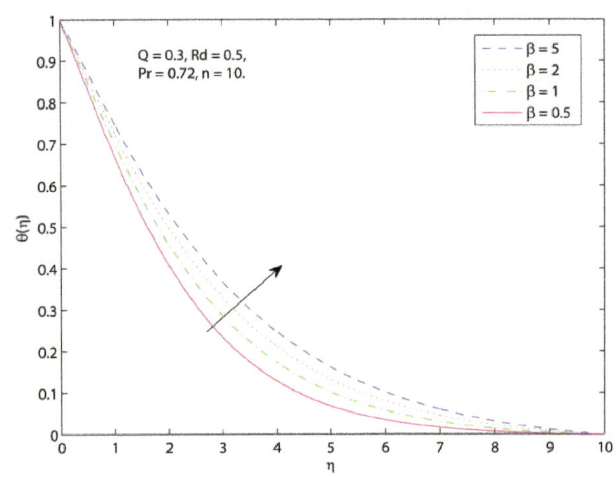

Figure 6. Temperature profile with variation in β.

the Prandtl number means an increase of fluid viscosity, which causes a decrease in the temperature distribution.

Figure 8 indicates that the increase of radiation parameter leads to increase in the temperature as well as thermal boundary layer thickness this is because the thermal radiation leads to heat transfer.

Figure 9 depicts the variation of temperature with respect to heat source/sink. It is observed that the increase in heat source/sink parameter increases the temperature profile. This is due to the fact that heat source can add

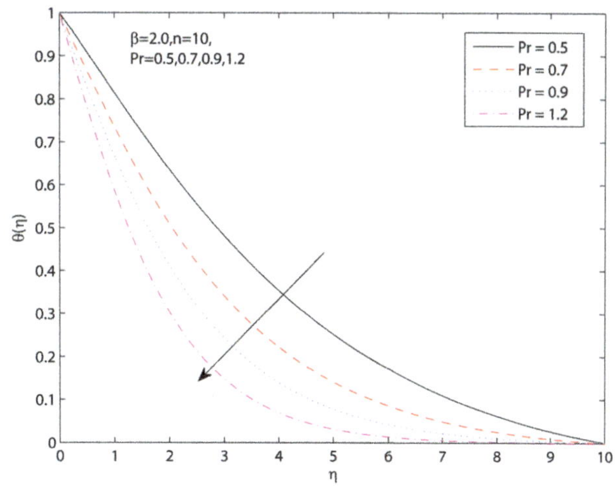

Figure 7. Temperature profile with variation in Pr.

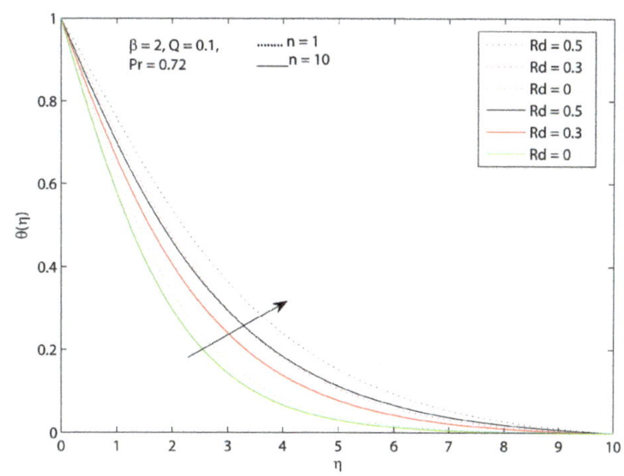

Figure 8. Temperature profile with variation in Rd.

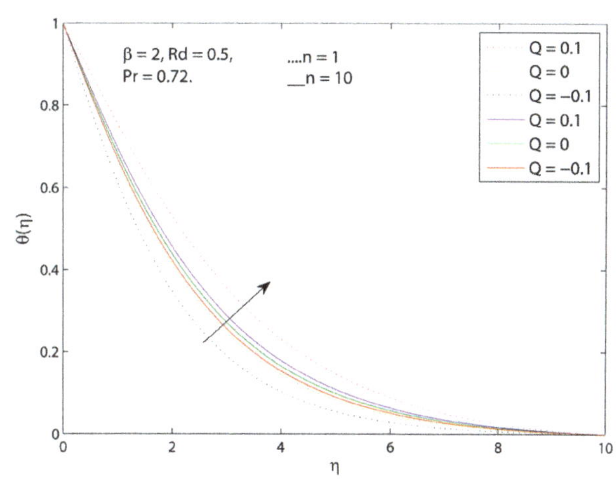

Figure 9. Temperature profile with variation in *Q*.

more heat to the stretching sheet which increases its temperature. This increases the thermal boundary layer thickness. In the flow of fluid which reduce the velocity of the flow. In addition to this it is also the velocity approaches to zero as the distance from the sheet increases.

4. Conclusions

The basic governing equations are transformed into coupled nonlinear ordinary differential equations. Keller Box method is used to perform the numerical computations. The effects of nonlinear stretching parameter, Cassson fluid parameter, Prandtl number, radiation parameter and heat source/sink parameters on the heat transfer characteristics are examined with the help of graphs. Finally got an excellent agreement with the previous paper [5]. Briefly the above discussions can be summarized as follows.

1) Both the skin friction coefficient and local Nusselt number increased with the increase in Casson parameter β. It is also noticed that the nonlinear stretching parameter is to suppress the velocity field.
2) The velocity of the fluid is found to be decrease with the increase in n where as the temperature is increased in this case.
3) The decreasing of momentum boundary layer thickness with the increase in β noted and the thermal boundary layer thickness increase with the increase in β.
4) The increasing effect of the Prandtl decreases the temperature.
5) An increment in the radiation parameter increases the temperature distribution.
6) The increase in the heat source increases the temperature of the fluid.

References

[1] Crane, L.J. (1970) Flow past a Stretching Plate. *Zeitschrift für Angewandte Mathematik und Physik*, **21**, 645-647. http://dx.doi.org/10.1007/BF01587695

[2] Wang, C.Y. (1984) The Three-Dimensional Flow Due to a Stretching Flat Surface. *Physics of Fluids*, **27**, 1915-1917. http://dx.doi.org/10.1063/1.864868

[3] Andersson, H.I. (1992) MHD Flow of a Viscoelastic Fluid past a Stretching Surface. *Acta Mechanica*, **95**, 227-230. http://dx.doi.org/10.1007/BF01170814

[4] Gupta, P.S. and Gupta, A.S. (1977) Heat and Mass Transfer on a Stretching Sheet with Suction or Blowing. *The Canadian Journal of Chemical Engineering*, **55**, 744-746. http://dx.doi.org/10.1002/cjce.5450550619

[5] Vajravelu, K. (2001) Viscous Flow over a Nonlinearly Stretching Sheet. *Applied Mathematics and Computation*, **124**, 281-288. http://dx.doi.org/10.1016/S0096-3003(00)00062-X

[6] Raptis, A. and Perdikis, C. (2006) Viscous Flow over a Non-Linearly Stretching Sheet in the Presence of a Chemical Reaction and Magnetic Field. *International Journal of Non-Linear Mechanics*, **41**, 527-529. http://dx.doi.org/10.1016/j.ijnonlinmec.2005.12.003

[7] Cortell, R. (2007) Viscous Flow and Heat Transfer over a Nonlinearly Stretching Sheet. *Applied Mathematics and Computation*, **184**, 864-873. http://dx.doi.org/10.1016/j.amc.2006.06.077

[8] Eerdunbuhe and Temuerchaolu (2012) Approximate Solution of the Magneto-Hydrodynamic Flow over a Nonlinear Stretching Sheet. *Chinese Physics B*, **21**, Article ID: 035201. http://dx.doi.org/10.1088/1674-1056/21/3/035201

[9] Ganji, D., Bararnia, H., Soleimani, S. and Ghasemi, E. (2009) Analytical Solution of the Magneto-Hydrodynamic Flow over a Nonlinear Stretching Sheet. *Modern Physics Letters B*, **23**, 2541. http://dx.doi.org/10.1142/S0217984909020692

[10] Mukhopadhyay, S. (2013) Casson Fluid Flow and Heat Transfer over a Nonlinearly Stretching Surface. *Chinese Physics B*, **22**, Article ID: 074701. http://dx.doi.org/10.1088/1674-1056/22/7/074701

[11] Pramanic, S. (2014) Casson Fluid Flow and Heat Transfer Past an Exponentially Porous Stretching Surface in Presence of Thermal Radiation. *Ain Shams Engineering Journal*, **5**, 205-212. http://dx.doi.org/10.1016/j.asej.2013.05.003

[12] Bhattacharyya, K. (2013) MHD Stagnation-Point Flow of Casson Fluid and Heat Transfer over a Stretching Sheet with Thermal Radiation. *Journal of Thermodynamics*, **2013**, Article ID: 169674. http://dx.doi.org/10.1155/2013/169674

[13] Mustafa, M., Hayat, T., Pop, I. and Aziz, A. (2011) Unsteady Boundary Layer Flow of a Casson Fluid Due to an Impulsively Started Moving Flat Plate. *Heat Transfer*, **40**, 563-576.

[14] Nadeem, S., Haq, R.U. and Lee, C. (2012) MHD Flow of a Casson Fluid over an Exponentially Shrinking Sheet. *Scientia Iranica*, **19**, 1550-1553. http://dx.doi.org/10.1016/j.scient.2012.10.021

[15] Bhattacharyya, K., Hayat, T. and Alsaedi, A. (2013) Analytic Solution For Magnetohydrodynamic Boundary Layer Flow of Casson Fluid over a Stretching/Shrinking Sheet with Wall Mass Transfer. *Chinese Physics B*, **22**, Article ID:

024702. http://dx.doi.org/10.1088/1674-1056/22/2/024702

[16] Mabood, F., Khan, W.A. and Ismail, A.I.M. (2015) MHD Boundary Layer Flow and Heat Transfer of Nanofluids over a Nonlinear Stretching Sheet: A Numerical Study. *Journal of Magnetism and Magnetic Materials*, **374**, 569-576. http://dx.doi.org/10.1016/j.jmmm.2014.09.013

[17] Behrouz, R., Syed Tauseef, M.-D. and Ahmet, Y. (2011) Solution to the MHD Flow over a Non-Linear Stretching Sheet Homotopy Perturbation Method. *Science China Physics*, *Mechanics & Astronomy*, **54**, 342-345. http://dx.doi.org/10.1007/s11433-010-4180-1

[18] Fredrickson, A.G. (1964) Principles and Applications of Rheology. Prentice-Hall, Englewood Cliffs.

[19] Mustafa, M., Hayat, T., Pop, I. and Aziz, A. (2011) Unsteady Boundary Layer Flow of a Casson Fluid Due to an Impulsively Started Moving Flat Plate. *Heat Transfer—Asian Research*, **40**, 563-576. http://dx.doi.org/10.1002/htj.20358

[20] Hayat, T., Hussain, Q. and Javed, T. (2009) The Modified Decomposition Method and Pade Approximation for the MHD Flow over a Non-Linear Stretching Sheet. *Nonlinear Analysis*: *Real World Applications*, **10**, 966-973. http://dx.doi.org/10.1016/j.nonrwa.2007.11.020

[21] Hady, F.M., Ibrahim, F.S., Abdel-Gaied, S.M. and Eid, M.R. (2012) Radiation Effect on Viscous Flow of a Nanofluid and Heat Transfer over a Nonlinearly Stretching Sheet. *Nanoscale Research Letters*, **7**, 229. http://dx.doi.org/10.1186/1556-276X-7-229

[22] Mukhopadhyay, S., Bhattacharyya, K. and Hayat, T. (2013) Exact Solutions for the Flow of Casson Fluid over a Stretching Surface with Transpiration and Heat Transfer Effects. *Chinese Physics B*, **22**, Article ID: 114701. http://dx.doi.org/10.1088/1674-1056/22/11/114701

[23] Hayat, T., Nawaz, M., Sajid, M. and Asghar, S. (2009) The Effect of Thermal Radiation on the Flow of a Second Grade Fluid. *Computers & Mathematics with Applications*, **58**, 369-379. http://dx.doi.org/10.1016/j.camwa.2009.01.040

[24] Liao, S.J. (1992) The Proposed Homotopy Analysis Technique for the Solution of Nonlinear Problems. PhD Thesis, Shanghai Jiao Tong University, Shanghai.

[25] Ibrahim, W. and Shanker, B. (2014) Magnetohydrodynamic Boundary Layer Flow and Heat Transfer of a Nano Fluid over Non-Isothermal Stretching Sheet. *Journal of Heat Transfer*, **136**, Article ID: 051701. http://dx.doi.org/10.1115/1.4026118

[26] Nandeppanavar, M.M., Vajravelu, K. and Subhas Abel, M. (2011) Heat Transfer in MHD Viscoelastic Boundary Layer Flow over a Stretching Sheet with Thermal Radiation and Non-Uniform Heat Source/Sink. *Communications in Nonlinear Science and Numerical Simulation*, **16**, 3578-3590. http://dx.doi.org/10.1016/j.cnsns.2010.12.033

[27] Bhattacharyya, K. (2011) Effects of Radiation and Heat Source/Sink on Unsteady MHD Boundary Layer Flow and Heat Transfer over a Shrinking Sheet with Suction/Injection. *Frontiers of Chemical Science and Engineering*, **5**, 376-384. http://dx.doi.org/10.1007/s11705-011-1121-0

[28] Abd El-Aziz, M. (2009) Radiation Effect on the Flow and Heat Transfer over an Unsteady Stretching Sheet. *International Communications in Heat and Mass Transfer*, **36**, 521-524. http://dx.doi.org/10.1016/j.icheatmasstransfer.2009.01.016

[29] Hossain, M.A., Alim, M.A. and Rees, D.A.S. (1999) The Effect of Radiation on Free Convection from a Porous Vertical Plate. *International Journal of Heat and Mass Transfer*, **42**, 181-191. http://dx.doi.org/10.1016/S0017-9310(98)00097-0

[30] Keller, H.B. (1971) A New Difference Scheme for Parabolic Problems. In: Hubbard, B., Ed., *Numerical Solutions of Partial Differential Equations*, Vol. II, Academic Press, New York, 327-350.

[31] Sarif, N.M., Salleha, M.Z. and Nazar, R. (2013) Numerical Solution of Flow and Heat Transfer over a Stretching Sheet with Newtonian Heating Using the Keller Box Method. *Procedia Engineering*, **53**, 542-554. http://dx.doi.org/10.1016/j.proeng.2013.02.070

23

Magneto Hydrodynamics Stagnation Point Flow of a Nano Fluid over an Exponentially Stretching Sheet with an Effect of Chemical Reaction, Heat Source and Suction/Injunction

Ch. Achi Reddy[1*], B. Shankar[2]

[1]M.L.R. Institute of Technology, Dundigal, Hyderabad, India
[2]Professor, Department of Mathematics, Osmania University, Hyderabad, India
Email: [*]achireddy.ch@gmail.com

Abstract

A numerical investigation is carried out on the effects of heat source suction and viscous dissipation on Magneto hydrodynamics boundary layer flow of a viscous, steady and incompressible fluid. The flow is assumed to be over on exponentially stretching sheet. The governing system of partial differential equations has been transformed into ordinary differential equation using similarity transformation. Keller box method is simulated on the dimensionless system of differential equations. The skin friction coefficient and the heat and mass transfer rates are very significant parameters that are computed, analysed discussed in detail.

Keywords

Boundary Layer Flow, Exponentially Stretching Sheet, Keller Box Method, Heat Source, Suction, Chemical Reaction and Viscous Dissipation

1. Introduction

Stagnation point is of considerable importance in most of industrial applications. The stagnation point is the one where fluid becomes stationary. In certain situations, flow is stagnated by a solid wall while in others; there is a line interior to a homogeneous fluid domain or the interface between two immiscible fluids [1]-[3]. A good

[*]Corresponding author.

amount of research is done drawing the attention of several researchers [4]-[14].

A new kind of fluid known as Nano fluid is first time proposed in Argonne National Laboratory in US by Chai at the time of investigations on coolants techniques and cooling processes. The Nano fluids are of nanometre sized particles that are metals, oxides and nanotubes.

The properties of Nano fluids are of special importance over the base fluid since thermal conductivity and convective properties of the Nano fluid predominant over the properties of the base fluid. Thermal conductivity is observed to be more effectively enhanced in the range of 15% - 40% over the base fluid. Stagnation point flow is most significant in various fields of science and technology. The study of stagnation point flow was carried out by Hiemenz in 1911 [15] who concentrated his studies on the two-dimensional stagnation point problem. Magyari and Keller [16] investigated the steady boundary layers flow on a stretching continuous surface with exponential temperature distribution while Partha *et al.* and M. Shakhaoath Khan [17] [18] analysed the effects of viscous dissipation on the mixed convection heat transfer from an exponentially stretching surface. However, researchers are getting interested in studying the boundary layer flow of Nano fluids with different aspects [19]-[36].

In the present study, we have investigated the viscous dissipation, heat source, suction for different values of velocity ratio parameter and observed that the nanoparticle volume decreases with the increase of chemical reaction parameter for $\gamma = 0.1$ and 2.1.

2. Mathematical Formulation

Consider a steady, two-dimensional boundary layer stagnation-point flow of a Nano fluid over an exponentially stretching sheet, the stretching and free stream velocities are assumed to be of the forms $u_w(x) = ae^{x/l}$, and $u_\infty(x) = be^{x/l}$, respectively. Where $a > 0$ and b are constants, x is the coordinate measure along the stretching surface and l is length of the sheet. A non-uniform transverse magnetic field of strength $B(x) = B_0 e^{x/2l}$ is imposed parallel to the y-axis, where Bo is a uniform magnetic field strength. It is assumed that the induced magnetic field due to the motion of an electrically conducting field is negligible. Further it is also assumed that the external electric field is zero and the electrical field due to polarization of charges is negligible [22].

Figure 1 shows that the temperature T and the Nano-particles fraction C take forms $T\omega(x)$ and $C\omega(x)$, respectively whereas the ambient values of temperature T_∞ and Nano-particle fraction C_∞ are obtained when y tends to infinity.

The governing boundary layer equations of the conservation Law of mass, momentum, energy and concentration in the flow as follows:

$$\frac{\partial u}{\partial x} + \frac{\partial v}{\partial y} = 0 \tag{1}$$

$$u\frac{\partial u}{\partial x} + v\frac{\partial u}{\partial y} = u_\infty \frac{du_\infty}{dx} + \frac{u}{\rho_f}\frac{\partial^2 u}{\partial y^2} + \frac{\sigma B^2(x)}{\rho_f}\left[u_\infty - u\right], \tag{2}$$

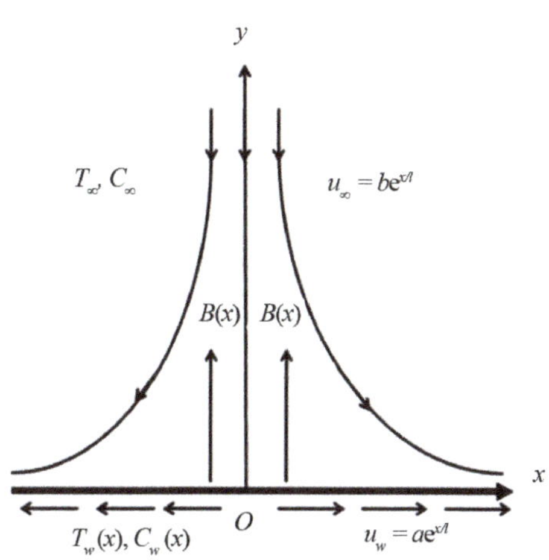

Figure 1. Physical flow model and coordinate system.

$$u\frac{\partial T}{\partial x}+v\frac{\partial T}{\partial y}=\alpha\frac{\partial^2 T}{\partial y^2}-\frac{1}{(\rho c)_f}\frac{\partial q_r}{\partial y}+\tau\left[D_B\frac{\partial C}{\partial y}\frac{\partial T}{\partial y}+\frac{D_T}{T_\infty}\left(\frac{\partial T}{\partial y}\right)^2\right]+\frac{\mu}{(\rho c_p)_f}\left(\frac{\partial u}{\partial y}\right)^2+\frac{Q}{(\rho C_p)_f}(T-T_\infty) \quad (3)$$

$$u\frac{\partial C}{\partial x}+v\frac{\partial C}{\partial y}=D_B\frac{\partial^2 C}{\partial y^2}+\frac{D_T}{T_\infty}\frac{\partial^2 T}{\partial y^2}-\kappa(C-C_\infty) \quad (4)$$

Here u and v are the velocity components in the x and y directions respectively, μ is the viscosity, ρ is the density of the base field and σ is the electrical conductivity. $\alpha=\frac{\kappa}{(\rho c)_f}$ where k is the thermal conductivity and $(\rho c)_f$ is the heat capacitance of the base fluid and $\tau=\frac{(\rho c)_p}{(\rho c)_f}$ where $(\rho c)_p$ is the heat capacitance of the nanoparticle. D_B is the Brownian diffusion coefficient, D_T is the thermophoresis diffusion coefficient and q_r is the radiation flux. The Rosseland approximation is defined as [23] [24];

$$q_r=-\frac{4\sigma^*}{3k^*}\frac{\partial T^4}{\partial y} \quad (5)$$

where σ^* is the Stefan-Boltzmann constant and k^* is the mean absorption coefficient. It is assumed that the temperature difference between the free stream T_∞ and local temperature T is small enough expanding T^4 in a Taylor series about T_∞ and neglecting higher order terms results;

$$T^4\cong 4T_\infty^3 T-3T_\infty^4. \quad (6)$$

After substituting Equations (5) and (6) in Equation (3), it will be reduced to

$$u\frac{\partial T}{\partial x}+v\frac{\partial T}{\partial y}=\left[\alpha+\frac{16\sigma^* T_\infty^3}{3\kappa^*(\rho c)_f}\right]\frac{\partial^2 T}{\partial y^2}+\tau\left[D_B\frac{\partial C}{\partial y}\frac{\partial T}{\partial y}+\frac{D_T}{T_\infty}\left(\frac{\partial T}{\partial y}\right)^2\right]+\frac{\mu}{(\rho c_p)_f}\left(\frac{\partial u}{\partial y}\right)^2 \quad (7)$$

the subjected boundary conditions are

$$\left.\begin{array}{lll} u=u_\omega(x)=ae^{x/l}, v=-v(x), T=T_\omega(x), C=C_\omega(x), & \text{at} & y=0 \\ u\to u_\infty(x)=be^{x/l}, T\to T_\infty, C\to C_\infty, & \text{as} & y\to\infty \end{array}\right\}. \quad (8)$$

The prescribed temperature and concentration on the surface of the sheet is assumed to be $T_\omega(x)=T_\infty+T_0e^{x/2l}$ and $C_\omega(x)=C_\infty+C_0e^{x/2l}$ where T_0, C_0 are the reference temperature and concentration respectively, now, the non-linear partial differential equations for the purpose of a stream function $\psi=\psi(x,y)$ is defined as

$$u=\frac{\partial\psi}{\partial y}, v=-\frac{\partial\psi}{\partial x} \quad (9)$$

where the continuity Equation (1) is satisfied identically. A similarity transformation is defined as [20] follows

$$\psi=\sqrt{2l\nu a}e^{x/2l}f(\eta), \qquad \theta(\eta)=\frac{T-T_\infty}{T_\omega-T_\infty}$$
$$\phi(\eta)=\frac{C-C_\infty}{C_\omega-C_\infty}, \quad \eta=y\sqrt{a/2\nu l} \quad e^{x/2l} \quad (10)$$

as such Equation (10), Equations (2), (4) and (7) reduce to the following system of nonlinear ordinary differential equations.

$$f'''+ff''-2f'^2+2B^2+M(B-f')=0 \quad (11)$$

$$p_{rN}\theta''+f\theta'-f'\theta+Nb\phi'\theta'+Nt\theta'^2+E_C f''^2+A\theta=0 \tag{12}$$

$$\phi''+Lef\phi'-Lef'\phi+Nt_b\theta''-\gamma\phi Le=0 \tag{13}$$

where

$$B=b/a,\ v=\frac{\mu}{p_f},\ M=\frac{2l\sigma B_0^2}{ap_f} \tag{14}$$

$$Nb=\frac{\tau D_B(C_\omega-C_\infty)}{v}$$

$$Nb=\frac{\tau D_T(T_\omega-T_\infty)}{vT_\infty},\ E_C=\frac{u_\omega^2}{(Cp)_f(T_\omega-T_\infty)},\ S=\frac{v_0}{\sqrt{\frac{v_0}{2l}}}$$

$$\gamma=\frac{2lk}{u_\omega}\quad A=\frac{2lQ_0}{a\rho c_p}$$

Here, prime denote the differentiation with respect to η, B is the velocity ratio parameter, v is the kinematic viscosity of the fluid, P_r is the Prandtl number, Le is the Lewis number, M is the magnetic parameter, A is the heat source parameter, S is the suction parameter.

$$P_{rN}=\frac{1}{P_r}\left(1+\frac{4}{3}N\right)$$

where $N=\frac{4\sigma^*T_\infty^3}{kk^*}$ is the radiation parameter.

$$Nt_b=\frac{Nt}{Nb}$$

where Nb is the Brownian motion parameter and Nt is the thermophoresis parameter, Ec Eckert number, γ is chemical reaction parameter; The corresponding boundary conditions Equation (8) are transformed into

$$\begin{aligned}&f(\eta)=s,f'(\eta)=1,\theta(\eta)=1,\phi(\eta)=1 \text{ at } \eta=0,\\&f'(\eta)\to\varepsilon,\theta(\eta)\to0,\phi(\eta)\to0 \qquad \text{as } \eta\to\infty\end{aligned}. \tag{15}$$

The parameters of practical interest in the formulated problem are velocity, heat and mass transfer respectively, which are presented in terms of Skin friction C_f, Nusselt number Nu and Sherwood numbers Sh. Using the transformed variables (10), the non-dimensional expressions for the Skin friction coefficient $C_{fx}(0)=f''(0)$, the reduced Nusselt number $-\theta'(0)$ and the reduced Sherwood number $-\phi'(0)$ respectively are defined as;

Skin friction coefficient:

The Skin friction coefficient C_f is defined by

$$C_f=\frac{T_\omega}{\frac{1}{2}pU_\omega^2}$$

where τ_ω is the local wall Shear stress, ρ is the fluid density.

$$\sqrt{\mathrm{Re}_x}\cdot C_f=f''(0)$$

Nusselt Number:

In heat transfer at a boundary within a fluid, the Nusselt number (Nu) is the ratio of convection to conductive heat transfer across the boundary

$$Nu_x = \frac{\text{convective heat transfer}}{\text{conductive heat transfer}} = \frac{xq_\omega}{K(T_\omega - T_\infty)}$$
$$-\theta'(0) = \frac{Nu_x}{\sqrt{\mathrm{Re}_x}} \tag{16}$$

Sherwood Number:

The Sherwood number (*Sh*) is also called the mass transfer Nusselt number. It represents the ratio of convective to diffusive mass transport

$$Sh_x = \frac{\text{Convective mass Transfer coefficient}}{\text{Diffusive mass Transfer coefficient}} = \frac{xJ_\omega}{D_B(C_\omega - C_\infty)}$$
$$\frac{Sh_x}{\sqrt{\mathrm{Re}_x}} = -\phi'(0)$$

where $\mathrm{Re}_x = u_\omega(x)\cdot\frac{x}{v}$ is the local Reynolds number based on the stretching velocity.

3. Numerical Procedure

The equations (11)-(14) subject to the boundary conditions (15) are solved numerically using an implicit finite-difference scheme known as Keller box method. The method has the following four basic steps.

1) Reduce equations (11)-(14) to first order equations;
2) Write the difference equations using central differences;
3) Linearize the resulting algebraic equations by Newton's method and write them in matrix-vector;
4) Use the Block-tridiagonal elimination technique to solve the linear system.

The system of ordinary differential equations (11)-(13) has been solved numerically using Keller-box method. From the numerical computation, the main physical quantities of interest namely the local skin friction coefficient, the local Nusselt number and the local Sherwood number are obtained and the results are presented in **Table 1**, **Table 2** and **Table 3**.

From **Table 2**, it is observed that with the increase in heat source parameter, there is a decrease in rate of heat transfer and increase in mass transfer.

From **Table 3**, it is observed that with increase in suction parameter, there is a increase in rate of heat transfer, mass transfer and skin friction coefficient.

Figure 2 shows the effects of the suction parameter "*s*" on the flow field velocity $f'(\eta)$ for three different values of the velocity ratio parameter *B*, *B* = 0.1, 1 and 2.

When *B* = 0.1 the velocity profile and the boundary layer thickness decrease with an increases in "*s*". When *s* = 2.0, higher value of the Lorentz force further reduces the velocity and consequently the thickness of boundary

Table 1. Comparison of the reduced Nusselt number $-\theta'(0)$ when $Nb = Nt = Le = B = 0$.

P_r	M	N	[19] $-\theta'(0)$	[20] $-\theta'(0)$	[21] $-\theta'(0)$	Present results $-\theta'(0)$
1	0	0	0.9548	0.9548	0.9548	0.9548
2	0	0	1.4714	1.4714	1.4714	1.4715
3	0	0	1.8691	1.8691	1.8691	1.8692
1	0	1.0	0.5315	0.5312	0.5312	0.5311
1	1.0	0	---	0.8611	0.8611	0.8611
1	1.0	1.0	---	0.4505	0.4505	0.4505

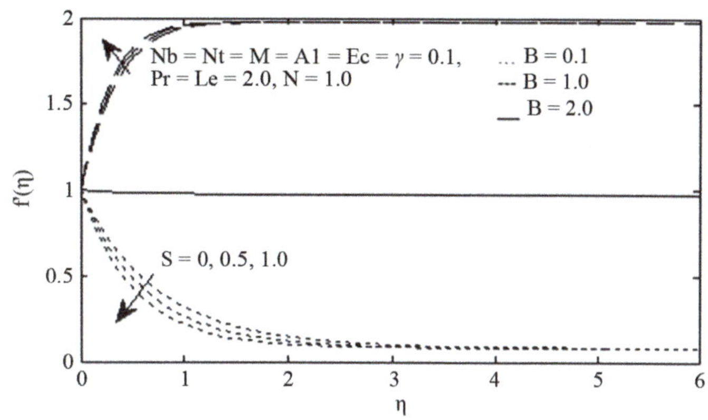

Figure 2. Velocity profile against η for different values of S.

Table 2. Values of the reduced Nusselt number $-\theta'(0)$, reduced Sherwood number $-\phi'(0)$, and the skin friction coefficient $C_{fx}(0)$ for various values of heat source parameter.

A	$Nb = Nt = M = B = Ec = \gamma = 0.1$			$-\theta'(0)$	$-\phi'(0)$	$C_{fx}(0)$
	P_r	Le	N			
0.1	1.00	10	1.0	0.4640	3.7137	1.2856
0.2	1.00	10	1.0	0.3978	3.7291	1.2856
0.3	1.00	10	1.0	0.3084	3.7470	1.2856
0.4	1.00	10	1.0	0.1727	3.7697	1.2856
0.5	1.00	10	1.0	0.0625	3.8035	1.2856

Table 3. Values of the reduced Nusselt number $-\theta'(0)$, reduced Sherwood number $-\phi'(0)$, and the skin friction coefficient $C_{fx}(0)$ for various values of suction parameter.

S	$Nb = Nt = M = B = E_C = \gamma = 0.1$			$-\theta'(0)$	$-\phi'(0)$	$C_{fx}(0)$
	Pr	Le	N			
0.1	1.00	10	1.0	0.4837	4.2958	1.3298
0.2	1.00	10	1.0	0.5042	4.9304	1.3757
0.3	1.00	10	1.0	0.5257	5.6106	1.4232
0.4	1.00	10	1.0	0.5479	6.3300	1.4723
0.5	1.00	10	1.0	0.5710	7.0825	1.5230

layer reduces. When $B = 1$, there is no influence of source on $f'(\eta)$ and $f'(\eta)$ attains a constant value of "1" for any values of η indicating that there is no boundary layer of fluid, as shown by a thick (-) line in **Figure 2**.

Figure 3 and **Figure 4** demonstrate the effects of suction parameter on temperature and concentration. It is clear that the fluid velocity decreases significantly with increasing values of suction parameter while it is found to enhance with blowing. The presence of suction would result in the reduction of the thickness of the boundary layer. So the Nano fluid temperature and concentration decreases with an increase in the suction parameter.

Figure 5 shows effects of the heat source parameter on temperature. Increasing the heat source parameter from −0.3 to 0.5 increases the temperature profile. Because the presence of source of heat enhances thermal energy, as a result of this, temperature profile increases. **Figure 6** shows the heat source parameter significantly decreases the concentration profile.

Figure 7 and **Figure 8** shows the effects of viscous dissipation parameter "*Ec*" on Temperature and concentration profile. Eckert number is the ratio of the kinetic energy of the flow to the boundary Layer enthalpy differ-

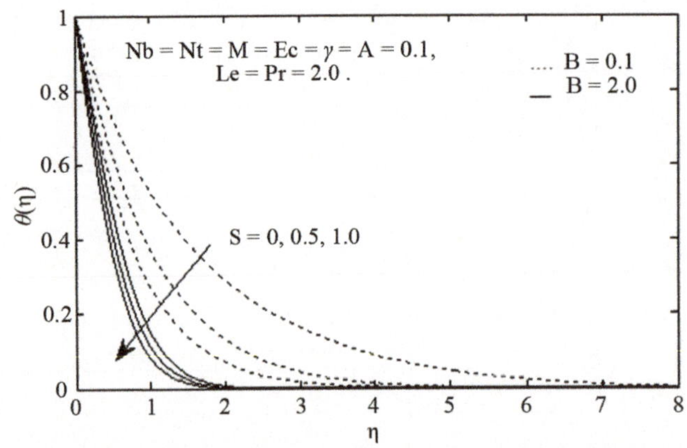

Figure 3. Temperature profile against η for different values of S.

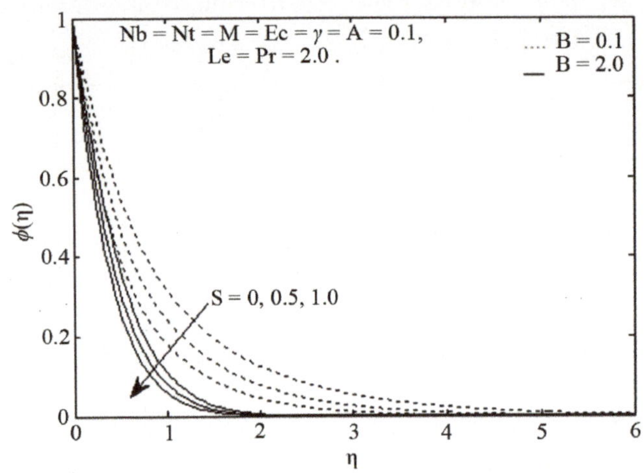

Figure 4. Concentration profile against η for different values of S.

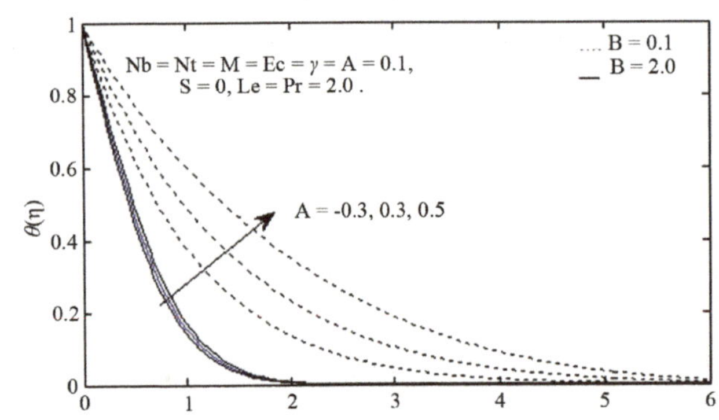

Figure 5. Temperature profile against η for different values of A.

ence. The effect of viscous dissipation on flow field is to increase the energy, yielding a greater fluid temperature and as consequence greater buoyancy force. The increase in the buoyancy force due to an increase in the dissipation parameter enhances the temperature. While concentration decreases with an increase in the viscous dissipation parameter.

The influence of chemical reaction parameter "γ" on concentration profile is shown in **Figure 9**. Concentration decreases with an increase in the chemical reaction parameter indicating that the nanoparticle volume fraction decreases with the increase of chemical reaction parameter, while effect of chemical reaction parameter γ is

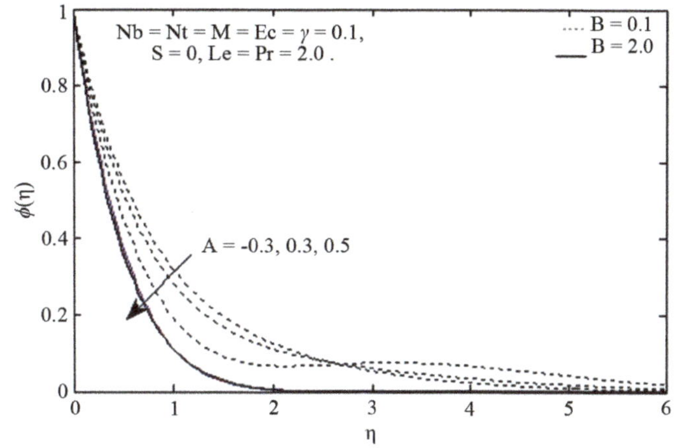

Figure 6. Concentration profile against η for different values of A.

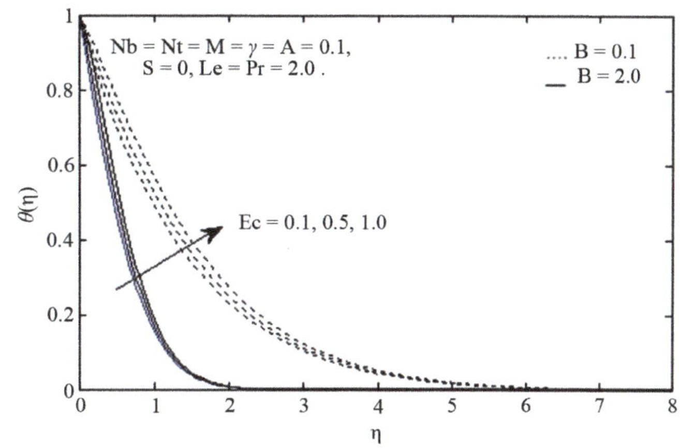

Figure 7. Temperature profile against η for different values of Ec.

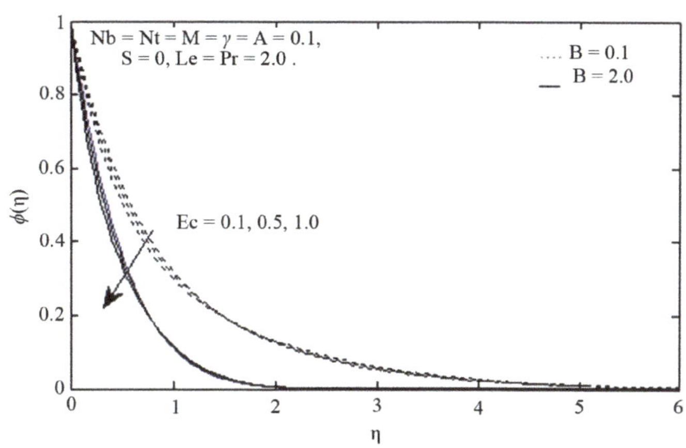

Figure 8. Concentration profile against η for different values of Ec.

not significant on the temperature profile.

4. Conclusions

A numerical study corresponding to the flow and heat transfer in a steady flow region of Nano fluid over an exponential stretching surface and effects of heat source, suction parameter and Eckert number are examined and

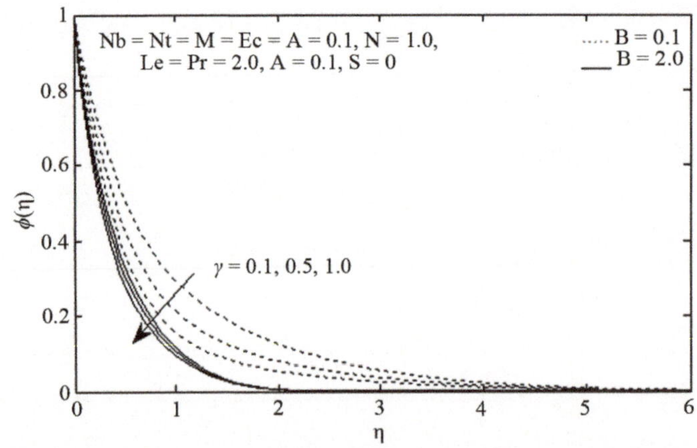

Figure 9. Concentration profile against η for different values of γ.

discussed in detail. The main observation of the present study as follows:

1) It is interesting to note that an increase in the suction parameter reduces the velocity profile and increase the temperature and concentration;

2) An increase in the heat source parameter increases the temperature profile while it reduces the concentration profile;

3) An increase in the Eckert number increases the temperature profile while it reduces the concentration profile.

References

[1] Hiemenz, K. (1911) Die Grenzschicht an einem inden gleichförmigen Flüssigkeitsstrom eingetauchten geraden Kreiszylinder, *Dingler's Polytech. J.*, **326**, 321-324.

[2] Wong, K.V. and Leon, O.D. (2010) Applications of Nanofluids: Current and Future. *Advances in Mechanical Engineering*, 1-12

[3] Mahapatra, T.R. and Gupta, A.S. (2002) Heat Transfer in Stagnation-Point Flow towards a Stretching Sheet. *Heat and Mass Transfer*, **38**, 517-521. http://dx.doi.org/10.1007/s002310100215

[4] Mahapatra, T.R. and Gupta, A.S. (2003) Stagnation-Point Flow towards a Stretching Surface. *The Canadian Journal of Chemical Engineering*, **81**, 258-263. http://dx.doi.org/10.1002/cjce.5450810210

[5] Nazar, R., Amin, N., Filip, D. and Pop, I. (2004) Stagnation Point Flow of a Micropolar Fluid towards a Stretching Sheet. *International Journal of Non-Linear Mechanics*, **39**, 1227-1235. http://dx.doi.org/10.1016/j.ijnonlinmec.2003.08.007

[6] Lok, Y.Y., Amin, N. and Pop, I. (2006) Non-Orthogonal Stagnation Point Flow towards a Stretching Sheet. *International Journal of Non-Linear Mechanics*, **41**, 622-627. http://dx.doi.org/10.1016/j.ijnonlinmec.2006.03.002

[7] Nadeem, S., Hussain, A. and Khan, M. (2010) HAM Solutions for Boundary Layer Flow in the Region of the Stagnation Point towards a Stretching Sheet. *Communications in Nonlinear Science and Numerical Simulation*, **15**, 475-481. http://dx.doi.org/10.1016/j.cnsns.2009.04.037

[8] Labropulu, F., Li, D. and Pop, I. (2010) Non-Orthogonal Stagnation-Point Flow towards a Stretching Surface in a Non-Newtonian Fluid with Heat Transfer. *International Journal of Thermal Sciences*, **49**, 1042-1050. http://dx.doi.org/10.1016/j.ijthermalsci.2009.12.005

[9] Ishak, A., Lok, Y.Y. and Pop, I. (2010) Stagnation-Point Flow over a Shrinking Sheet in a Micropolar Fluid. *Chemical Engineering Communications*, **197**, 1417-1427. http://dx.doi.org/10.1080/00986441003626169

[10] Bachok, N., Ishak, A. and Pop, I. (2010) Boundary-Layer Flow of Nanofluid over a Moving Surface in a Flowing Fluid. *International Journal of Thermal Sciences*, **49**, 1663-1668. http://dx.doi.org/10.1016/j.ijthermalsci.2010.01.026

[11] Patel, M. and Timol, M.G. (2011) Orthogonal Stagnation Point Flow of a Power Law Fluid towards a Stretching Surface. *International Journal of Applied Mathematics and Mechanics*, **7**, 31-37.

[12] Bachok, N., Ishak, A. and Pop, I. (2011) On the Stagnation Point Flow towards a Stretching Sheet with Homogeneous-Heterogeneous Reactions Effects. *Communications in Nonlinear Science and Numerical Simulation*, **16**, 4296-4302. http://dx.doi.org/10.1016/j.cnsns.2011.01.008

[13] Choi, S.U.S., Zhang, Z.-G., Yu, W., Lockwood, F.E. and Grullce, E.A. (2001) Anomalous Thermal Conductivity Enhancement in Nanotube Suspensions. *Applied Physics Letters*, **79**, 2252. http://dx.doi.org/10.1063/1.1408272

[14] Yu, W., France, D.M., Routbort, J.L. and Choi, S.U.S. (2008) Review and Comparison of Nanofluid Thermal Conductivity and Heat Transfer Enhancements. *Heat Transfer Engineering*, **29**, 432-460. http://dx.doi.org/10.1080/01457630701850851

[15] Bachok. N., Ishak, A. and Pop, I. (2010) Melting Heat Transfer in Boundary Layer Stagnation—Point Flow Towards a Stretching/Shrinking Sheet. *Physics Letters A*, **374**, 4075-4079.

[16] Magyari, E. and Keller, B. (1999) Heat and Mass Transfer in the Boundary Layers on an Exponentially Stretching Continuous Surface. *Journal of Physics D*, **32**, 577-585. http://dx.doi.org/10.1088/0022-3727/32/5/012

[17] Partha, M.K., Murthy, P.V.S.N. and Rajasekhar, G.P. (2005) Effect of Viscous Dissipation on the Mixed Convection Heat Transfer from an Exponentially Stretching Surface. *Heat and Mass Transfer*, **11**, 360-366. http://dx.doi.org/10.1007/s00231-004-0552-2

[18] Khan, M.S., Karim, I. and Islam, M.S. (2014) Possessions of Chemical Reaction on MHD Heat and Mass Transfer Nanofluid Flow on a Continuously Moving Surface. *The American Chemical Science Journal*, **4**, 401-415. http://dx.doi.org/10.9734/ACSJ/2014/5422

[19] Bidin, B. and Nazar, R. (2009) Numerical Solution of the Boundary Layer Flow over an Exponentially Stretching Sheet with Thermal Radiation. *European Journal of Scientific Research*, **33**, 710-717.

[20] Ishak, A. (2011) MHD Boundary Layer Flow Due to an Exponentially Stretching Sheet with Radiation Effect. *Sains Malaysiana*, **40**, 391-395.

[21] Anwar, I., Shafie, S. and Salleh, M.Z. (2014) Radiation Effect on MHD Stagnation Point Flow of a Nanofluid over an Exponentially Stretching Sheet. *Walailak Journal of Science and Technology*, **11**, 569-591.

[22] Chen, C.H. (2008) Effects of Magnetic Field and Suction/Injection on Convection Heat Transfer of Non-Newtonian Power-Law Fluids Past a Power-Law Stretched Sheet with Surface Heat Flux. *International Journal of Thermal Sciences*, **47**, 954-961. http://dx.doi.org/10.1016/j.ijthermalsci.2007.06.003

[23] Cess, R.D. (1966) The Interaction of Thermal Radiation with Free Convection Heat Transfer. *International Journal of Heat and Mass Transfer*, **9**, 1269-1277. http://dx.doi.org/10.1016/0017-9310(66)90119-0

[24] Hady, F.M., Ibrahim, F.S., Abdel-Gaied, S.M. and Eid, M.R. (2012) Radiation Effect on Viscous Flow of a Nanofluid and Heat Transfer over a Nonlinearly Stretching Sheet. *Nanoscale Research Letters*, **7**, 229. http://dx.doi.org/10.1186/1556-276X-7-229

[25] El-Aziz, M.A. (2009) Viscous Dissipation Effect on Mixed Convection Flow of a Micropolar Fluid over an Exponentially Stretching Sheet. *Canadian Journal of Physics*, **87**, 359-368. http://dx.doi.org/10.1139/P09-047

[26] Partha, M.K., Murthy, P.V.S.N. and Rajasekhar, G.P. (2005) Effect of Viscous Dissipation on the Mixed Convection Heat Transfer from an Exponentially Stretching Surface. *Heat and Mass Transfer*, **41**, 360-366. http://dx.doi.org/10.1007/s00231-004-0552-2

[27] Khan, M.S., Karim, I., Ali, L.E. and Ismail, A. (2012) Unsteady MHD Free Convection Boundary-Layer Flow of a Nanofluid along a Stretching Sheet with Thermal Radiation and Viscous Dissipation Effects. *International Nano Letters*, **2**, 24. http://dx.doi.org/10.1186/2228-5326-2-24

[28] Hamad, M.A., Pop, I. and Ismail, A.I. (2011) Magnetic Field Effects on Free Convection Flow of a Nanofluid Past a Vertical Semi-Infinite Flat Plate. *Nonlinear Analysis*: *Real World Applications*, **12**, 1338-1346. http://dx.doi.org/10.1016/j.nonrwa.2010.09.014

[29] Jafar, K., Nazar, R., Ishak, A. and Pop, I. (2011) MHD Flow and Heat Transfer over Stretching/Shrinking Sheets with External Magnetic Field, Viscous Dissipation and Joule Effects. *The Canadian Journal of Chemical Engineering*, **90**, 1336-1346. http://dx.doi.org/10.1002/cjce.20609

[30] Reddy, M.G. (2011) Influence of the Thermal Radiation, Viscous Dissipation and Hall Current on MHD Convection Flow over a Stretched Vertical Flat Plate. *Ain Shams Engineering Journal*, **3**, 169-175.

[31] Vajravelu, K. and Hadjinicolaou, A. (1993) Heat Transfer in a Viscous Fluid over a Stretching Sheet with Viscous Dissipation and Internal Heat Generation. *International Communications in Heat and Mass Transfer*, **20**, 417-430. http://dx.doi.org/10.1016/0735-1933(93)90026-R

[32] Postelnicu, A. (2007) Influence of Chemical Reaction on Heat and Mass Transfer by Natural Convection from Vertical Surfaces in Porous Media Considering Soret and Dufour Effects. *Heat and Mass Transfer*, **43**, 595-602. http://dx.doi.org/10.1007/s00231-006-0132-8

[33] Goyal, M. and Bhargava, R. (2013) Numerical Solution of MHD Viscoelastic Nanofluid Flow over a Stretching Sheet with Partial Slip and Heat Source/Sink. *ISRN Nanotechnology*, **2013**, Article ID: 931021.

[34] Cebeci, T. and Bradshaw, P. (1988) Physical and Computational Aspects of Convective Heat Transfer. Springer, New York.

[35] Vendabai, K. and Sarojamma, G. (2014) Unsteady Convective Boundary Layer Flow of a Nanofluid over a Stretching Surface in the Presence of a Magnetic Field and Heat Generation. *International Journal of Emerging Trends in Engineering and Development*, **3**, 214-230.

[36] Dulal, P. and Hiranmoy, M. (2012) Non-Darcian Buoyancy Driven Heat and Mass Transfer over a Stretching Sheet in a Porous Medium with Radiation and Ohmic Heating. *International Journal of Nonlinear Science*, **14**, 115-123.

24

Rotational Oscillation Effect on Flow Characteristics of a Circular Cylinder at Low Reynolds Number

Alice Rosa da Silva[1*], Aristeu da Silveira-Neto[2], Antônio Marcos Gonçalves de Lima[2]

[1]School of Civil Engineering, Federal University of Uberlândia, Uberlândia, Brazil
[2]School of Mechanical Engineering, Federal University of Uberlândia, Brazil
Email: [*]alicers@feciv.ufu.br, aristeus@mecanica.ufu.br, amglima@mecanica.ufu.br

Abstract

Two dimensional numerical simulations of flow around a rotationally oscillating circular cylinder were performed at $Re = 1000$. A wide range of forcing frequencies, f_r, and three values of oscillation amplitudes, A, are considered. Different vortex shedding modes are observed for a fixed A at several values of f_r, as well as for a fixed f_r at different values of A. The $2C$ mode of vortex shedding was obtained in the present study. It is important to point out that this mode has not been observed by other investigators for rotationally oscillating case. Also, it is verified that this mechanism has great influence on the drag coefficient for high frequency values. Furthermore, the lift and pressure coefficients and the power spectra density are also analyzed.

Keywords

Rotationally Oscillating Cylinder, Vortex Shedding Modes, Drag Coefficient

1. Introduction

It is widely known that the vibrations induced by vortex shedding process may have negative effects in engineering systems, such as economic loss, damage of installations, and very frequently with environment-related consequences. Thus, although the fluid flow around circular cylinder is a classical problem in fluid dynamic due to its simple geometry, this is a reason for which in the last decades, a great deal of effort has been devoted to the development of numerical and computational procedures for dealing with the problem of vortex shedding phenomena. Comprehensive studies on this subject have been reported in books [1]-[3].

*Corresponding author.

In applications in which a circular cylinder under rotational oscillations is involved, the flow dynamic is different from those observed for stationary cylinders and has fascinated researchers for a long time [4]-[7]. In this case, two important parameters related to the prescribed motion are the forcing frequency and amplitude. Hence, the characterization of those parameters on the flow structure becomes of capital importance.

Among some studies over rotationally oscillating cylinders, the so-named Hybrid Vortex Method and the Discrete Vortex Method have been proposed [8] [9] to investigate the process of vortex formation for a set of forcing frequency at Reynolds number 200 and 1000. Srinivas and Fujisawa [10] have used the unsteady form of Reynolds-averaged Navier-Stokes equations combined with the *k*-ε model of turbulence in order to determine the effects of several parameters on the flow structure over a rotationally oscillating circular cylinder for Reynolds in the range 2000 - 3.0×10^4. Ray and Christofides [11] studied a control system based on the open-loop simulations to reduce the effects of drag exerted on a circular cylinder subjected to rotational oscillations for Reynolds number in the range 100 - 500.

The present study focuses on two-dimensional, incompressible viscous flow over a rotationally oscillating circular cylinder by using the Immersed Boundary Methodology (*IBM*) [12] in order to investigate the oscillation effect in the wake structure behind the cylinder, at different forcing frequencies and amplitudes. In addition, the influence of those parameters on the drag, lift and pressure coefficients and the power spectra density are also analyzed. The simulations were carried out for flows at $Re = 1000$, amplitudes equals to 1, 2 and 3 and for various frequency ratios. The used methodology has showed a promising tool to simulate mobile bodies and in the present study, it captures the vortex shedding mode, named 2*C*, which has not been found for the case of rotational oscillation by other researchers.

2. General Aspects of the Numerical Methodology

Mathematical Formulation for the Fluid and for the Fluid-Solid Interface

One approach, which is not very common to solve the Navier-Stokes equations, is the so called velocity-vorticity formulation [13]. Flows can also be modeled by these equations in primitive variables [14]. In this section, the *IBM* combined with the Virtual Physical Model (*VPM*) [15] is summarized. For a viscous and incompressible flow, the *IBM* is based on the Navier-Stokes Equation (1) with an added force source term which acts on the fluid so that a particle perceives the existence of the solid interface. The continuity Equation (2) is also expressed.

$$\frac{\partial u_i}{\partial t} + \frac{\partial (u_i u_j)}{\partial x_j} = -\frac{1}{\rho}\frac{\partial p}{\partial x_i} + \frac{\partial}{\partial x_j}\left[\nu\left(\frac{\partial u_i}{\partial x_j} + \frac{\partial u_j}{\partial x_i}\right)\right] + f_i \tag{1}$$

$$\frac{\partial u_i}{\partial x_i} = 0 \tag{2}$$

where ρ [kg/m^3] and ν [m^2/s] are the specific mass and the kinematic viscosity, respectively; u_i [m/s] and p [N/m^2] are, respectively, the *i*-th velocity component and the pressure; f_i [N/m^3] is the *i*-th component of the Eulerian force calculated as follows:

$$\boldsymbol{f}(\boldsymbol{x}) = \sum_k D_{ij}(\boldsymbol{x} - \boldsymbol{x}_k)\boldsymbol{F}(\boldsymbol{x}_k)\Delta S^2(\boldsymbol{x}_k). \tag{3}$$

In Equation (3) $\boldsymbol{x}$ [m] and $\boldsymbol{x}_k$ [m] are the position vectors of a Eulerian and a Lagrangian point, respectively, and ΔS [m] is the arc length centered in each Lagrangian point, which is evenly spaced as shown in **Figure 1(a)**. $\boldsymbol{F}(\boldsymbol{x}_k)$ [N] is the Lagrangian force over the interface, and D_{ij} [m^{-2}] is the distribution function [16].

At this point, the mixed Eulerian-Lagrangian formulation is retained, in which the Eulerian fixed grid describes the flow and the Lagrangian grid (which can be fixed or not) describes the immersed body. These meshes are geometrically independent from each other, and this fact enables to study the flows around simple, complex, movable and deformable geometries, without any remeshing process. These two formulations are coupled by a force field obtained at the Lagrangian points and then distributed over the Eulerian nodes in the body neighborhood. By this strategy, one can use a simple Cartesian grid and it is not necessary to move the grids. According to the *VPM* [15] the Lagrangian force field is calculated based on the momentum balance over a fluid

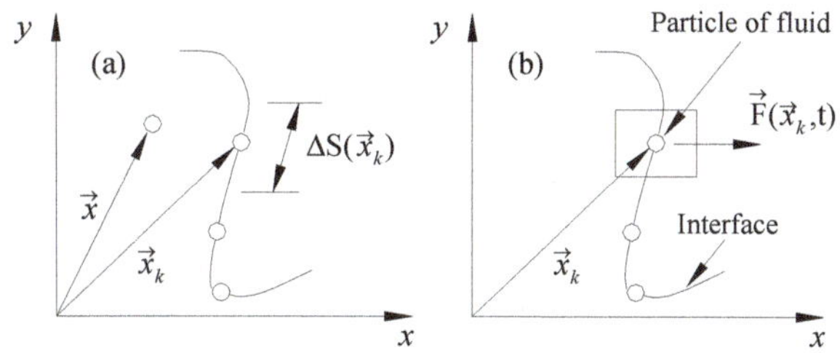

Figure 1. (a) Illustration of the distance between two Lagrangian points; (b) particle of fluid on the interface.

particle placed on the Lagrangian points.

By considering a particle of fluid placed on the fluid-solid interface as illustrated in **Figure 1(b)**, the Lagrangian force field is given as follows:

$$F(x_k,t) = \underbrace{\rho\frac{\partial V(x_k,t)}{\partial t}}_{F_a} + \underbrace{\rho\nabla\left[V(x_k,t)V(x_k,t)\right]}_{F_i} - \underbrace{\mu\nabla\left[\nabla V(x_k,t)+\nabla^T V(x_k,t)\right]}_{F_v} + \underbrace{\nabla p(x_k,t)}_{F_p} \tag{4}$$

where F_a [N] is the acceleration force, F_i [N] is the inertial force, F_v [N] is the viscous force, and F_p [N] is the pressure force. After obtaining the force field given by Equation (4), its values are distributed over the Eulerian nodes by using Equation (3) to generate the Eulerian force field that models the immersed body.

3. Numerical Method

A number of mesh-free methods have been developed in recent years [17] to circumvent the polygonisation problem found in the classical numerical methods. Here, the momentum and continuity equations are numerically solved using the finite difference method through the fractional step scheme based on the pressure correction concept [18]. Given the initial velocity, the pressure and the force fields, an estimated velocity field is obtained. This velocity field is used to calculate the pressure correction, by solving a linear system of algebraic equations, for which the Modified Strongly Implicit Procedure (*MSI*) [19] is used. The Poisson equation gives the coupling between Equation (1) and Equation (2). Also, it provides values of pressure that allow that the velocities components, obtained by using the Navier-Stokes equations, satisfy the mass conservation condition. The time discretization is done by the second order Runge-Kutta method [20]. The estimation of the velocity is calculated as:

$$\frac{\tilde{u}_i^{n+1}-u_i^n}{\Delta t} = -\frac{1}{\rho}\frac{\partial p^n}{\partial x_i} - \left[\frac{\partial(u_i u_j)}{\partial x_j}\right]^n + \frac{\partial}{\partial x_j}\left[\nu\left(\frac{\partial u_i}{\partial x_j}+\frac{\partial u_j}{\partial x_i}\right)\right]^n + f_i^n \tag{5}$$

where $\tilde{u}_i$ [m/s] is the estimated velocity component, Δt [s] is the computational time step and n is the substep index. The Poisson equation for pressure correction, p'^{n+1}, with the source term given by the divergent of the estimated velocity, is expressed by:

$$\nabla^2 p'^{n+1} = \frac{\rho\nabla\tilde{\boldsymbol{u}}^{n+1}}{\Delta t} \tag{6}$$

and the velocity field is updated by solving the algebraic equation $u_i^{n+1} = \tilde{u}_i^{n+1} - (\Delta t/\rho)(\partial p'^{n+1}/\partial x_i)$. The previous pressure field p^n and the correction pressure p'^{n+1} are used to calculate the updated values of the pressure field, according to the expression $p^{n+1} = p^n + p'^{n+1}$.

4. Results and Discussions

In this section, numerical simulations are performed to investigate the effects of the oscillating amplitude and forcing frequency on the flow structure of a circular cylinder. **Figure 2(a)** shows the computational domain which dimensions are 40*d* (*Lu* = 16.5*d; Ld* = 23.5*d*), in the streamwise and 15*d* (*H*) in the cross-stream direction, where *d* [m] is the cylinder diameter. The upper and lower boundaries are placed at 7.5*d*. The flow direction is from the

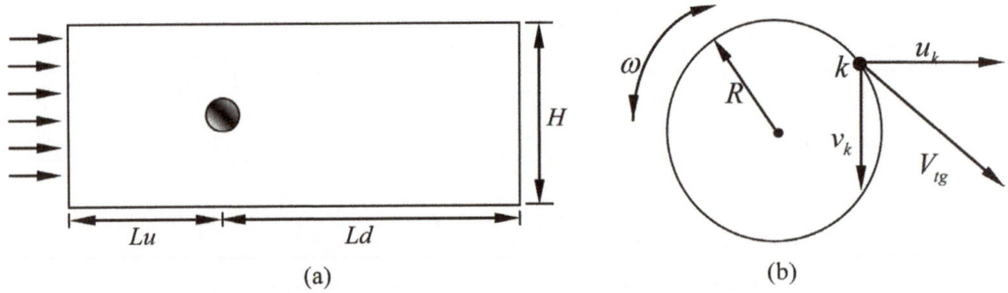

Figure 2. (a) Computational domain with a single rotating-oscillating cylinder; (b) Illustration of the velocities components and the angular velocity.

left to the right side of the domain and at the inlet, a uniform velocity profile $(u = 1, \upsilon = 0)$ is imposed. A Neumann boundary condition is used at the outlet and lateral boundaries, ($\partial u/\partial x = \partial \upsilon/\partial x = 0$, $\partial u/\partial y = \partial \upsilon/\partial y = 0$), respectively. For the pressure, the Neumann condition is used at the inlet $(\partial p/\partial x = 0)$ and the Dirichlet condition is used on the outlet and in the lateral boundaries $(p = 0)$. On the surface of the cylinder, no-slip boundary condition $(u = \upsilon = 0)$ is virtually employed using the *IBM* method. The time spent for each numerical simulation was about one hour. Also, the numerical code used herein was developed in Visual C.

Rotational oscillations at a prescribed set of frequency ratios and amplitudes are then imposed on the cylinder, where the tangential velocity over each Lagrangian point *k*, as shown in **Figure 2(b)**, is defined as $V_{tg} = \omega R = A\sin(2\pi f_c t)R$, where *A* [m] is the oscillating amplitude, f_c [Hz] is the forcing frequency, and *t* [s] is the physical time.

The simulations are performed at Reynolds number 1000, with a time step chosen arbitrarily of $\Delta t = 1\times10^{-5}$ s for the first iteration. After that, and during the first 100 iterations, the time step size is increased gradually until $\Delta t = 1\times10^{-3}$ s, for which the geometry of the immersed cylinder is completely defined according to the immersed boundary methodology. The grid is composed by 400 × 125 points in *x* and *y* directions, respectively; the oscillating amplitude, *A*, ranges from 1 to 3, while the frequency ratio, $f_r = f_c/f_o$, (f_o is the stationary cylinder frequency) varies from 0 to 6. Many cases have been performed, however, for the purposes of this work, only selected cases are presented.

4.1. Vortex Shedding Modes

Figure 3 shows the vorticity contours for $A = 1$ for different frequency ratios. When $f_r = 0$ (**Figure 3(a)**), it is seen that the vortex street is aligned and symmetric with respect to the central axis of the flow, showing two single vortices shedding per cycle. It is so called 2*S* mode of vortex shedding. In the frequency range $0.2 \le f_r \le 0.6$ (**Figure 3(b)**) to (**Figure 3(d)**) the patterns of vortex shedding seem to be similar, but the sizes appear uneven. When $f_r = 0.9$ (**Figure 3(e)**), one observes that the process of vortex formation is completely different from those of previous cases, and at the distance 12.5*d* downstream of the cylinder, the 2*S* mode appears again. Nevertheless, the transversal spacing is greater than that of the stationary case shown in **Figure 3(a)**. When $f_r = 1.05$ (**Figure 3(f)**), the 2*S* mode reappear. **Figure 3(g)** shows the *P* + *S* mode in which one pair and a single vortex are shed in each cycle. As frequency ratio further increases, **Figure 3(h)** shows that the wake structure remains essentially unchanged and similar to the stationary case. The no synchronized flow with the cylinder movement resembles to stationary cylinder flow, with some additional instabilities, due to the cylinder movement [21]. In other words, the instability caused by cylinder oscillation is limited to a region near to the cylinder, while far from immersed body, the vortices redirect to form the stable Von Kármán Street. This implies that occurs vortex-vortex interaction of the same sign, near to the cylinder, resulting in large scale vortices, whose values frequencies are close to the stationary cylinder frequency ($f_o = 0.23$).

Figure 4 shows visualizations of the flow structure obtained for *A* = 2 in the same frequency range as shown at **Figure 3**. In the range $0.2 \le f_r \le 0.5$, one can note the presence of two vortex pairs shed per cycle, named as 2*P* mode. When $f_r = 0.6$, the wake vortices are similar to the classical Von Kármán Street shown in **Figure 3(a)**, named 2*S* mode. However, the longitudinal and transversal spaces are larger than those corresponding to the stationary condition. In the frequency range $0.9 \le f_r \le 1.2$, the evolution of the patterns remains unchanged, as shown from **Figure 4(d)** to **Figure 4(f)**. In addition, it is observed that the transversal spacing is greater near

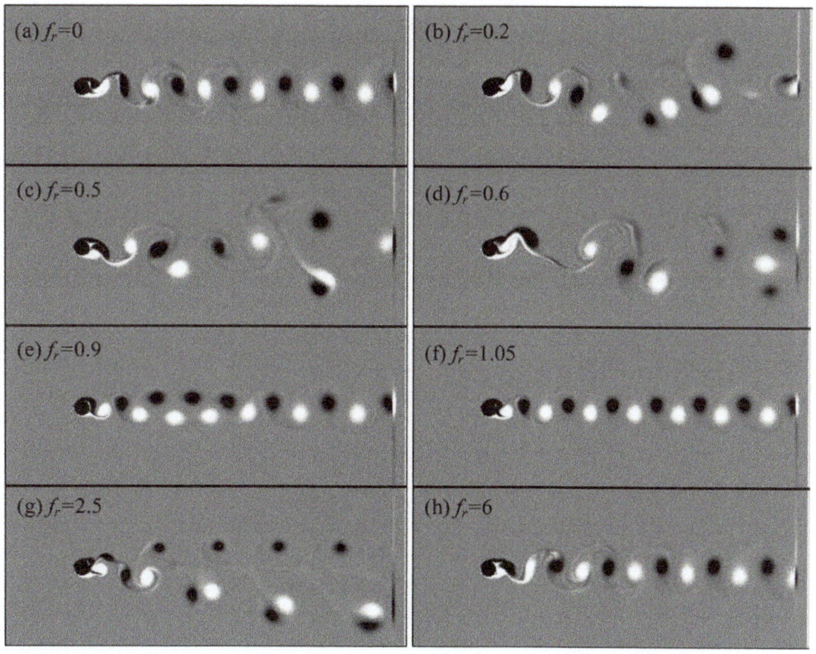

Figure 3. Vorticity contours for Re = 1000 and A = 1 and several values of f_r.

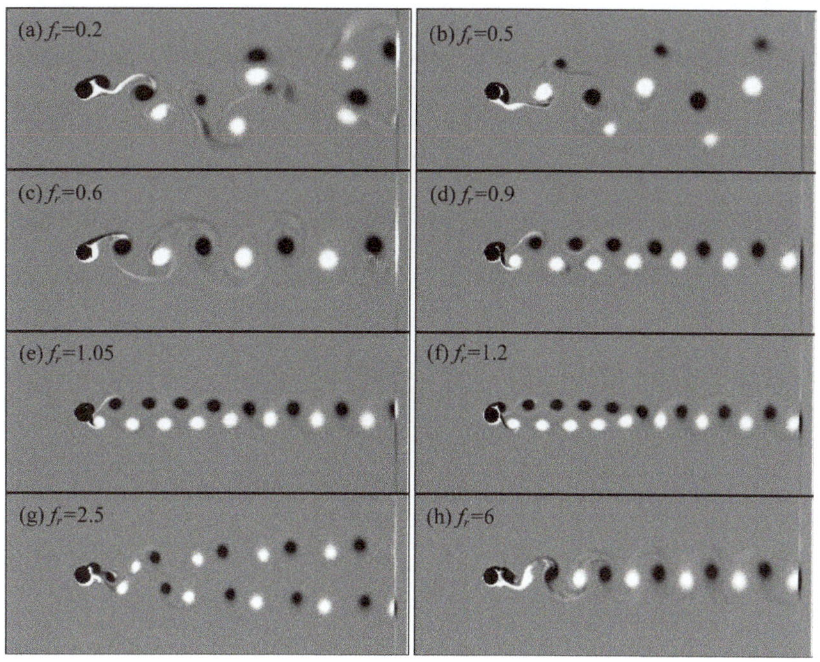

Figure 4. Vorticity contours for Re = 1000 and A = 2 and several values of f_r.

the cylinder and decreases away from it. When $f_r = 2.5$ (**Figure 4(g)**) it can be noted the presence of two vortex pairs in each side of the central line of the flow, resulting in a conical wake structure. When f_r is higher, it is observed that the vortex pattern is the same as the Kármán vortex street, indicating that for higher frequency values, the effect of rotational oscillation is confined to the flow near the cylinder, and its influence in the far field of the vortex structure is insignificant, as shown in **Figure 4(h)**.

The patterns of vortex shedding from the cylinder in the near and far wakes are shown in **Figure 5**, for A = 3 and various values of f_r. In **Figure 5(a)** one can note that the vortex street is not aligned with respect to the central axis of the flow and there are few vortex pairs in the wake. In addition, **Figure 5(b)** shows that the vortex

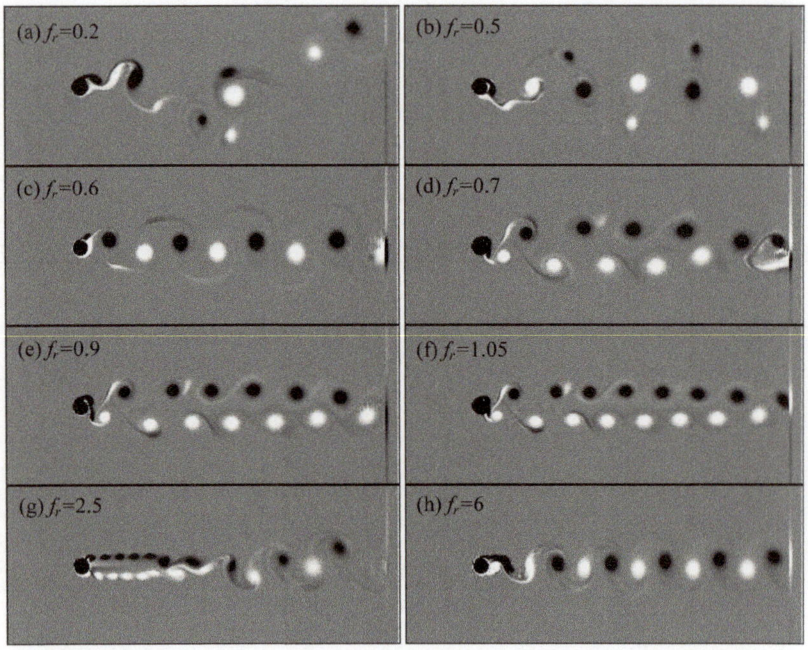

Figure 5. Vorticity contours for Re = 1000 and A = 3 and several values of f_r.

formation process is completely different from that of the previous cases, in which a new mode of vortex shedding, named as 2C mode, appears. This mode indicates that two vortex pairs of the same signal are shed per cycle. It is important to mention that this mode was not observed by the authors cited in the references for the case of circular cylinders subjected to rotational oscillations. However, it is noted in the case of a pivoted cylinder as reported in Williamson and Jauvtis [22].

In the frequency range $0.6 \le f_r \le 1.05$, one can note that the vortex wakes are found to be basically the same as those obtained for A = 2 in the same frequency range. The result shown in **Figure 5(g)** demonstrates a new pattern of vortex shedding for which the double wake near the cylinder, composed by vortex of the same signal in each row coalesce at the final double street to form a single wake. As frequency ratio further increases, the pattern of vortex shedding tends to be the same from the non-oscillating case.

4.2. Time Histories of the Fluid Dynamics Coefficients and Power Spectra Density

It is known that the vortex shedding process causes fluctuations in the dynamic coefficients and affects the behavior of the flow structure. **Figure 6** presents the time histories of the lift (C_l) and drag (C_d) coefficients as well as the power spectra of the lift coefficients for A = 1. **Figure 6(a)** shows that the transient behavior takes approximately 40 dimensionless time units (nine cycles in the present case) before reaching the periodic regime. From **Figure 6(b)** to **Figure 6(h)** one can note that the steady regime is reached more quickly than in the previous case due to fact that the cylinder's oscillations accelerate the vortex shedding process. As can be seen in **Figure 6(b)**, there is no harmonic behavior as observed for the previous stationary case, and the lift curve clearly resembles the shape of a signal with frequency f_c that is beaten by another signal at f_o, indicating that there is a significant interaction between these two frequencies. Also, from **Figure 6(b)** to **Figure 6(d)** it is verified that the amplitudes of the lift coefficient are greater than that for the stationary case. Moreover, it is observed a transition between the two vortex shedding modes when the frequency ratio is increased from $f_r = 1.05$ (**Figure 6(f)**) (2S mode) to $f_r = 2.5$ (**Figure 6(g)**) (P + S mode), but the oscillation amplitudes for the P + S mode are larger than that those corresponding to the 2S mode. This means that the change on the vortex modes has a strong influence on the time histories of the dynamics coefficients. As frequency further increases, the oscillation amplitudes tend to be the same of the stationary case, as can be noted by the comparison between **Figure 6(h)** and **Figure 6(a)**.

When $f_r = 0$, **Figure 6(a)** shows that the lift spectrum for the stationary case is composed by one peak at the dimensionless frequency $St_o = 0.23$. When $f_r = 0.2$ there are two prominent frequencies in the lift spectrum

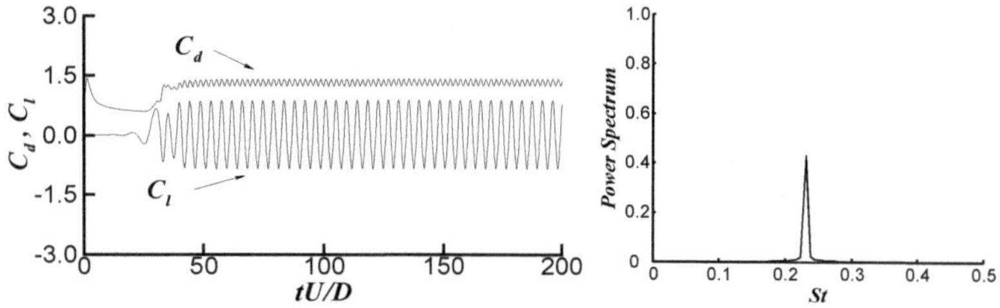
C_d
C_l
C_d, C_l
tU/D
Power Spectrum
St

(a)

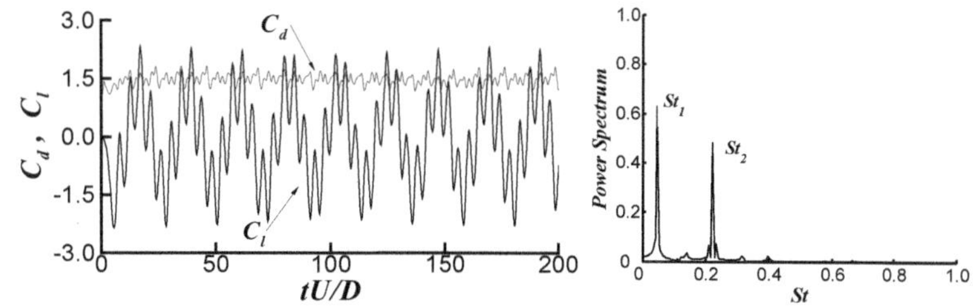
C_d
C_l
C_d, C_l
tU/D
Power Spectrum
St_1
St_2
St

(b)

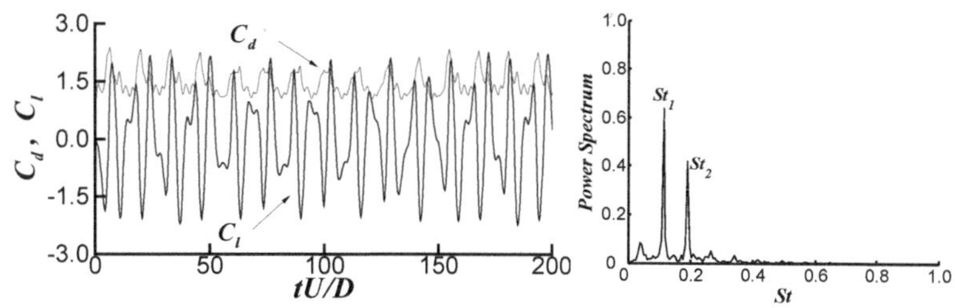
C_d
C_l
C_d, C_l
tU/D
Power Spectrum
St_1
St_2
St

(c)

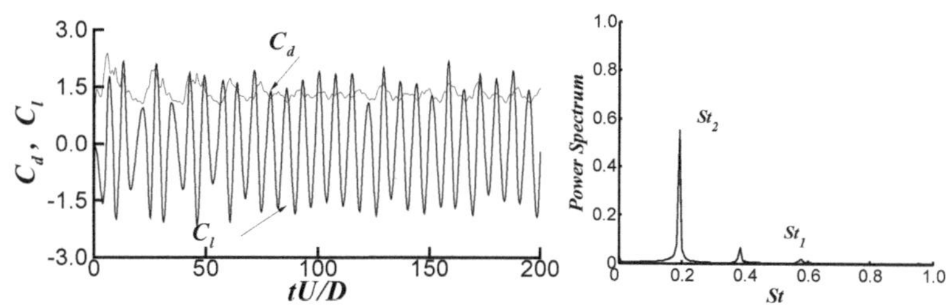
C_d
C_l
C_d, C_l
tU/D
Power Spectrum
St_2
St_1
St

(d)

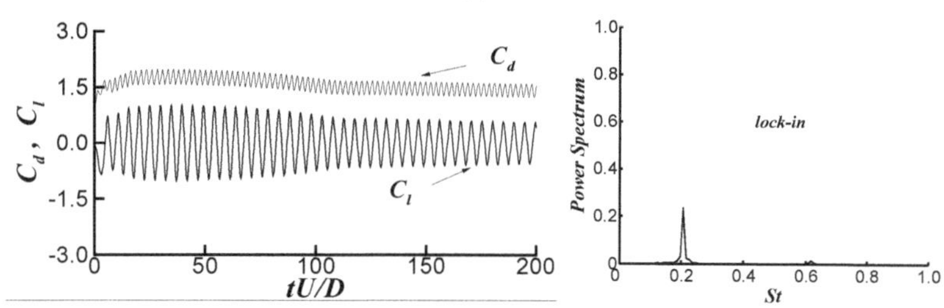
C_d
C_l
C_d, C_l
tU/D
Power Spectrum
lock-in
St

(e)

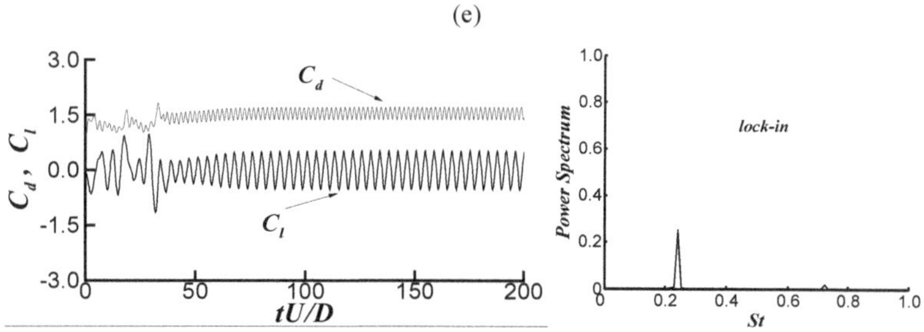
C_d
C_l
C_d, C_l
tU/D
Power Spectrum
lock-in
St

(f)

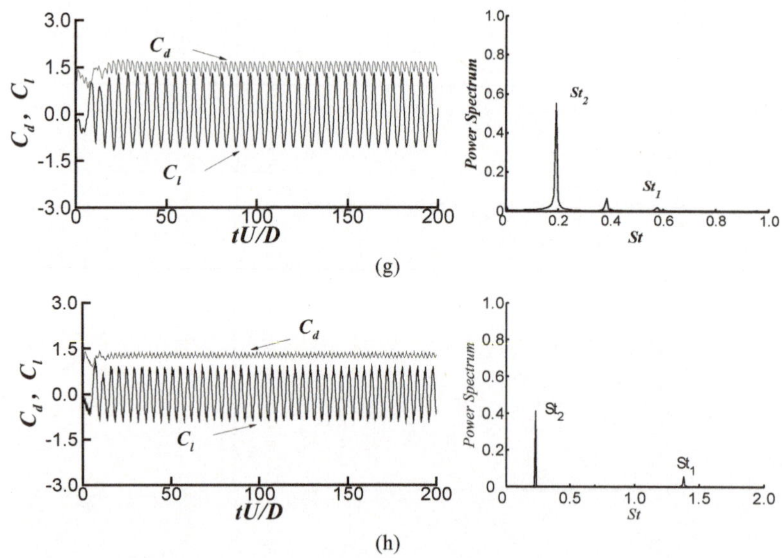

Figure 6. Time histories of the dynamic coefficients and power spectra for A = 1: (a) $f_r = 0$; (b) $f_r = 0.2$; (c) $f_r = 0.5$; (d) $f_r = 0.6$; (e) $f_r = 0.9$; (f) $f_r = 1.05$; (g) $f_r = 2.5$; and (h) $f_r = 6.0$.

shown in **Figure 6(b)**. They correspond to the vortex shedding frequency $St_2 = 0.22$, which is closed to St_o and the forcing frequency St_1. As the frequency ratio further increases from $f_r = 0.2$ to $f_r = 0.5$ (**Figure 6(c)**) shows two frequencies in the power spectrum, but there is a discernible drift of the vortex shedding frequency, St_2, towards the forcing frequency, St_1. The energy level of St_2 (secondary peak) is reduced and the frequency value is 83% of $St_o = 0.23$.

When $0.6 \le f_r \le 1.05$, the frequency of vortex shedding, St_2, locks-in to the forcing frequency, St_1, resulting in only one dominating component in the lift frequency spectrum, as shown from **Figure 6(d)** to **Figure 6(f)**. During the lock-in regime the wake structure is synchronized with the oscillatory motion of the cylinder and the interaction becomes strong. Also, it is observed that as the frequency ratio increases, the amplitudes of lift coefficient curve decrease and then increase again outside the lock-in regime. **Figure 6(d)** shows an increase in energy level, compared with the previous, $f_r = 0.5$, followed by a reduction when the frequency ratio is increased, as shown in **Figure 6(e)** and **Figure 6(f)**. Consequently, this increase and decrease in the frequency peak also affect the mean drag coefficient that reaches a maximum value at $f_r = 0.8$ ($C_d = 1.97$). As frequency ratio further increases, the lock-in regime no longer exists as shown in **Figure 6(g)** and **Figure 6(h)**. Also, it is interesting to note that when $f_r = 2.5$ (corresponding to the $P + S$ mode), the smallest magnitude of the energy peak corresponds to St_1. On the other hand, as f_r further increases, the magnitude of St_2 tends to be constant, indicating that the large-scale Kármán vortex street has reached some stable state, corresponding to stationary condition. It is worth mentioning that St_2 is equal to the stationary case $St_o = 0.23$, as shown in **Figure 6(h)**.

Figure 7 shows the time evolution of the dynamics coefficients as well as the power spectra density of the lift coefficient for $A = 2$. When $f_r = 0.2$, the drag and lift coefficients present behaviors similar to those shown in **Figure 6(b)**. Given that each peak (both positive and negative) in the lift curve is related to the shedding of one vortex, then **Figure 7(a)** means that there should be eighteen vortices in the wake when the dimensionless time reaches 200. By regarding the lift curve, it is also interesting to observe a reduction in the interaction between the two frequencies, from the comparison between **Figure 7(a)** and **Figure 6(b)**. In the range $0.9 \le f_r \le 1.2$, from **Figure 7(d)** to **Figure 7(f)** it is verified that the amplitudes decrease when compared with the previous frequency ratio. When $f_r = 2.5$, the fluctuation of the lift coefficient increases again due to the change of the vortex shedding mode. For higher frequency ratios, the behavior is similar to that observed to $A = 1$, in such a way that the flow tends to the same vortex shedding mode as the one observed for the stationary cylinder. As oscillation amplitude increases, the lock-in regime increases ($0.5 \le f_r \le 1.2$) as shown in **Figure 7**. The amplitude peak before the lock-in regime for St_1 ($f_r = 0.2$), are larger than those observed in **Figure 6(b)** and **Figure 6(c)** for $A = 1$. As f_r increases from $f_r = 0.5$ to $f_r = 1.2$, it is observed a reduction in the magnitude

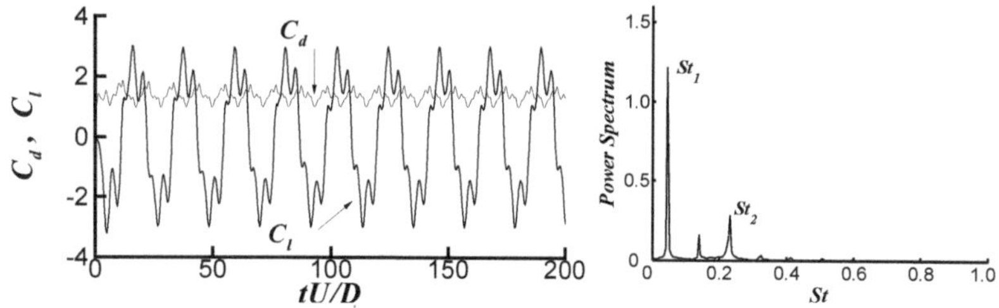
C_d, C_l
tU/D
Power Spectrum
St
St_1
St_2

(a)

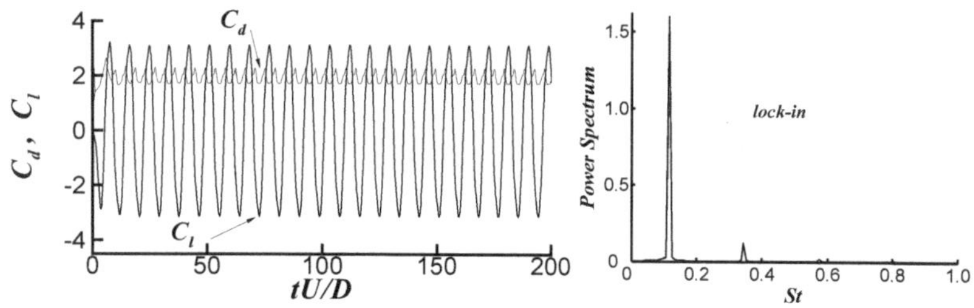
C_d, C_l
tU/D
Power Spectrum
St
lock-in

(b)

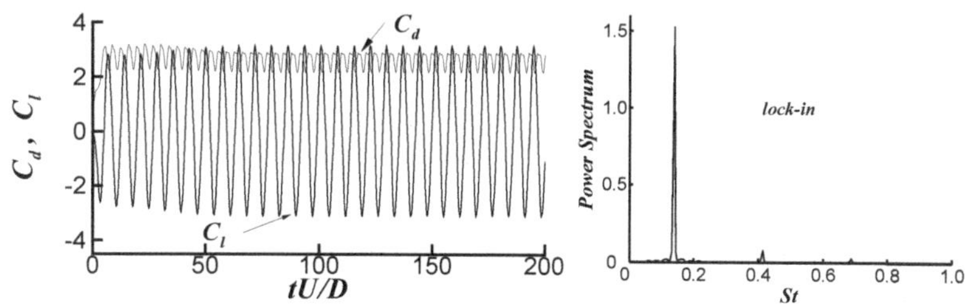
C_d, C_l
tU/D
Power Spectrum
St
lock-in

(c)

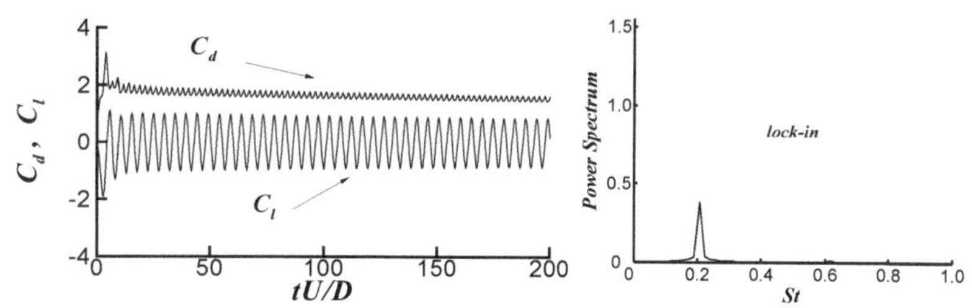
C_d, C_l
tU/D
Power Spectrum
St
lock-in

(d)

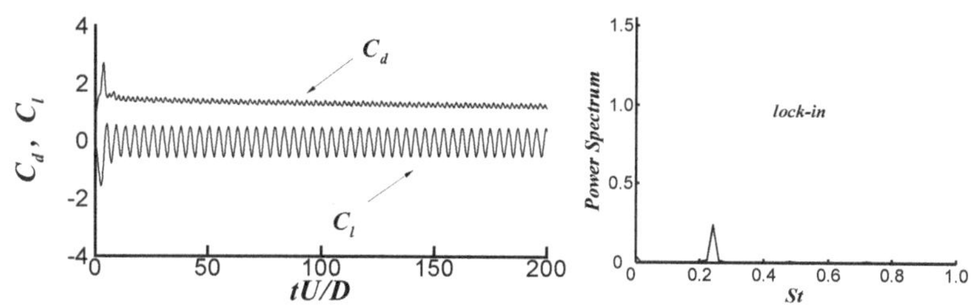
C_d, C_l
tU/D
Power Spectrum
St
lock-in

(e)

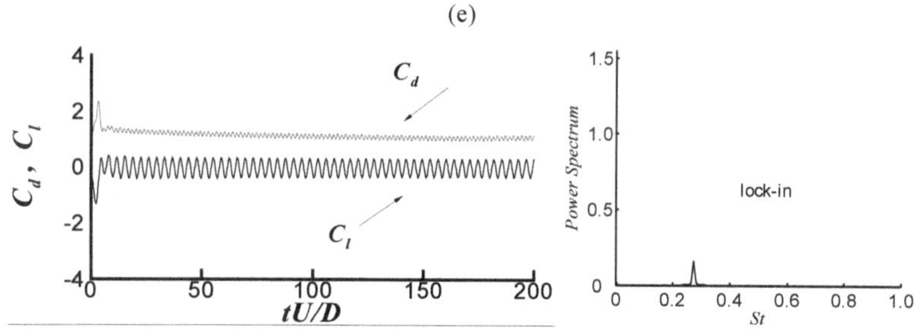
C_d, C_l
tU/D
Power Spectrum
St
lock-in

(f)

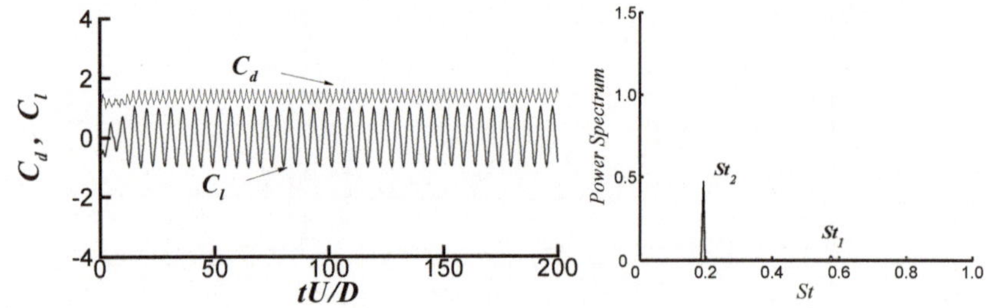

(g)

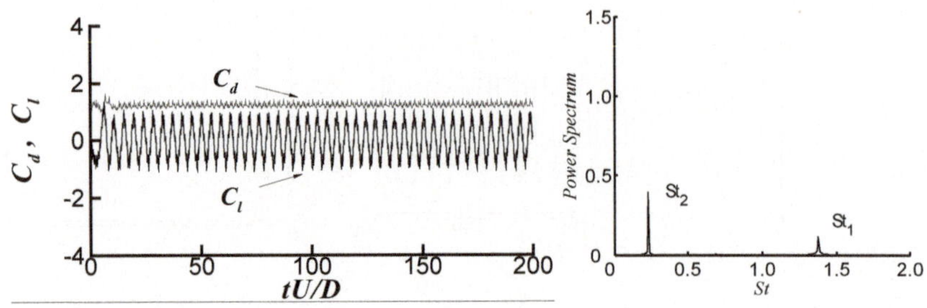

(h)

Figure 7. Time histories of the dynamic coefficients and power spectra for $A = 2$: (a) $f_r = 0.2$; (b) $f_r = 0.5$; (c) $f_r = 0.6$; (d) $f_r = 0.9$; (e) $f_r = 1.05$; (f) $f_r = 1.2$; (g) $f_r = 2.5$; and (h) $f_r = 6.0$.

of the peak, as shown in **Figure 7(b)** to **Figure 7(f)**. During the lock-in regime, the fluctuation amplitude of lift coefficient decreases, as the frequency ratio increases, and increases again outside the lock-in range. It is interesting to point out that in the lock-in regime, for the cases for which has occurred reduction in the magnitude of the energy peak, the wake structure has greater transversal space near the cylinder, while, far away from the cylinder, it decreases (see **Figure 4(d)** to **Figure 4(f)**). When $f_r = 2.5$, two spectral peaks can be seen again in the lift spectrum, indicating that the lock-in regime no longer exists. The energy peak associated to St_2 is greater than St_1 and its frequency value ($St_2 = 0.19$) is smaller than that of the non-oscillating case. As f_r further increases, it can be seen in **Figure 7(h)** that the wake configuration tends to be the same of classical Von Kármán Street corresponding to $St_2 = 0.23$.

Figure 8 shows the time histories of the lift and drag coefficients and the power spectra of the lift coefficient for $A = 3$. When $f_r = 0.2$, still there are "kink" on the lift curve. This kink phenomenon occurs due to the fact that as the oscillation amplitude increases, a main vortex is formed and shed on one side of the cylinder at the same time as an adjacent secondary vortex having opposite sign is formed and annihilated later by the main vortex. When $f_r = 0.6$, it can be observed a reduction in the amplitude of the lift curve when compared to the previous frequency ratios, $f_r = 0.2$ and $f_r = 0.5$. In the frequency range $0.6 \le f_r \le 2.5$ (**Figure 8(c)** to **Figure 8(g)**) it is observed that as the frequency ratio increases, the fluctuation amplitude of the lift coefficient is reduced. This reduction is due to the vortex wake structure as shown from **Figure 5(c)** to **Figure 5(g)**, where it was verified the 2*S* mode for $f_r = 0.6$, the elliptical wake for $0.7 \le f_r \le 1.05$, and a wake with two rows of vortices near the cylinder for $f_r = 2.5$. At higher frequency ratios, the amplitude of the lift curves approaches to that observed for the stationary case.

For $A = 3$, the lock-in regime is located at the interval $0.2 \le f_r \le 2.5$, for which the maximum mean drag coefficient is obtained at $f_r = 0.6$ ($C_d = 2.93$). Also, as the frequency ratio increases from $f_r = 0.2$ to $f_r = 0.5$, the energy level increases. After that, the energy level reduces from $f_r = 0.5$ to $f_r = 2.5$. Finally, another aspect to be pointed out is that as the oscillation amplitude increases, the frequency range in which the lock-in regime occurs is enlarged: $[0.6-1.05]$ for $A = 1$; $[0.5-1.2]$ for $A = 2$ and $[0.2-2.5]$ for $A = 3$.

Another relevant aspect to be investigated is the energy level of the power spectra, as shown in **Figure 9**. The most immediate conclusion is that as the frequency ratio increases, the energy level increases and after decreases keeping approximately unchanged for all analyzed amplitudes, for frequency ratio equal to 2. Also, higher energy levels are obtained in the lock-in regime for all amplitudes. Moreover, as the oscillating amplitude increase the energy level in the lower boundary of the lock-in regime also increase.

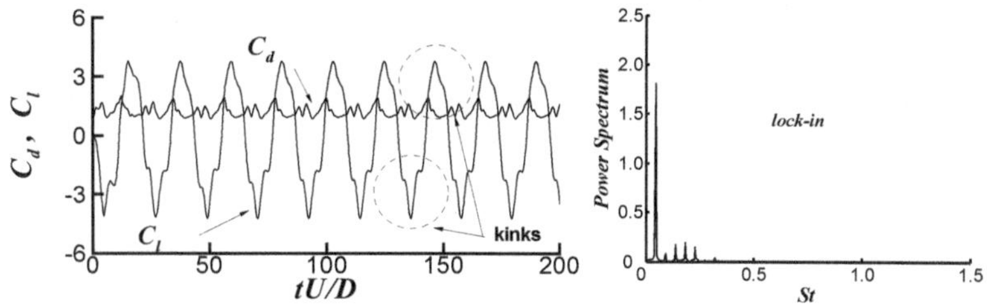

C_d
C_l
kinks
C_d, C_l
tU/D
Power Spectrum
lock-in
St

(a)

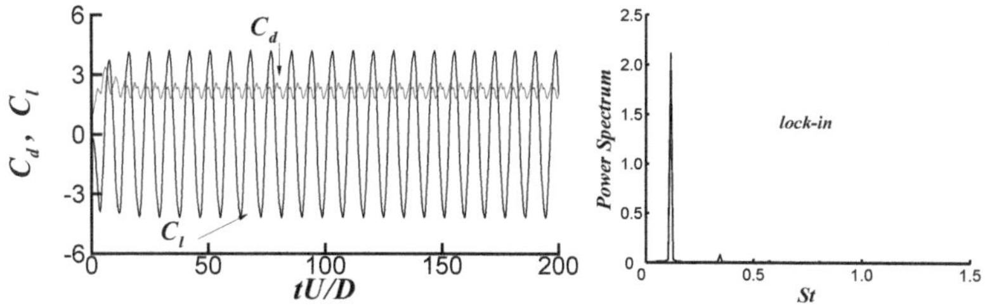

C_d
C_l
C_d, C_l
tU/D
Power Spectrum
lock-in
St

(b)

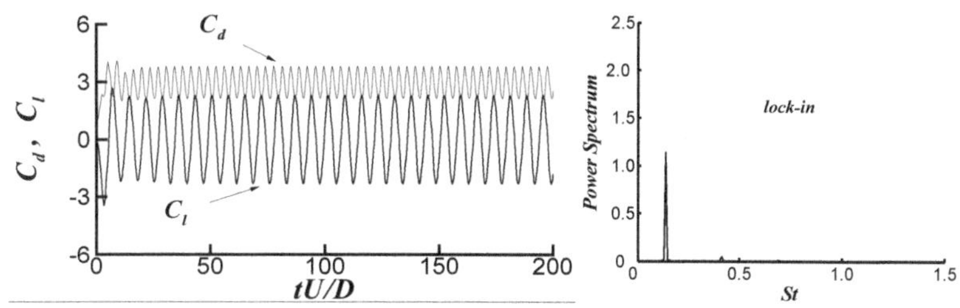

C_d
C_l
C_d, C_l
tU/D
Power Spectrum
lock-in
St

(c)

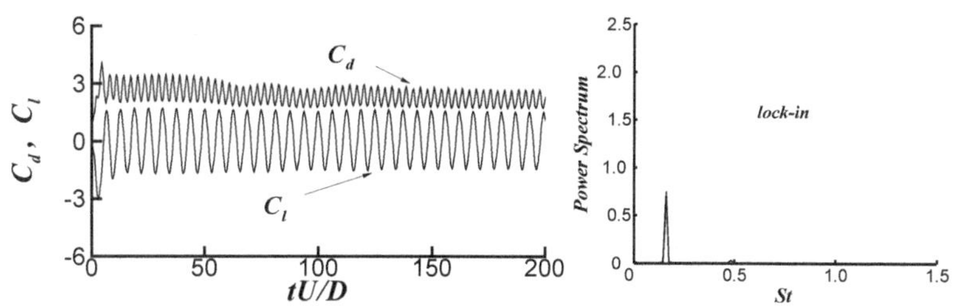

C_d
C_l
C_d, C_l
tU/D
Power Spectrum
lock-in
St

(d)

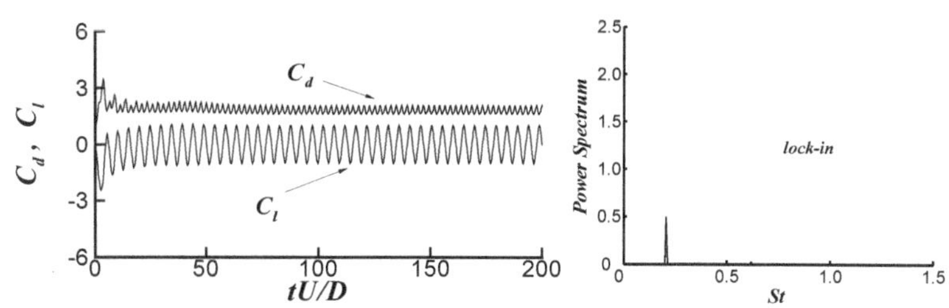

C_d
C_l
C_d, C_l
tU/D
Power Spectrum
lock-in
St

(e)

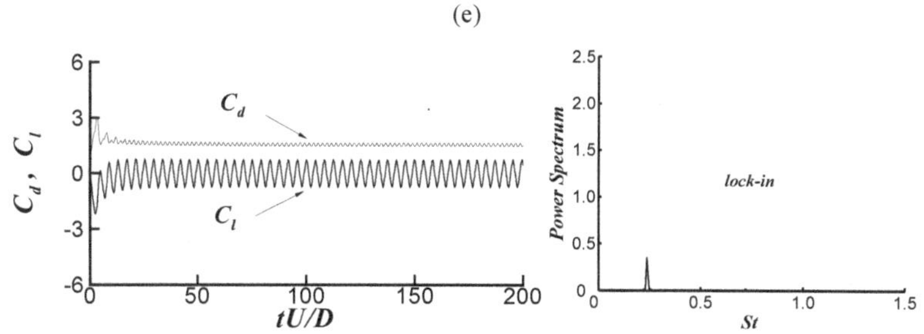

C_d
C_l
C_d, C_l
tU/D
Power Spectrum
lock-in
St

(f)

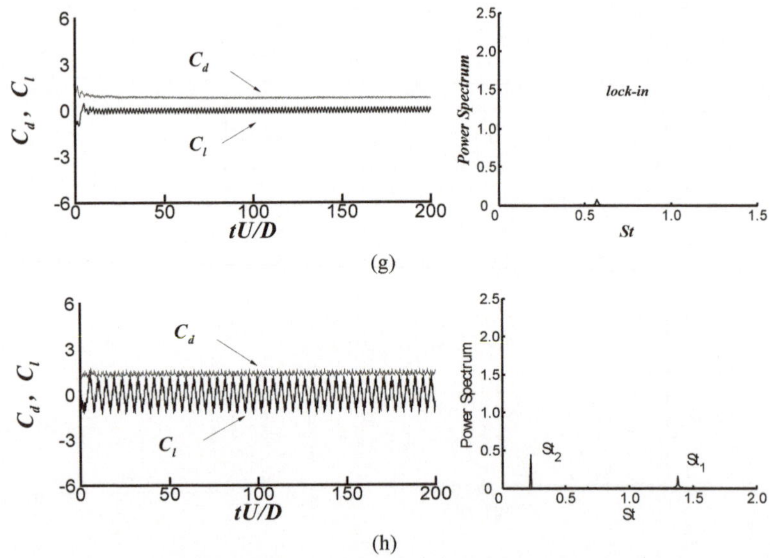

Figure 8. Time histories of the dynamic coefficients and power spectra for $A = 3$: (a) $f_r = 0.2$; (b) $f_r = 0.5$; (c) $f_r = 0.6$; (d) $f_r = 0.7$; (e) $f_r = 0.9$; (f) $f_r = 1.05$; (g) $f_r = 2.5$; and (h) $f_r = 6.0$.

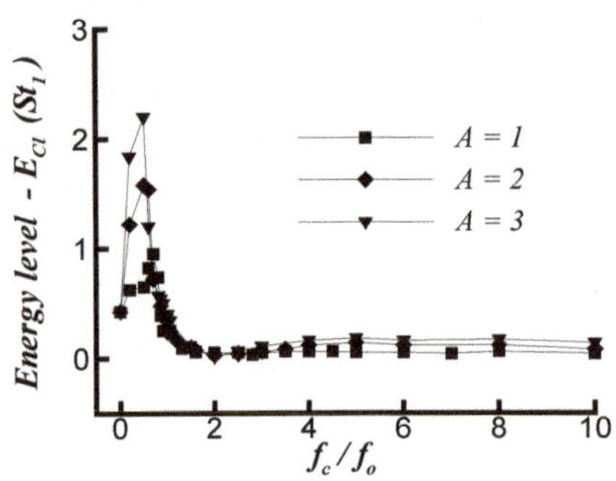

Figure 9. Energy level of St_l in function of frequency ratio fr for $A = 1$, $A = 2$ and $A = 3$.

4.3. Comparison with other Previous Studies

It is desirable to compare the present results with some numerical results from previous studies reported by other investigators. **Figure 10** (left) shows a plot of the mean drag coefficient as a function of the frequency ratio obtained for the present numerical simulations and the literature results [4] [9], for $A = 3$ at Reynolds number 1000. It can be noted that all the results indicate a sharp peak of the C_d curve at low frequency ratios, and small values of C_d at high frequencies ratio This behavior has been also observed for other Reynolds number and oscillation amplitudes [10] [23]. It can be verified that the present computational results are in good agreement with the results by references for low frequency ratios, but overestimates C_d in high frequencies, which may reflect the assumptions of the numerical methodology used in the present calculations. It is important to mention that according to Srinivas and Fujisawa [10], there is a great discrepancy in the literature concerning the behavior of the mean drag coefficient in the Reynolds numbers range [1000-3000].

Another aspect to be pointed out is that the maximum mean drag coefficients were obtained for the *2S* mode for all the analyzed amplitudes, in which the longitudinal and transversal spacing for the oscillating cylinder is greater than those of the stationary cylinder. This fact can be observed by analyzing the vorticity contours corresponding to the maximum C_d, as shown in **Figure 11**, where **a** is the transversal spacing and **b** indicates the longitudinal spacing. It should be noted that as the oscillating amplitude increases, **b** increases and **a** takes a constant value. This enables to conclude that the mean drag coefficient is more strongly dependent on the longitudinal spacing than the transversal spacing.

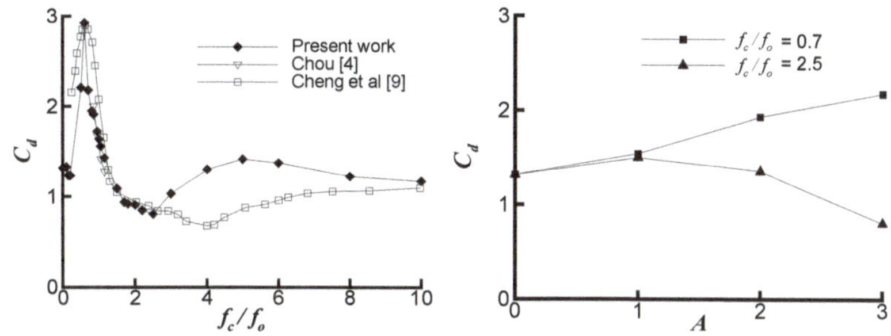

Figure 10. (left) Mean drag coefficient in function of f_r for A = 3 and (right) in function of oscillation amplitude. Re = 1000.

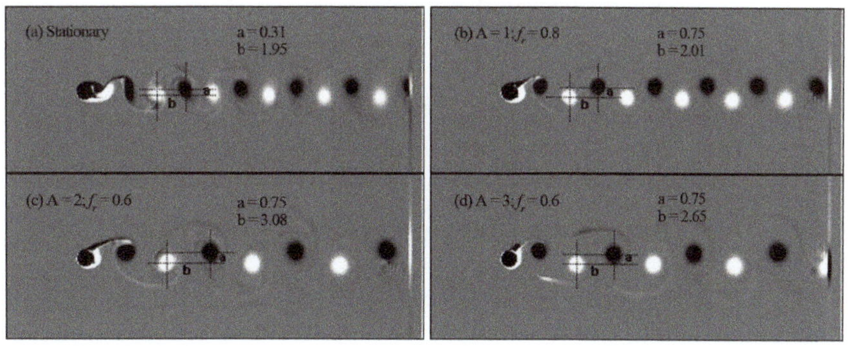

Figure 11. Presentation of longitudinal and transversal spacing between the vortices.

4.4. Pressure Distribution

The mechanism of increase and reduction of C_d for low and high frequency ratios, respectively, can also be analyzed through the mean pressure distribution along the surface of the stationary and oscillating cylinders. **Figure 12(a)** shows an illustrative scheme of the angle θ along the surface of the immersed body, which is defined as zero at the downstream point of the cylinder.

The pressure coefficient is obtained by the relation $C_p = (p_k - p_\infty)/(0.5\rho U_\infty^2)$, where p_k is the static pressure in the Lagrangian points, p_∞ is the static pressure of the undisturbed flow at the inlet of the domain, and ($0.5\rho U_\infty^2$) is the dynamic pressure taken as the reference. **Figure 12(b)** and **Figure 12(c)** show the distribution of the mean pressure coefficient C_p for all the analyzed amplitudes and two frequency ratios. It can be noted that C_p at the maximum local pressure ($\theta = 0°$) is almost unit, as expected. Also, it can be verified how the mean pressure coefficient downstream of the cylinder ($\theta = 180°$) varies with the cylinder oscillation amplitude. When $f_r = 0.6$, it can be observed that for A = 1 the mean pressure coefficient downstream of the cylinder ($\theta = 180°$) presents a small decrease (in absolute values) when compared with the stationary cylinder. Nevertheless, it is noted a considerable increase in the mean pressure coefficient when the oscillation amplitude increases from A = 1 to A = 2. As amplitude further increases from A = 2 to A = 3, C_p tends to be the same for $\theta = 180°$. When $f_r = 1.5$, it can be observed a little decrease in C_p at $\theta = 180°$ for A = 1, while for A = 2, a greater reduction in C_p value is observed from $C_p = -2.84$ (f_r = 0.6) to $C_p = -0.25$ (f_r = 1.5).

By comparing **Figure 12(b)** and **Figure 12(c)** for A = 3, it can be noted an inversion of the mean pressure coefficient signal, from $C_p = -2.8$ (f_r = 0.6) to $C_p = 0.04$ (f_r = 1.5). Thus, the forcing frequency plays an important role on the pressure distribution, enabling to verify that for low frequency ratios, the flow downstream of the cylinder is very dissipative due to viscous effects. This effect contributes to reduce the pressure and consequently to increase the drag over the cylinder. However, at high frequency ratios, the wake dynamic downstream of the cylinder is not so strong when compared to the corresponding ones obtained for low frequencies, which contribute to increase the pressure and, consequently, to reduce the drag coefficient.

5. Concluding Remarks

The numerical simulations of the flow over the rotationally-oscillating circular cylinder by using the Immersed

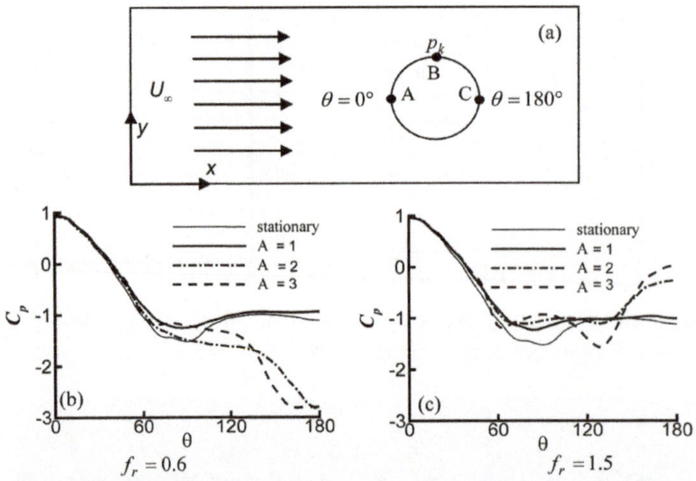

Figure 12. Illustrative scheme of the angle θ along the cylinder surface (a) and the distribution of the mean pressure coefficient in function of the angle (b) and (c).

Boundary Method combined with the Virtual Physical Model have been addressed. Given the strong influence of the oscillation amplitude and frequency on the flow around the cylinder, the influence of these parameters on the lift and drag coefficients, on the pressure distribution, as well as on the vortex shedding frequency has been investigated.

The simulations shown that the flow structure in the near wake was strongly dependent on the oscillation frequency. Also, it observed different vortex shedding modes ("2*C*", "2*P*", "2*S*", and "*P* + *S*") for a fixed oscillation amplitude at different values of frequency ratios, and also for a fixed frequency ratio at different values of oscillation amplitudes, in addition to the conical wakes. Another important feature of the numerical methodology is its capability to identify the 2*C* mode of the vortex shedding, which has not been observed by other investigators for the case of rotationally-oscillating cylinders.

The study reported herein enabled to observe a number of important features that should be mentioned:

- The range of resonance increases as the oscillation amplitude increases;
- For high frequency ratios, the wake structure configuration is similar to the classical Von Kármán Street, and the values of the vortex shedding frequency take the values corresponding to the stationary cylinder case;
- The pressure distribution over the cylinder is influenced by forcing frequency, and consequently affects the drag over the cylinder. It implies that drag control can be done by the rotational oscillations mechanism;
- The Immersed Boundary Method combined with the Virtual Physical Model can be easily employed in the case of moving bodies, being a very useful tool to simulate problems involving prescribed motion.

Acknowledgements

The author is grateful to the following organizations: Agency of the Ministry of Science, Technology and Innovation—CNPq for the continued support to their research work, and the Minas Gerais State Agency FAPEMIG.

References

[1] Anagnostopoulos, P. (2002) Flow-Induced Vibrations in Engineering Practice. WIT Press, Southampton, Boston.

[2] Naudascher, E. and Rockwell, D. (1994) Flow-Induced Vibrations: An Engineering Guide. Dover Publications, Inc., Mineola, New York.

[3] Païdoussis, M.P. (2004) Fluid-Structure Interactions: Slender Structures and Axial Flow. Vol. 2, Elsevier Academic Press, San Diego.

[4] Chou, M.H. (1997) Synchronization of Vortex Shedding from a Cylinder under Rotary Oscillation. *Computers & Fluids*, **36**, 755-774. http://dx.doi.org/10.1016/S0045-7930(97)00028-5

[5] He, J.W., Glowinski, R., Metcalfe, R., Nordlander, A. and Periaux, J. (2000) Active Control and Drag Optimization for Flow Past a Circular Cylinder I. Oscillatory Cylinder Rotation. *Journal of Computational Physics*, **163**, 83-117.

http://dx.doi.org/10.1006/jcph.2000.6556

[6] Lee, S.-J. and Lee, J.-Y. (2006) Flow Structure of Wake behind a Rotationally Oscillating Circular Cylinder. *Journal of Fluids and Structures*, **22**, 1097-1112. http://dx.doi.org/10.1016/j.jfluidstructs.2006.07.008

[7] Du, L. and Dalton, C. (2013) LES Calculation for Uniform Flow past Rotationally Oscillating Cylinder. *Journal of Fluids and Structures*, **42**, 40-54. http://dx.doi.org/10.1016/j.jfluidstructs.2013.05.008

[8] Cheng, M., Liu, G.R. and Lam, K.Y. (2001) Numerical Simulation of Flow Past a Rotationally Oscillating Cylinder. *Computers & Fluids*, **30**, 365-392. http://dx.doi.org/10.1016/S0045-7930(00)00012-8

[9] Cheng, M., Chew, Y.T. and Luo, S.C. (2001) Numerical Investigation of a Rotationally Oscillating Cylinder in Mean Flow. *Journal of Fluids and Structures*, **15**, 981-1007. http://dx.doi.org/10.1006/jfls.2001.0387

[10] Srinivas, K. and Fujisawa, N. (2003) Effect of Rotational Oscillation upon Fluid Forces about a Circular Cylinder. *Journal of Wind Engineering and Industrial Aerodynamics*, **91**, 637-652. http://dx.doi.org/10.1016/S0167-6105(02)00460-9

[11] Ray, P. and Christofides, P.D. (2005) Control of Flow over a Cylinder Using Rotational Oscillations. *Computers and Chemical Engineering*, **29**, 877-885. http://dx.doi.org/10.1016/j.compchemeng.2004.09.014

[12] Peskin, C.S. (1977) Numerical Analysis of Blood Flow in the Heart. *Journal of Computational Physics*, **25**, 220-252. http://dx.doi.org/10.1016/0021-9991(77)90100-0

[13] Nicolás, A. and Bermúdez, B. (2007) Viscous Incompressible Flows by the Velocity-Vorticity Navier-Stokes Equations. *CMES*: *Computer Modeling in Engineering & Sciences*, **20**, 73-83.

[14] Báez, E. and Nicolás, A. (2009) Recirculation of Viscous Incompressible Flows in Enclosures. *CMES*: *Computer Modeling in Engineering & Sciences*, **41**, 107-130.

[15] Lima e Silva, A.L.F., Silva, A.R. and Silveira-Neto, A. (2007) Numerical Simulation of Two-Dimensional Complex Flows around Bluff Bodies Using the Immersed Boundary Method. *Journal of the Brazilian Society of Mechanical Sciences and Engineering*, **XXIX**, 378-386. http://dx.doi.org/10.1590/s1678-58782007000400006

[16] Peskin, C.S. and McQueen, D.M. (1994) A General Method for the Computer Simulation of Biological Systems Interacting with Fluids. *SEB Symposium on Biological Fluid Dynamics*, Leeds, England, 5-8 July 1994.

[17] Vertnik, R. and Šarler, B. (2009) Solution of Incompressible Turbulent Flow by a Mesh-Free Method. *CMES*: *Computer Modeling in Engineering & Sciences*, **44**, 65-95.

[18] Chorin, A. (1968) Numerical Solution of the Navier-Stokes Equations. *Mathematics of Computations*, **22**, 745-762. http://dx.doi.org/10.1090/S0025-5718-1968-0242392-2

[19] Schneider, G.E. and Zedan, M.A. (1981) Modified Strongly Implicit Procedure for the Numerical Solution of Field Problems. *Numerical Heat Transfer*, **4**, 1-19. http://dx.doi.org/10.1080/01495728108961775

[20] Ferziger, J.H. and Peric, M. (2002) Computational Methods for Fluid Dynamics. 3rd Edition, Springer-Verlag, Berlin, 423 p. http://dx.doi.org/10.1007/978-3-642-56026-2

[21] Tuszynski, J. and Löhner, R. (1998) Control of a Kármán Vortex Flow by Rotational Oscillations of a Cylinder. George Mason University, USA, 1-12.

[22] Williamson, C.H.K. and Jauvtis, N. (2004) A High-Amplitude 2T Mode of Vortex-Induced Vibration for a Light Body in X-Y Motion. *European Journal of Mechanics—B/Fluids*, **23**, 107-114. http://dx.doi.org/10.1016/j.euromechflu.2003.09.008

[23] Fujisawa, N., Asano, Y., Arakawa, C. and Hashimoto, T. (2005) Computational and Experimental Study on Flow around a Rotationally Oscillating Circular Cylinder in a Uniform Flow. *Journal of Wind Engineering and Industrial Aerodynamics*, **93**, 137-153. http://dx.doi.org/10.1016/j.jweia.2004.11.002

25

Elastic Stress Predictor for Stochastic Finite Element Problems

Drakos Stefanos

International Center for Computational Engineering, Rhodes, Greece
Email: stefanos.drakos@gmail.com

Abstract

The paper presents a new algorithm of elastic stress predictor in non linear stochastic finite element method using the Generalized Polynomial Chaos. The statistical moments of strains calculated based on the displacement Polynomial Chaos expansion. To descretise the stochastic process of material the Karhunen-Loeve Expansion was used and it is presented. Using the strains and the material Karhunen-Loeve Expansion the stress components are calculated. A numerical example of shallow foundation was carried out and the results of stress and strain of the new algorithm were compared with those raised from Monte Carlo method which is treated as the exact solution. A great accuracy was presented.

Keywords

Polynomial Chaos, Stochastic Finite Element, Karhunen-Loeve Expansion, Quantification of Uncertainty

1. Introduction

The analysis and design in structural and geotechnical engineering problems requires the calculation of stress and strain which is generally a difficult task because of the uncertainty and spatial variability of the materials' properties. Various forms of uncertainties arise which depend on the nature of geological formation or construction method, the site investigation, the type and the accuracy of design calculations etc. In recent years there has been considerable interest amongst engineers and researchers in the issues related to quantification of uncertainty as it affects safety, design as well as the cost of projects [1]-[4].

A number of approaches using statistical concepts have been proposed in engineering in the past 25 years or so. These include the Stochastic Finite Element Method (SFEM) [5]-[7], and the Random Finite Element Method (RFEM) [8]-[12]. The RFEM involves generating a random field of soil or structure properties with controlled mean, standard deviation and spatial correlation length, which is then mapped onto a finite element mesh. However the number of works on the stochastic stress and strain calculation and their statistical moments are limited. An essential paper on the field is presented by Ghosh & Farhat [13] where the constitutive relation of stress and strain calculated by different approaches.

In this paper we present SFEM [14]-[18] using the method of Generalized Polynomial Chaos (GPC) [19]. To descretise the stochastic process of material the Karhunen-Loeve Expansion was used and it is presented. The constitutive relation of stress and strain calculated using the Generalized Polynomial Chaos and verified against Monte Carlo simulation which is treated as the exact solution based on a series of computational experiment. In order to solve an elastoplastic problem the invariants of stress are also needed. In the current work the stochastic stress invariants are given also.

A numerical example of shallow foundation is given in the last part of the paper. The results of the two methods of stress and strains calculation are compared and presented.

2. Probability Density Definition

Considering an arbitrary body and a the sample space $(\Omega,\mathcal{F},\mathbb{P})$ where $\mathcal{F}$ is the σ-algebra and is considered to contain all the information that is available, $\mathbb{P}$ is the probability measure and the spatial domain of the body is $D\subset\mathbb{R}^3$. Assuming that the parameters of the body $v(x,\xi(\omega))$ for $x\in D$ and $\omega\in\Omega$ dependent on a finite number M of random variables $\xi=[\xi_1,\xi_2,\cdots,\xi_M]:\Omega\to\Gamma\subset\mathbb{R}^M$ and $\Gamma=\Gamma_1\times\Gamma_1\times\cdots\times\Gamma_M$. To compute the statistical moments of the results we perform a change of variable $y_k:=\xi_k(\omega)$ and $y=[y_1,y_2,\cdots,y_M]$ [20]. If ρ is the joint density and random variables are independent and ρ_i denote the density of ξ_i then:

$$\rho(y)=\rho_1(y_1)\rho_2(y_2)\cdots\rho_M(y_M) \tag{1}$$

The expected value of a quantity of the problem is given by the following norm:

$$\|\cdot\|_{L^2(\Omega,L^2(D))}=\int_\Omega\int_D|\cdot|^2(x,\omega)\,\mathrm{d}x\mathrm{d}\mathbb{P}=\mathbb{E}\left(\|\cdot\|_{L^2(D)}\right)<\infty \tag{2}$$

3. Computation of Strains

The author has presented a stochastic finite element procedure to solve boundary problems using polynomial chaos [15]-[18]. The outcome displacement of the problem is given by the polynomial chaos expansion as:

$$u(x,y)=\sum_{k=1}^{Q}u_k(x)\psi_\kappa(y) \tag{3}$$

where the order Q and the formula ψ of Polynomial Chaos are given in **Appendix A**.

In order to propagate the uncertainties from input parameters to the results for an elastoplastic problem throw the constitutive equation the strains must be calculated first.

In an elastostatic problem of homogeneous isotropic body one of the field equations that must be satisfied at all interior points of the body is the Strain-Displacement relations:

$$\varepsilon_{ij}(x,y)=\frac{1}{2}\left(u(x,y)_{i,j}+u(x,y)_{j,i}\right)\quad i,j=1,2,3 \tag{4}$$

Using the displacement polynomial chaos expansion the Equation (4) leads to:

$$\begin{aligned}\varepsilon_{ij}(x,y)&=\frac{1}{2}\left(\left(\sum_{k=0}^{Q}u_i^{(k)}(x)\psi_\kappa(y)\right)_{,j}+\left(\sum_{k=0}^{Q}u_j^{(k)}(x)\psi_\kappa(y)\right)_{,i}\right)\\ \Rightarrow\varepsilon_{ij}(x,y)&=\frac{1}{2}\left(\sum_{k=0}^{Q}u_{i,j}^{(k)}(x)\psi_\kappa(y)+\sum_{k=0}^{Q}u_{j,i}^{(k)}(x)\psi_\kappa(y)\right)=\sum_{k=0}^{Q}\varepsilon_{ij}^{(k)}\psi_\kappa(y)\end{aligned} \tag{5}$$

4. Integration Algorithm

Solving for each increment the boundary problem the strain Polynomial Chaos Expansion can be calculated as before. At each increment $n+1$ they are also known the stress from the previous state σ^n the plastic strain

$d\varepsilon_{kl}^{p,n}$. The basic steps in computing the new state of stress are as follows:

The mean value of elastic predictor and of the trial stress are given:

$$\left\langle d\sigma_{ij}^{n+1}\right\rangle = \left\langle C_{ijkl} d\varepsilon_{kl}^{n+1} - p_f^n d\varepsilon_{kl}^{p,n}\right\rangle \quad (6)$$

$$\left\langle \sigma_{ij}^{tr,n+1}\right\rangle = \left\langle \sigma_{ij}^{n}\right\rangle + \left\langle d\sigma_{ij}^{n+1}\right\rangle \quad (7)$$

The 4th-order stochastic elasticity tensor of elastic module is given by the equation:

$$C_{ijkl}(x,y) = E(x)C_{ijkl}^{*}(x), \quad i,j,k,l = 1,2,3 \quad (8)$$

$C_{ijkl}^{*}(x)$: is expressed in terms of (deterministic) Poisson's ratio as

$$C_{ijkl}^{*}(x) = \frac{v}{(1+v)}\delta_{ij}\delta_{kl} + \frac{1}{2(1+v)}\left(\delta_{ik}\delta_{jl} + \delta_{ll}\delta_{jk}\right) \quad (9)$$

Based on that the stochastic process of Young modulus over the spatial domain with a known mean value $\tilde{E}(x)$ and covariance matrix $Cov(x_1,x_2)$ assuming lognormal distribution the Karhunen-Loeve expansion has been used which is the most efficient method for the discretization of a random field. Thus:

$$E(x,y) = \exp\left(\tilde{E}(x) + \sum_{\kappa}^{Q}\sqrt{\lambda_\kappa}\varphi_\kappa(x)y_\kappa\right) \quad (10)$$

where:

λ_κ: is the eigenvalues of the covariance function;

$\varphi_\kappa(x)$: is the eigenfunctions of the covariance function $Cov(x_1,x_2)$.

5. Constitutive Equations for Plain Strain Condition

Assuming one dimension and 3rd order polynomial chaos and plain strain conditions:

$$\begin{bmatrix}\sigma_{11}\\ \sigma_{22}\\ \sigma_{12}\end{bmatrix} = \exp\left(mu_{lnE} + \sigma_{lnE}y\right)\frac{1}{(1+v)(1-2v)}\begin{bmatrix}1-v & v & 0\\ v & 1-v & 0\\ 0 & 0 & (1-2v)/v\end{bmatrix}\begin{bmatrix}\varepsilon_{11}\\ \varepsilon_{22}\\ \varepsilon_{12}\end{bmatrix} \quad (11)$$

Considering as μ_k, σ_k and $v_k = \frac{\sigma_k}{\mu_k}$ the mean value the standard deviation and the coefficient of variation of Elasticity modulus , the mean values and the variance of the lognormal distribution are equal to:

$$\begin{cases}\sigma_{lnk}^2 = \ln\left(1+v_k^2\right)\\ \mu_{lnk} = \ln(\mu_k) - \frac{1}{2}\sigma_{lnk}^2\end{cases} \quad (12)$$

Using the Chaos polynomial expansion the stochastic equation of each component of stress is given:

$$\begin{cases}\sigma_{11} = \frac{\exp\left(mu_{lnE}+\sigma_{lnE}y\right)}{(1+v)(1-2v)}\left[(1-v)\sum_{k=1}^{Q}\varepsilon_{11}^{(k)}\psi_k(y) + v\sum_{k=1}^{Q}\varepsilon_{22}^{(k)}\psi_k(y)\right]\\ \sigma_{22} = \frac{\exp\left(mu_{lnE}+\sigma_{lnE}y\right)}{(1+v)(1-2v)}\left[(v)\sum_{k=1}^{Q}\varepsilon_{11}^{(k)}\psi_k(y) + (1-v)\sum_{k=1}^{Q}\varepsilon_{22}^{(k)}\psi_k(y)\right]\\ \sigma_{12} = \frac{\exp\left(mu_{lnE}+\sigma_{lnE}y\right)}{(1+v)(1-2v)}\left[\frac{1-2v}{v}\sum_{k=1}^{Q}\varepsilon_{12}^{(k)}\psi_k(y)\right]\\ \sigma_{3} = \frac{\exp\left(mu_{lnE}+\sigma_{lnE}y\right)}{(1+v)}\left(\frac{v}{1-2v}\right)\left(\sum_{k=1}^{Q}\varepsilon_{11}^{(k)}\psi_k(y) + \sum_{k=1}^{Q}\varepsilon_{22}^{(k)}\psi_k(y)\right)\end{cases} \quad (13)$$

The expected value after some algebra gives:

$$\left\langle\begin{bmatrix}\sigma_{11}\\ \sigma_{22}\\ \sigma_{12}\end{bmatrix}\right\rangle = \underbrace{\frac{e^{mu_{lnE}}}{(1+v)(1-2v)}\begin{bmatrix}1-v & v & 0\\ v & 1-v & 0\\ 0 & 0 & (1-2v)/v\end{bmatrix}}_{\text{deterministic}}\left\langle\underbrace{\begin{bmatrix}e^{\sigma_{lnE}y}\cdot\varepsilon_{11}\\ e^{\sigma_{lnE}y}\cdot\varepsilon_{22}\\ e^{\sigma_{lnE}y}\cdot\varepsilon_{12}\end{bmatrix}}_{\text{stochastic}}\right\rangle$$

$$\Rightarrow\left\langle\begin{bmatrix}\sigma_{11}\\ \sigma_{22}\\ \sigma_{12}\end{bmatrix}\right\rangle = \frac{e^{mu_{lnE}}}{(1+v)(1-2v)}\begin{bmatrix}1-v & 0 & 0\\ v & 1-v & 0\\ 0 & 0 & (1-2v)/v\end{bmatrix}\begin{bmatrix}\sum_{k=0}^{Q}\varepsilon_{11}^{(k)}\int_\Gamma\rho(y)e^{\sigma_{lnE}y}\psi_k(y)dy\\ \sum_{k=0}^{Q}\varepsilon_{22}^{(k)}\int_\Gamma\rho(y)e^{\sigma_{lnE}y}\psi_k(y)dy\\ \sum_{k=0}^{Q}\varepsilon_{12}^{(k)}\int_\Gamma\rho(y)e^{\sigma_{lnE}y}\psi_k(y)dy\end{bmatrix} \quad (14)$$

The variance of stress:

$$Var(\sigma_{11}) = E\left[\sigma_{11}^2\right]-\left(E[\sigma_{11}]\right)^2 \quad (15)$$

where:

$$\begin{cases} E\left(\sigma_{11}^2\right) = E\left[\left(\dfrac{e^{mu_{lnE}}e^{\sigma_{lnE}y}}{(1+v)(1-2v)}(1-v)\sum_{k=1}^{Q}\varepsilon_{11}^{(\kappa)}\psi_\kappa(y)+v\sum_{k=1}^{Q}\varepsilon_{22}^{(\kappa)}\psi_\kappa(y)\right)^2\right]\\ \left(E[\sigma_{11}]\right)^2 = \left[\dfrac{e^{mu_{lnE}}}{(1+v)(1-2v)}\left((1-v)\sum_{k=1}^{Q}\varepsilon_{11}^{(\kappa)}\int_\Gamma\rho(y)e^{\sigma_{lnE}y}\psi_\kappa(y)dy+v\sum_{k=1}^{Q}\varepsilon_{22}^{(\kappa)}\int_\Gamma\rho(y)e^{\sigma_{lnE}y}\psi_\kappa(y)dy\right)\right]\end{cases} \quad (16)$$

Similarly for the other components.
Expected value of I_2:

$$E[I_2] = E[\sigma_{11}\sigma_{22}]+E[\sigma_{22}\sigma_{33}]+E[\sigma_{33}\sigma_{11}]-E\left[\sigma_{12}^2\right] \quad (17)$$

Based on the stochastic Equation (13) of σ_{ij} components the expected value of the invariant I_2 is given as following:

$$\begin{aligned} E[\sigma_{11}\sigma_{22}] = E\Bigg[&\left(\frac{\exp(mu_{lnE}+\sigma_{lnE}y)}{(1+v)(1-2v)}\left[(1-v)\sum_{k=1}^{Q}\varepsilon_{11}^{(k)}\psi_k(y)+v\sum_{k=1}^{Q}\varepsilon_{22}^{(k)}\psi_k(y)\right]\right)\\ &\times\left(\frac{\exp(mu_{lnE}+\sigma_{lnE}y)}{(1+v)(1-2v)}\left[(v)\sum_{k=1}^{Q}\varepsilon_{11}^{(k)}\psi_k(y)+(1-v)\sum_{k=1}^{Q}\varepsilon_{22}^{(k)}\psi_k(y)\right]\right)\Bigg]\end{aligned} \quad (18)$$

$$\begin{aligned} E[\sigma_{22}\sigma_{33}] = E\Bigg[&\left(\frac{\exp(mu_{lnE}+\sigma_{lnE}y)}{(1+v)(1-2v)}\left[(v)\sum_{k=1}^{Q}\varepsilon_{11}^{(k)}\psi_k(y)+(1-v)\sum_{k=1}^{Q}\varepsilon_{22}^{(k)}\psi_k(y)\right]\right)\\ &\times\left(\frac{\exp(mu_{lnE}+\sigma_{lnE}y)}{(1+v)}\left(\frac{v}{1-2v}\right)\left[\sum_{k=1}^{Q}\varepsilon_{11}^{(k)}\psi_k(y)+\sum_{k=1}^{Q}\varepsilon_{22}^{(k)}\psi_k(y)\right]\right)\Bigg]\end{aligned} \quad (19)$$

$$\begin{aligned} E[\sigma_{33}\sigma_{11}] = E\Bigg[&\left(\frac{\exp(mu_{lnE}+\sigma_{lnE}y)}{(1+v)}\left(\frac{v}{1-2v}\right)\left[\sum_{k=1}^{Q}\varepsilon_{11}^{(k)}\psi_k(y)+\sum_{k=1}^{Q}\varepsilon_{22}^{(k)}\psi_k(y)\right]\right)\\ &\times\left(\frac{\exp(mu_{lnE}+\sigma_{lnE}y)}{(1+v)(1-2v)}\left[(1-v)\sum_{k=1}^{Q}\varepsilon_{11}^{(k)}\psi_k(y)+v\sum_{k=1}^{Q}\varepsilon_{22}^{(k)}\psi_k(y)\right]\right)\Bigg]\end{aligned} \quad (20)$$

$$E\left[\sigma_{12}^2\right] = E\left[\left(\frac{\exp(mu_{lnE}+\sigma_{lnE}y)}{(1+v)(1-2v)}\left[\frac{1-2v}{v}\sum_{k=1}^{Q}\varepsilon_{12}^{(k)}\psi_k(y)\right]\right)\times\left(\frac{\exp(mu_{lnE}+\sigma_{lnE}y)}{(1+v)(1-2v)}\left[\frac{1-2v}{v}\sum_{k=1}^{Q}\varepsilon_{12}^{(k)}\psi_k(y)\right]\right)\right]$$

Variance value of I_2:

Using again the stochastic Equation (13) of σ_{ij} components the variance *of* I_2:

$$Var[I_2] = Var[\sigma_{11}\sigma_{22}] + Var[\sigma_{22}\sigma_{33}] + Var[\sigma_{33}\sigma_{11}] - Var[\sigma_{12}^2] \tag{21}$$

where:

$$\begin{aligned} Var[\sigma_{11}\sigma_{22}] &= E\left[(\sigma_{11}\sigma_{22})^2\right] - \left(E[\sigma_{11}\sigma_{22}]\right)^2 \\ Var[\sigma_{22}\sigma_{33}] &= E\left[(\sigma_{22}\sigma_{33})^2\right] - \left(E[\sigma_{22}\sigma_{33}]\right)^2 \\ Var[\sigma_{11}\sigma_{33}] &= E\left[(\sigma_{11}\sigma_{33})^2\right] - \left(E[\sigma_{11}\sigma_{33}]\right)^2 \end{aligned} \tag{22}$$

As an example the variance of $\sigma_{11}\sigma_{22}$ is given:

$$\begin{cases} E\left[(\sigma_{11}\sigma_{22})^2\right] = E\left[\left(\left(\dfrac{\exp(mu_{lnE} + \sigma_{lnE}y)}{(1+v)(1-2v)}\left[(1-v)\sum_{k=1}^{Q}\varepsilon_{11}^{(k)}\psi_k(y) + v\sum_{k=1}^{Q}\varepsilon_{22}^{(k)}\psi_k(y)\right]\right)\right.\right. \\ \qquad \left.\left.\times\left(\dfrac{\exp(mu_{lnE} + \sigma_{lnE}y)}{(1+v)(1-2v)}\left[(v)\sum_{k=1}^{Q}\varepsilon_{11}^{(k)}\psi_k(y) + (1-v)\sum_{k=1}^{Q}\varepsilon_{22}^{(k)}\psi_k(y)\right]\right)\right)^2\right] \\ \left(E[\sigma_{11}\sigma_{22}]\right)^2 = \left(E\left[\left(\dfrac{\exp(mu_{lnE} + \sigma_{lnE}y)}{(1+v)(1-2v)}\left[(1-v)\sum_{k=1}^{Q}\varepsilon_{11}^{(k)}\psi_k(y) + v\sum_{k=1}^{Q}\varepsilon_{22}^{(k)}\psi_k(y)\right]\right)\right.\right. \\ \qquad \left.\left.\times\left(\dfrac{\exp(mu_{lnE} + \sigma_{lnE}y)}{(1+v)(1-2v)}\left[(v)\sum_{k=1}^{Q}\varepsilon_{11}^{(k)}\psi_k(y) + (1-v)\sum_{k=1}^{Q}\varepsilon_{22}^{(k)}\psi_k(y)\right]\right)\right]\right)^2 \end{cases} \tag{23}$$

The expected value of I_3:

$$E[I_3] = E[\sigma_{11}\sigma_{22}\sigma_{33}] - E[\sigma_{33}\sigma_{12}^2] \tag{24}$$

The variance value of I_3:

$$Var[I_3] = Var[\sigma_{11}\sigma_{22}\sigma_{33}] - Var[\sigma_{33}\sigma_{12}^2] \tag{25}$$

The analysis are carried out similar as the invariant of I_2.

6. Numerical Example

A shallow foundation problem for various values of variation's coefficient v_e is solved taken to account the randomness of the ground. To estimate the statistical moments of the soil deformation the numerical algorithm of SFEM using the Generalized Polynomial Chaos as described in the previous paragraphs is applied. In [16] the author compared the displacement results to those obtained by the closed form solution.

The geometry of the finite elements used for the simulation of the problem presented in **Figure 1**. The input data of the problem is the random field modulus with a constant average value equal to 100 Mpa and a fixed Poisson ratio equal to 0.25. Calculations have been made for ten different coefficients $v_e = \dfrac{\sigma_E}{\mu_E}$ of the elastic modulus with a minimum value of 0.1 and then with step 0.1 to a maximum value equal to 1. For SFEM one dimensional Hermite GPC with order 5 [19] were used. In the **Figures B1-B10 (Appendix B)**, the strains and stress components, and the stress tensor invariants are presented as resulted by the Chaos Polynomial expansion (**Appendix A**) and compared with those raised by the Monte Carlo Method. The convergence of the outcomes decreases as the number of Monte Carlo simulations increases.

7. Conclusions

To propagate the uncertainties of input parameters to constitutive relations of strain and stress where arises due

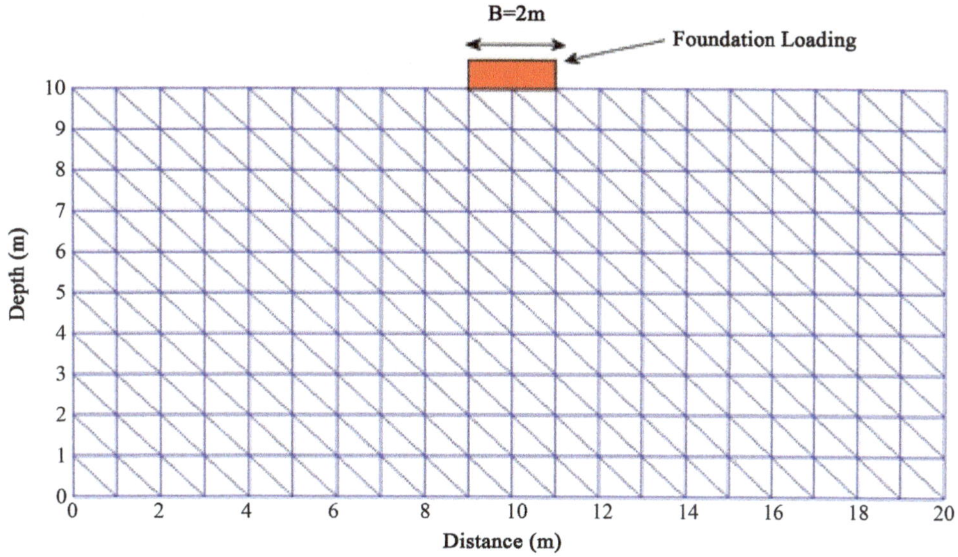

Figure 1. Finite element mesh.

to spatial variability of mechanical parameters in engineering problems, a new algorithm of Stochastic Finite Element Method has been presented.

An algorithm of Stochastic Finite Element using Polynomial Chaos has been developed and the elastic predictor of stress in a non linear problem is calculated.

A numerical example of shallow foundation was carried out and the results of stress and strain of the new algorithm were compared with those raised from Monte Carlo method which is treated as the exact solution. A great accuracy was presented.

The main advantage in using the proposed methodology is that a large number of realizations which have to be made for (Random Finite Element Method) avoided, thus making the procedure viable for realistic practical problems.

References

[1] Hasselman, T. (2001) Quantification of Uncertainty in Structural Dynamic Models. *Journal of Aerospace Engineering*, **14**, 158-165. http://dx.doi.org/10.1061/(ASCE)0893-1321(2001)14:4(158)

[2] Preis, A., Perelman, L. and Ostfeld, A. (2008) Uncertainty Quantification of Contamination Source Identification. *World Environmental and Water Resources Congress* 2008, 1-6. http://dx.doi.org/10.1061/40976(316)505

[3] Keitel, H. (2013) Quantifying Sources of Uncertainty for Creep Models under Varying Stresses. *Journal of Structural Engineering*, **139**, 949-956. http://dx.doi.org/10.1061/(ASCE)ST.1943-541X.0000716

[4] Karim, A., Ahmed, I., Boutton, T., Strom, K. and Fox, J. (2015) Quantification of Uncertainty in a Sediment Provenance Model. *World Environmental and Water Resources Congress* 2015, 1790-1799. http://dx.doi.org/10.1061/9780784479162.175

[5] Phoon, K., Quek, S., Chow, Y. and Lee, S. (1990) Reliability Analysis of Pile Settlements. *Journal of Geotechnical Engineering, ASCE*, **116**, 1717-35. http://dx.doi.org/10.1061/(ASCE)0733-9410(1990)116:11(1717)

[6] Mellah, R., Auvinet, G. and Masrouri, F. (2000) Stochastic Finite Element Method Applied to Non-Linear Analysis of Embankments. *Probabilistic Engineering Mechanics*, **15**, 251-259. http://dx.doi.org/10.1016/S0266-8920(99)00024-7

[7] Eloseily, K., Ayyub, B. and Patev, R. (2002) Reliability Assessment of Pile Groups in Sands. *Journal of Structural Engineering, ASCE*, **128**, 1346-53. http://dx.doi.org/10.1061/(ASCE)0733-9445(2002)128:10(1346)

[8] Fenton, G.A. and Vanmarcke, E. H. (1990) Simulation of Random Fields via Local Average Subdivision. *Journal of Engineering Mechanics*, **116**, 1733-1749. http://dx.doi.org/10.1061/(ASCE)0733-9399(1990)116:8(1733)

[9] Paice, G.M., Griffiths, D.V. and Fenton, G.A. (1996) Finite Element Modeling of Settlements on Spatially Random Soil. *Journal of Geotechnical Engineering*, **122**, 777-779. http://dx.doi.org/10.1061/(ASCE)0733-9410(1996)122:9(777)

[10] Fenton, G.A. and Griffiths, D.V. (2002) Probabilistic Foundation Settlement on Spatially Random Soil. *Journal of*

Geotechnical and Geoenvironmental Engineering, **128**, 381-390.
http://dx.doi.org/10.1061/(ASCE)1090-0241(2002)128:5(381)

[11] Fenton, G.A. and Griffiths, D.V. (2005) Three-Dimensional Probabilistic Foundation Settlement. *Journal of Geotechnical and Geoenvironmental Engineering*, **131**, 232-239.
http://dx.doi.org/10.1061/(ASCE)1090-0241(2005)131:2(232)

[12] Fenton, G.A. and Griffiths, D.V. (2008) Risk Assessment in Geotechnical Engineering. Wiley, Hoboken.
http://dx.doi.org/10.1002/9780470284704

[13] Debraj, G. and Charbel, F. (2008) Strain and Stress Computations in Stochastic Finite Element Methods. *International Journal for Numerical Method in Engineering*, **74**, 1219-1239. http://dx.doi.org/10.1002/nme.2206

[14] Ghanem, R.G. and Spanos, P.D. (1991) Stochastic Finite Elements: A Spectral Approach. Springer-Verlag, New York.
http://dx.doi.org/10.1007/978-1-4612-3094-6

[15] Drakos, I.S. and Pande, G.N. (2015) Stochastic Finite Element Analysis for Transport Phenomena in Geomechanics Using Polynomial Chaos. Civil and Structural Engineering, GJRE-E, Vol. 15, Issue 2, Version 1.0.

[16] Drakos, I.S. and Pande, G.N. (2015) Quantitative of Uncertainties in Earth Structures. *World Congress on Advances in Structural Engineering and Mechanics*, Incheon, 25-29 August 2015, 25-29.

[17] Drakos, I.S. and Pande, G.N. (2015) Stochastic Finite Element Analysis Using Polynomial Chaos. *Studia Geotechnica et Mechanica Journal*. (under review)

[18] Drakos, I.S. and Pande, G.N. (2015) Stochastic Finite Element Consolidation Analysis Based on Polynomial Chaos Expansion. *Georisk Journal*. (under review)

[19] Xiu, D.B. and Em Karniadakis, G. (2003) Modeling Uncertainty in Steady State Diffusion Problems via Generalized Polynomial Chaos. *Computer Methods in Applied Mechanics and Engineering*, **191**, 4927-4948.
http://dx.doi.org/10.1016/S0045-7825(02)00421-8

[20] Lord, G., Powel, C. and Shardlow, T. (2014) An Introduction to Computational Stochastic PDEs. Cambridge Texts in Applied Mathematics. Cambridge University Press, Cambridge. http://dx.doi.org/10.1017/cbo9781139017329

Apendix A

Galerkin Approximation and Generalized Polynomial of Chaos

In order to solve the problem we have to create the new space $L_p^2\left(\Gamma, \mathrm{H}_0^1(D)\right)$. For that reason the subspace $S^k \subset L_p^2(\Gamma)$ is considered as [20].

$$S^k = span\{\psi_1, \psi_2, \cdots, \psi_\kappa\} \tag{A.1}$$

Assuming that the S_i^k represents a space of univariate orthonormal polynomial of variable $y_i \subset \Gamma_i \subset \mathbb{R}$ with order k or lower and:

$$S_i^k = span\left\{P_{a_i}^i(y_i), a_i = 0,1,2,\cdots,k\right\}, i = 1,\cdots,M \tag{A.2}$$

The tensor product of the M S_i^k subspace results the space of the *Generalized Polynomial Chaos*:

$$S^k = S_1 \otimes S_2 \otimes \cdots \otimes S_M \tag{A.3}$$

And using (A2)

$$S^k = span\left\{\prod_{i=1}^M P_{a_i}^i(y_i) : a_i = 0,1,\cdots,k, i = 1,\cdots,M, |a| \le k\right\} \tag{A.4}$$

where $|a| = \sum_{i=1}^M a_i$

And

$$Q = \dim\left(S^k\right) = \frac{(M+k)!}{M!k!} \tag{A.5}$$

Xiu & Karniadakis [19] show the application of the method for different kind of orthonormal polynomials and in the current paper the Hermite polynomial was used with the following characteristics:

$$P_0 = 1, \langle P_i \rangle = 0, i > 0$$

$$\langle P_m P_n \rangle = \int_\Gamma P_m(y) P_n(y) \rho(y) \mathrm{d}y = \gamma_n \delta_{mn} \tag{A.6}$$

where:

$\gamma_n = \left\langle P_n^2 \right\rangle$: are the normalization factors, δ_{mn} is the *Kronecker* delta;

$\rho(y) = \frac{1}{\sqrt{2\pi}} \mathrm{e}^{-\frac{y}{2}}$: is the density function and $P_n = (-1)^n \mathrm{e}^{\frac{y}{2}} \frac{\mathrm{d}^n}{\mathrm{d}y^n} \mathrm{e}^{-\frac{y}{2}}$.

For a 3rd order of one dimension of uncertainty the Hermite Polynomial Chaos is given by:

$$\psi_0(y) = P_0(y) = 1, \psi_1(y) = P_1(y) = y, \psi_2(y) = P_2(y) = y^2 - 1 \tag{A.7}$$

Apendix B

Results of Numerical Example

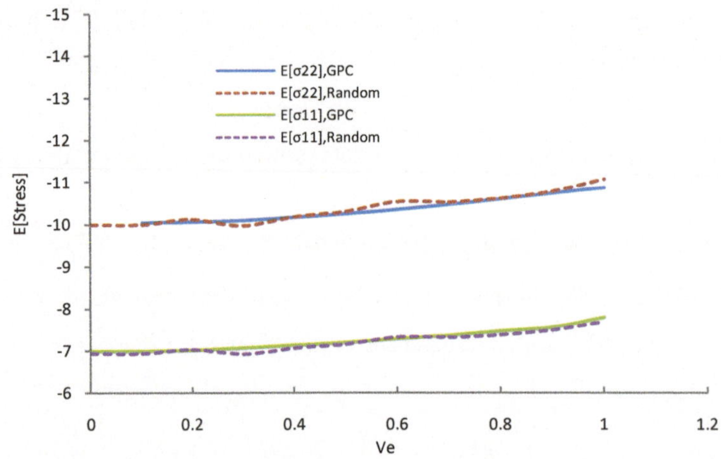

Figure B1. Expected value of stress tensor complements.

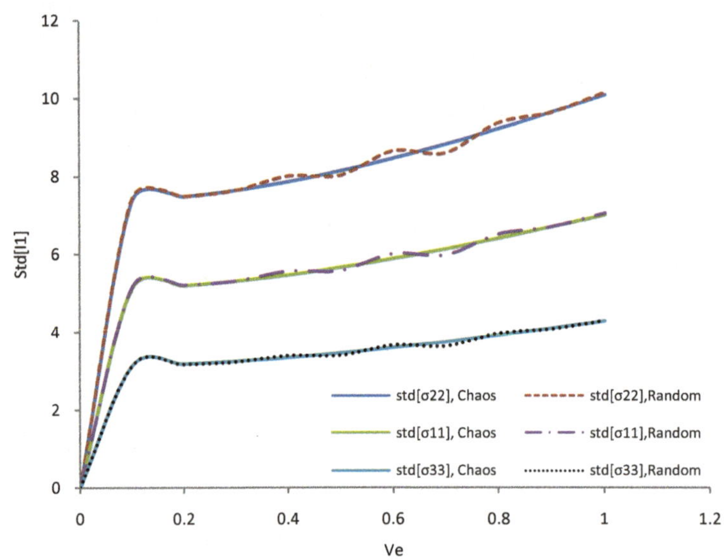

Figure B2. Standard deviation of stress tensor complements.

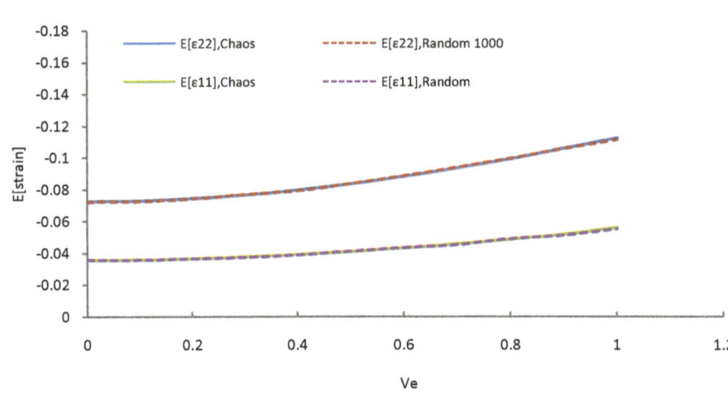

Figure B3. Expected value of strain tensor complements.

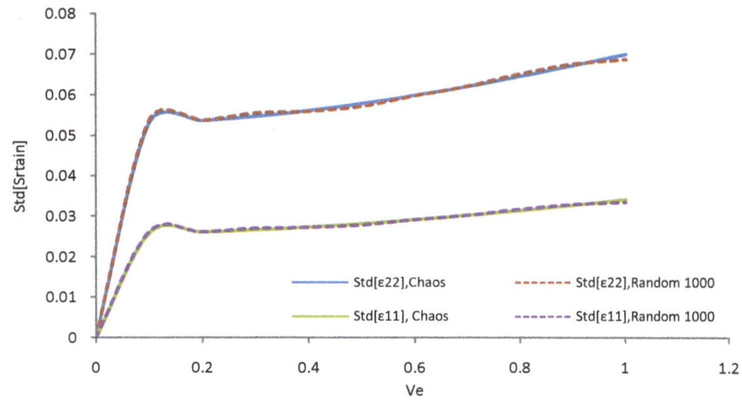

Figure B4. Standard deviation of strain tensor complements.

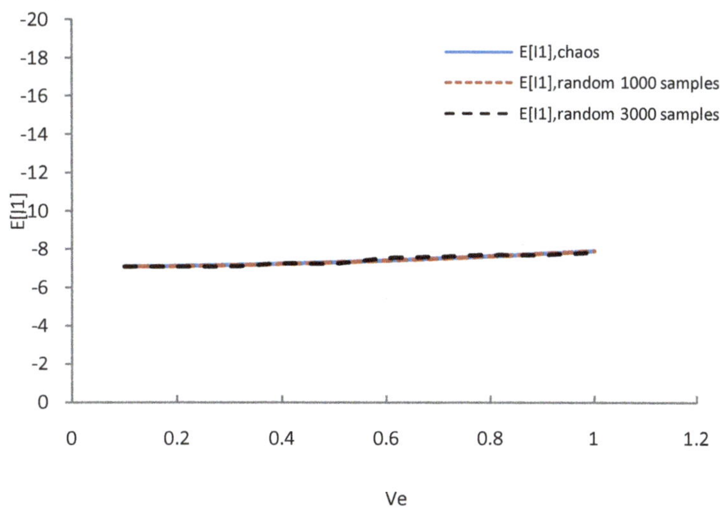

Figure B5. Expected value of stress tensor invariant I_1.

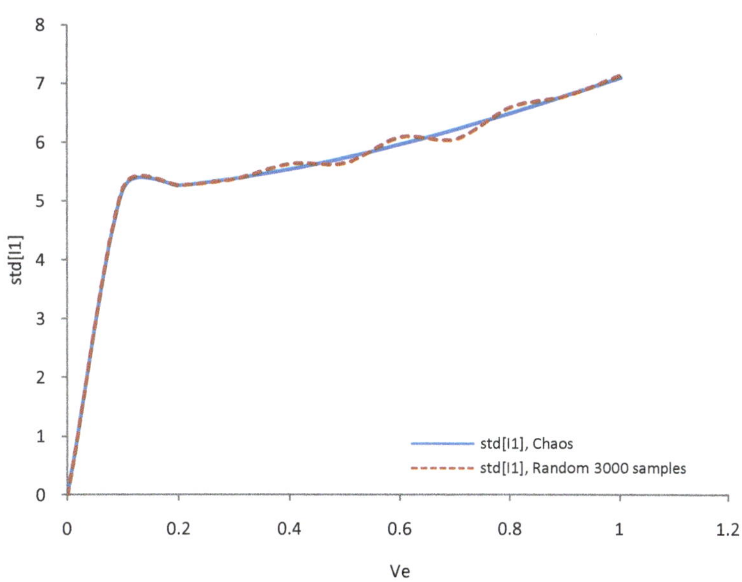

Figure B6. Standard deviation of stress tensor invariant I_1.

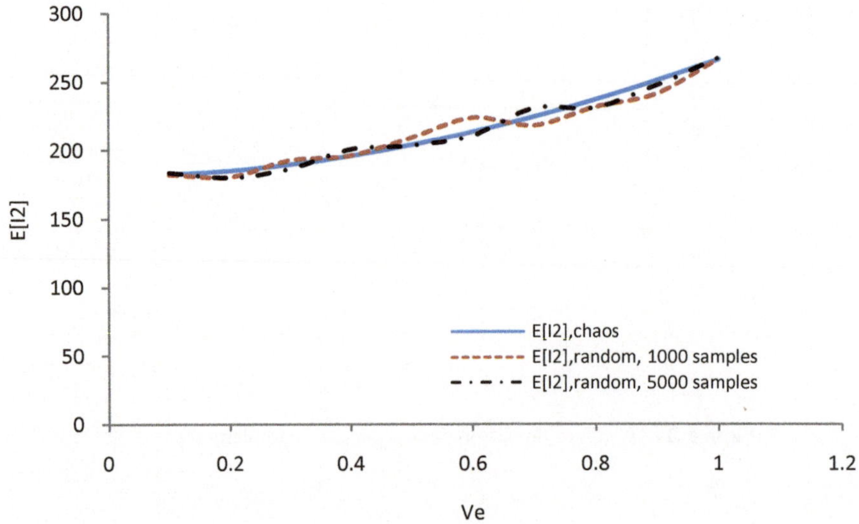

Figure B7. Expected value of stress tensor invariant I_2.

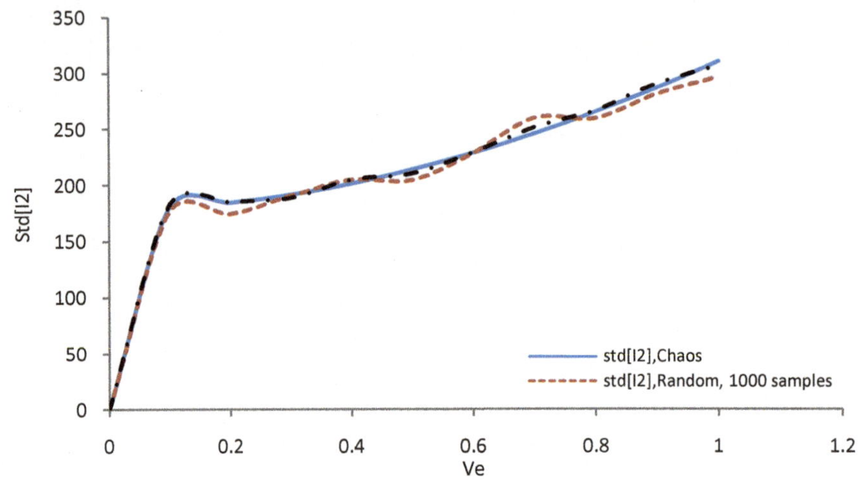

Figure B8. Standard deviation of stress tensor invariant I_2.

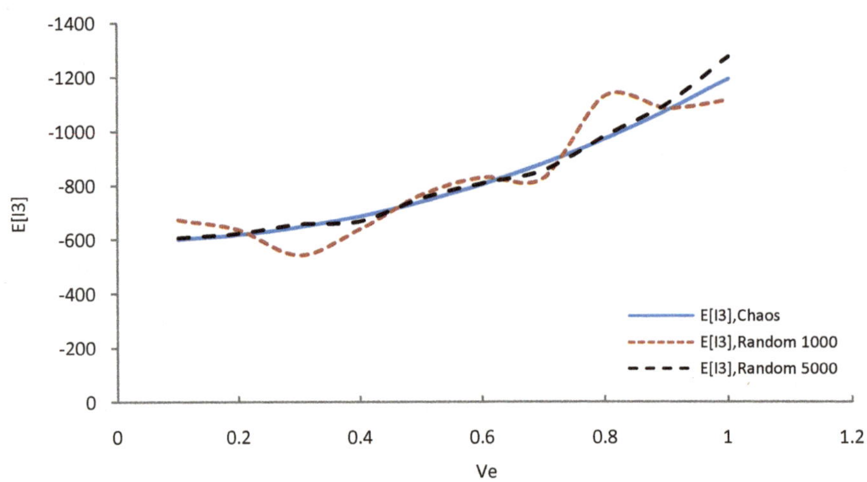

Figure B9. Expected value of stress tensor invariant I_3.

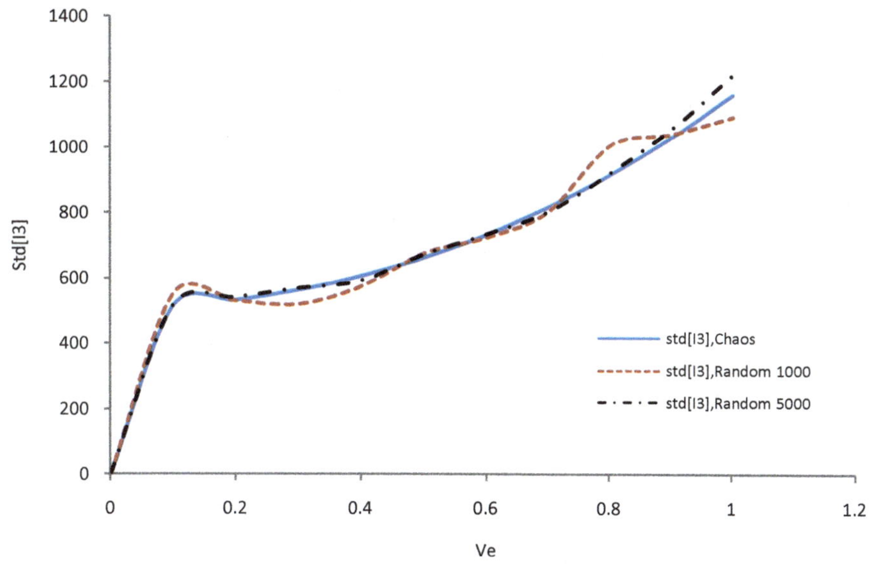

Figure B10. Standard deviation of stress tensor invariant I_3.

Permissions

All chapters in this book were first published in WJM, by Scientific Research Publishing; hereby published with permission under the Creative Commons Attribution License or equivalent. Every chapter published in this book has been scrutinized by our experts. Their significance has been extensively debated. The topics covered herein carry significant findings which will fuel the growth of the discipline. They may even be implemented as practical applications or may be referred to as a beginning point for another development.

The contributors of this book come from diverse backgrounds, making this book a truly international effort. This book will bring forth new frontiers with its revolutionizing research information and detailed analysis of the nascent developments around the world.

We would like to thank all the contributing authors for lending their expertise to make the book truly unique. They have played a crucial role in the development of this book. Without their invaluable contributions this book wouldn't have been possible. They have made vital efforts to compile up to date information on the varied aspects of this subject to make this book a valuable addition to the collection of many professionals and students.

This book was conceptualized with the vision of imparting up-to-date information and advanced data in this field. To ensure the same, a matchless editorial board was set up. Every individual on the board went through rigorous rounds of assessment to prove their worth. After which they invested a large part of their time researching and compiling the most relevant data for our readers.

The editorial board has been involved in producing this book since its inception. They have spent rigorous hours researching and exploring the diverse topics which have resulted in the successful publishing of this book. They have passed on their knowledge of decades through this book. To expedite this challenging task, the publisher supported the team at every step. A small team of assistant editors was also appointed to further simplify the editing procedure and attain best results for the readers.

Apart from the editorial board, the designing team has also invested a significant amount of their time in understanding the subject and creating the most relevant covers. They scrutinized every image to scout for the most suitable representation of the subject and create an appropriate cover for the book.

The publishing team has been an ardent support to the editorial, designing and production team. Their endless efforts to recruit the best for this project, has resulted in the accomplishment of this book. They are a veteran in the field of academics and their pool of knowledge is as vast as their experience in printing. Their expertise and guidance has proved useful at every step. Their uncompromising quality standards have made this book an exceptional effort. Their encouragement from time to time has been an inspiration for everyone.

The publisher and the editorial board hope that this book will prove to be a valuable piece of knowledge for researchers, students, practitioners and scholars across the globe.

List of Contributors

Emil Edipovich Lin
Russian Federal Nuclear Center, All-Russia Research Institute of Experimental Physics, Sarov, Russia

Yoshihiro Kubota
Faculty of Science and Engineering, Toyo University, Saitama, Japan

Osamu Mochizuki
Department of Biomedical Engineering, Toyo University, Saitama, Japan

Haiduke Sarafian
The Pennsylvania State University, University College, York, PA, USA

L. A. Sosnovskiy
S&P Group "TRIBO-FATIGUE", Gomel, Belarus

A. V. Bogdanovich and S. S. Sherbakov
Department of Theoretical and Applied Mechanics, Belarusian State University, Minsk, Belarus

Gaibnazar Khodievich Khojmetov, Alinazar Irismetovich Khodjimetov and Anvar Sayfullaevich Yuvmitov
Tashkent Automobile and Road Institute, Tashkent Architectural-Building Institute, Branch of I. M. Gubkin Russian State University of Oil-Gas, Tashkent, The Republic of Uzbekistan

Zakaria Zergoune and Bilal Harras
Laboratoire de Génie Mécanique, Faculté des Sciences et Techniques, Fez, Morocco

Rhali Benamar
Laboratoire d'Études et de Recherches en Simulation, Université Med V Rabat, École Mohammadia d'Ingénieurs, Rabat, Morocco

Seifedine Kadry
American University of the Middle East, Egaila, Kuwait

Abdelkhalak El Hami
INSA of Rouen Institute, Normandy University, St. Etienne du Rouvray, France

Mehdi Eshaghi and Rama Bhat
Department of Mechanical and Industrial Engineering, Concordia University, Montreal, Canada

Jakkula Anand Rao and Gandamalla Vasumathi
Department of Mathematics, Osmania University, Hyderabad, Telangana

Jakkula Mounica
Department of Mathematics, NGRI, Hyderabad, Telangana

Yukio Fujimoto, Chen Liu, Yoshikazu Tanaka, Eiji Shintaku and Toshiki Nakanishi
Department of Transportation and Environmental Engineering, Graduate School of Engineering, Hiroshima University, Hiroshima, Japan

I. A. Ciureanu
Department of Medical Informatics and Biostatistics, University of Medicine and Pharmacy "Gr. T. Popa", Iasi, Romania

D. Condurache
Department of Theoretical Mechanics, Technical University of Iasi, Iasi, Romania

Kankanala Sharada and Bandari Shankar
Department of Mathematics, Osmania University, Hyderabad, India

Ioan-Adrian Ciureanu
Department of Medical Informatics and Biostatistics, University of Medicine and Pharmacy "Gr.T. Popa", Iasi, Romania

Daniel Condurache
Department of Theoretical Mechanics, Technical University of Iasi, Iasi, Romania

Chen Liu, Yoshikazu Tanaka and Yukio Fujimoto
Department of Transportation and Environmental Engineering, Graduate School of Engineering Science, Hiroshima University, Higashi-Hiroshima, Japan

Yoshihiro Kubota
Faculty of Science and Engineering, Toyo University, Kawagoe, Japan

Osamu Mochizuki
Department of Biomedical Engineering, Toyo University, Kawagoe, Japan

Roberto Capata and Leone Martellucci
Department of Mechanical and Aerospace Engineering, University of Roma "Sapienza", Rome, Italy

Francisco Casanova-del-Angel
National Polytechnic Institute, Mexico City, Mexico

Moisés Gaytán-López
SACM SMA, Government of Mexico City, Mexico City, Mexico

Yuri Menshikov
Department of Mechanics and Mathematics, Dnepropetrovsk University, Dnepropetrovsk, Ukraine

Dimitrios S. Sophianopoulos and Vasiliki S. Pantazi
Department of Civil Engineering, University of Thessaly, Volos, Greece

Chenna Sumalatha and Shankar Bandari
Department of Mathematics, Osmania University, Hyderabad, India

Ch. Achi Reddy
M.L.R. Institute of Technology, Dundigal, Hyderabad, India

B. Shankar
Professor, Department of Mathematics, Osmania University, Hyderabad, India

Alice Rosa da Silva
School of Civil Engineering, Federal University of Uberlândia, Uberlândia, Brazil

Aristeu da Silveira-Neto and Antônio Marcos Gonçalves de Lima
School of Mechanical Engineering, Federal University of Uberlândia, Brazil

Drakos Stefanos
International Center for Computational Engineering, Rhodes, Greece